Structure and Applications of Microstrip Antennas

Structure and Applications of Microstrip Antennas

Edited by **Joe Myers**

$\mathcal{CL}$LANRYE
INTERNATIONAL

New Jersey

Published by Clanrye International,
55 Van Reypen Street,
Jersey City, NJ 07306, USA
www.clanryeinternational.com

Structure and Applications of Microstrip Antennas
Edited by Joe Myers

International Standard Book Number: 978-1-63240-472-5 (Hardback)

Printed in the United States of America.

Contents

Preface   VII

Section 1   Design Techniques   1

Chapter 1   Design Techniques for Conformal Microstrip Antennas and
Their Arrays   3
Daniel B. Ferreira, Cristiano B. de Paula and Daniel C. Nascimento

Chapter 2   Full-Wave Spectral Analysis of Resonant Characteristics and
Radiation Patterns of High Tc Superconducting Circular and
Annular Ring Microstrip Antennas   32
Ouarda Barkat

Chapter 3   Bandwidth Optimization of Aperture-Coupled Stacked
Patch Antenna   48
Marek Bugaj and Marian Wnuk

Section 2   Multiband Planar Antennas   72

Chapter 4   Shared-Aperture Multi-Band Dual-Polarized SAR Microstrip
Array Design   74
Shun-Shi Zhong and Zhu Sun

Chapter 5   Compact Planar Multiband Antennas for Mobile
Applications   98
Ahmad Rashidy Razali, Amin M Abbosh and Marco A Antoniades

Section 3   UWB Printed Antennas   121

Chapter 6   Recent Trends in Printed Ultra-Wideband (UWB)
Antennas   123
Mohammad Tariqul Islam and Rezaul Azim

Chapter 7   **Printed Wide Slot Ultra-Wideband Antenna**   **152**
Rezaul Azim and Mohammad Tariqul Islam

Chapter 8   **UWB Antennas for Wireless Applications**   **171**
Osama Haraz and Abdel-Razik Sebak

Chapter 9   **Dual Port Ultra Wideband Antennas for Cognitive Radio and
Diversity Applications**   **199**
Gijo Augustin, Bybi P. Chacko and Tayeb A. Denidni

Section 4   **Circular Polarization**   **223**

Chapter 10   **Axial Ratio Bandwidth of a Circularly Polarized
Microstrip Antenna**   **225**
Li Sun, Gang Ou, Yilong Lu and Shusen Tan

Section 5   **Recent Advanced Applications**   **243**

Chapter 11   **Drooped Microstrip Antennas for GPS Marine and
Aerospace Navigation**   **245**
Ken G. Clark, Hussain M. Al-Rizzo,
James M. Tranquilla, Haider Khaleel and Ayman Abbosh

Chapter 12   **Planar Microstrip-To-Waveguide Transition in
Millimeter-Wave Band**   **270**
Kazuyuki Seo

Chapter 13   **Wearable Antennas for Medical Applications**   **299**
Albert Sabban

Chapter 14   **Design, Fabrication, and Testing of Flexible Antennas**   **330**
Haider R. Khaleel, Hussain M. Al-Rizzo and
Ayman I. Abbosh

Chapter 15   **Reconfigurable Microstrip Antennas for Cognitive Radio**   **351**
Mohammed Al-Husseini,
Karim Y. Kabalan, Ali El-Hajj and Christos G. Christodoulou

Permissions

List of Contributors

Preface

Every book is initially just a concept; it takes months of research and hard work to give it the final shape in which the readers receive it. In its early stages, this book also went through rigorous reviewing. The notable contributions made by experts from across the globe were first molded into patterned chapters and then arranged in a sensibly sequential manner to bring out the best results.

The structure and applications of microstrip antennas are explained in this profound book. This book presents fundamental and enhanced theories of microstrip antennas comprising of design process and current functions. It discusses arrays, spectral domain, high Tc superconducting microstrip antennas, optimization, multiband, double and spherical polarization, microstrip to waveguide transitions and developing bandwidth and quality occurrence. Antenna fusion substance, microstrip circuits, spectral domain and efficiency are another few additional topics covered in this book. Planar UWB antennas have been extensively discussed. Plan of UWB antennas with solitary or multi notch bands has also been measured. This book intends to guide students and experts as well as those interested in dealing with microstrip antennas.

It has been my immense pleasure to be a part of this project and to contribute my years of learning in such a meaningful form. I would like to take this opportunity to thank all the people who have been associated with the completion of this book at any step.

Editor

Design Techniques

Design Techniques for Conformal Microstrip Antennas and Their Arrays

Daniel B. Ferreira, Cristiano B. de Paula and
Daniel C. Nascimento

Additional information is available at the end of the chapter

1. Introduction

Owing to their electrical and mechanical attractive characteristics, conformal microstrip antennas and their arrays are suitable for installation in a wide variety of structures such as aircrafts, missiles, satellites, ships, vehicles, base stations, etc. Specifically, these radiators can become integrated with the structures where they are mounted on and, consequently, do not cause extra drag and are less visible to the human eye; moreover they are low-weight, easy to fabricate and can be integrated with microwave and millimetre-wave circuits [1,2]. Nonetheless, there are few algorithms available in the literature to assist their design. The purpose of this chapter is to present accurate design techniques for conformal microstrip antennas and arrays composed of these radiators that can bring, among other things, significant reductions in design time.

The development of efficient design techniques for conformal microstrip radiators, assisted by state-of-the-art computational electromagnetic tools, is desirable in order to establish clear procedures that bring about reductions in computational time, along with high accuracy results. Nowadays, the commercial availability of high performance three-dimensional electromagnetic tools allows computer-aided analysis and optimization that replace the design process based on iterative experimental modification of the initial prototype. Software such as CST®, which uses the Finite Integration Technique (FIT), and HFSS®, based on the Finite Element Method (FEM), are two examples of analysis tools available in the market [3]. But, since they are only capable of performing the analysis of the structures, the synthesis of an antenna needs to be guided by an algorithm whereby iterative process of simulations, result analysis and model's parameters modification are conducted until a set of goals is satisfied [4].

Generally, the design of a probe-fed microstrip antenna starts from an initial geometry determined by means of an approximate method such as the Transmission-line Model [5-7] or the Cavity Model [8]. Despite their numerical efficiency, i.e., they are not time-consuming and do not require a powerful computer to run on, these methods are not accurate enough for the design of probe-fed conformal microstrip antennas, leading to the need of antenna model optimization through the use of full-wave electromagnetic solvers in an iterative process. However, the full-wave simulations demand high computational efforts. Therefore, it is advantageous to have a design technique that employs full-wave electromagnetic solvers for accuracy purposes, but requires a small number of simulations to accomplish the design. Unfortunately, the approximated methods mentioned before provide no means for using the full-wave solution data in a feedback scheme, what precludes their integration in an iterative design process, hence restricting them just to the initial design step. In this chapter, in order to overcome this drawback and to reduce the number of full-wave simulations required to synthesize a probe-fed conformal microstrip antenna with quasi-rectangular patch, a circuital model able to predict the antenna impedance locus calculated in the full-wave electromagnetic solver is developed with the aim of replacing the full-wave simulations for the probe positioning. This is accomplished by the use of a transmission-line model with a set of parameters derived to fit its impedance locus to the one obtained in the full-wave simulation [4]. Since this transmission line model adapts its input impedance to fit the one from the full-wave simulation, at each algorithm iteration, it is an adaptive model per nature, so it was named ATLM – Adaptive Transmission Line Model. In Section 2, the ATLM is described in detail and some design examples are given to demonstrate its applicability.

Similar to what occurs with conformal microstrip antennas, the literature does not provide a great number of techniques to guide the design of conformal microstrip arrays. Among these design techniques, there are, for example, the Dolph-Chebyshev design and the Genetic Algorithms [9]. However, the results provided by the Dolph-Chebyshev design are not accurate for beam steering [10], once it does not take the radiation patterns of the array elements into account in its calculations, i.e., for this pattern synthesis technique, the array is composed of only isotropic radiators; hence it implies errors in the main beam position and sidelobes levels when the real patterns of the array elements are considered. On the other hand, the Genetic Algorithms can handle well the radiation patterns of the array elements and guarantee that the sidelobes assume a level better than a given specification R [9]. Nonetheless, to control the array directivity [11], it is important that all these sidelobes have the same level R, but to obtain this type of result Genetic Algorithms frequently requires a high number of iterations which increases the design time. Thus, in Section 3, an elegant procedure is employed, based on the solution of linearly constrained least squares problems [12], to the design of conformal microstrip arrays. Not only does this algorithm take the radiation pattern of each array element into account, but it also assures that a determined number of sidelobes levels have the same value, so to get optimized array directivity. And, to obtain more accurate results, the radiation patterns of the array elements, which feed the developed procedure, are evaluated from the array full-wave simulation data. In this work, the CST® Version 2012 was used to get these data. The proposed design technique was coded in the Mathematica® package [13] to create a computer program capable of assisting the design of conformal microstrip arrays. Some examples are given in this section to illustrate the use and effectiveness of this computer program.

Another concern for designing conformal microstrip arrays is how to implement a feed network that can impose appropriate excitations (amplitude and phase) on the array elements to synthesize a desired radiation pattern. Some microstrip arrays used in tracking systems, for example, employ the Butler Matrix [11] as a feed network. Nevertheless, this solution can just accomplish a limited set of look directions and cannot control the sidelobes levels. Hence, in this work, in order not to limit the number of radiation patterns that can be synthesized, an active circuit, composed of phase shifters and variable gain amplifiers, is adopted to feed the array elements. Expressions for calculating the phase shifts and the gains of these components are addressed in Section 4, as well as some design examples are provided to demonstrate their applicability.

2. Algorithm for conformal microstrip antennas design

The main property of the proposed ATLM is to allow the prediction of the impedance locus determined in the antenna full-wave analysis when one of its geometric parameters is modified, for instance, the probe position, thereby replacing full-wave simulations in probe position optimization. It results in a dramatic computational time saving, since a circuital simulation is usually at least 1000 times faster than a full-wave one. In this section, the ATLM is described in detail and some design examples are provided to highlight its advantages.

2.1. Algorithm description

In order to describe the algorithm for the design of conformal microstrip antennas, for the sake of simplicity, let us first consider a probe-fed planar microstrip antenna with a gular patch of length L_{pa} and width W_{pa}, mounted on a dielectric substrate of thickness h_s, relative permittivity ε_r, and loss tangent tanδ, such as the one shown in Figure 1(a). The antenna feed probe is positioned d_p apart from the patch centre. For the following analysis, it is adopted that the antenna resonant frequency f_r is controlled by the length L_{pa} and once the probe is located along the x-axis, it excites the TM_{10} mode, whose main fringing field is also represented in Figure 1(a). Despite this geometry being of planar type, the same model parameters are used to describe the conformal quasi-rectangular microstrip antennas illustrated in Figure 1(b), 1(c) and 1(d), and consequently the algorithm is valid as well.

(a) Planar microstrip antenna

(b) Cylindrical microstrip antenna

(c) Spherical microstrip antenna

(d) Conical microstrip antenna

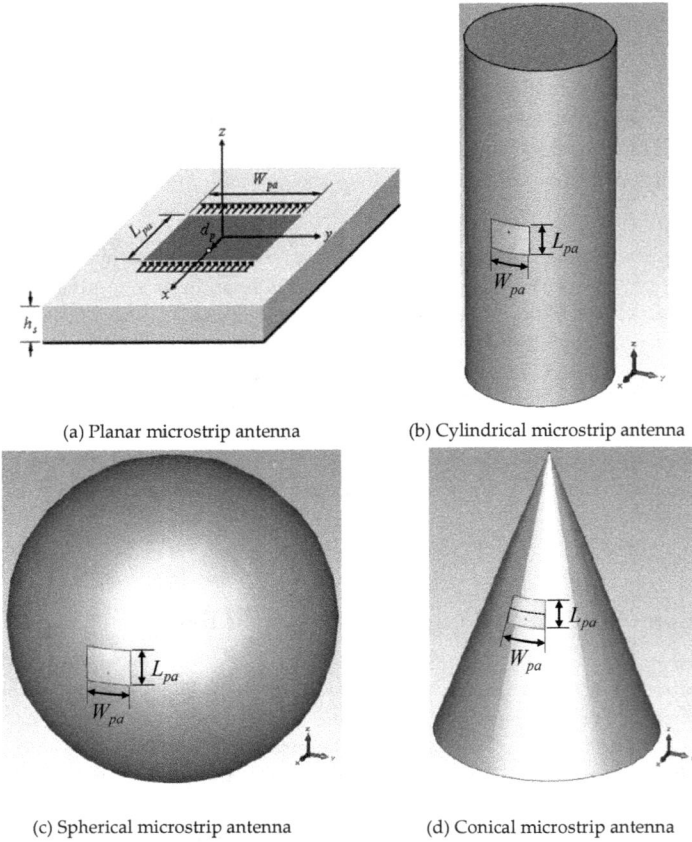

Figure 1. Microstrip antennas studied in this chapter

It is convenient to write both the probe position d_p and patch width W_{pa} as functions of the patch length L_{pa}, to establish a standard set of control variables. Hence, the probe position is written as

$$d_p = R_p L_{pa}, \ 0 < R_p \leq 0.5,$$

(1)

and the patch width as follows

$$W_{pa} = R L_{pa}, \ R \geq 1.$$

(2)

Therefore, the standard set of control variables is composed of L_{pa}, R(patch width to patch length ratio) and R_p (probe position to patch length ratio). The variables L_{pa} and R_p will be used in the algorithm to control its convergence and the variable R will be defined by specification, based on the desired geometry (rectangular, square). Usually, W_{pa} is made 30% higher than L_{pa}, i.e., R=1.3 [14].

In this work, it is considered that the resonant frequency f_r occurs when the magnitude of the antenna reflection coefficient reaches its minimum value. Under this assumption,

$$\left|\Gamma_a(f_r)\right| = \min_f \left|\Gamma_a(f)\right|, \text{ for } f \in [f_1, f_2],\tag{3}$$

in which $\Gamma_a(f)$ is the reflection coefficient determined in the antenna full-wave analysis, f_1 and f_2 are the minimum and maximum frequencies that define the simulation domain $[f_1, f_2]$. For electrically thin radiators it is usually enough to choose $f_1 = 0.95f_0$ and $f_2 = 1.05f_0$, where f_0 is the desired operating frequency, and whether the microstrip antenna is electrically thick, then $f_1 = 0.80f_0$ and $f_2 = 1.20f_0$, in order to locate f_r between f_1 and f_2 in the first algorithm iteration.

Since the antennas design will be conducted in an iterative manner, the optimization process of the model needs to be evaluated against optimization goals in order to set a stop criterion. Therefore, let the frequency error be defined as

$$e = \left|\frac{f_r}{f_0} - 1\right|\tag{4}$$

and its maximum value specified as e_{max}. It leads to the first optimization goal, that is,

$$e \le e_{max}.\tag{5}$$

The second optimization goal is expressed by means of

$$\left|\Gamma_a(f_r)\right| < \Gamma_{min},\tag{6}$$

where Γ_{min} is a positive real number defined by specification. So, the maximum reflection coefficient magnitude observed at the resonant frequency needs to be lower than Γ_{min}.

Now that the main parameters of the design algorithm have been derived, let us focus on the Adaptive Transmission Line Model, depicted in Figure 2. As can be seen, this circuital model is composed of two microstrip lines, μS_1 and μS_2, whose widths are equal to W_{pa}, an ideal transmission line TL_p – with characteristic impedance Z_p and electrical length $\angle E_l$ (in degrees) given by

$$\angle E_l = 360 h_s \left(\frac{c_0}{f \sqrt{\varepsilon_r}} \right)^{-1} , \qquad (7)$$

where c_0 is the speed of the light in free-space –, a capacitor C, and two load terminations L_s. The ideal transmission line together with the capacitor C were included in the model to account for the impedance frequency shift due to the feed probe. In order to fit the input impedance of this model to the one determined in the antenna full-wave analysis, the reflection coefficients at the terminals of the loads L_s are written as

$$\Gamma_f(f) = (a_0 + a_1 f) e^{-j(b_0 + b_1 f)} , \qquad (8)$$

in which $\Gamma_f(f)$ is the reflection coefficient of the equivalent slot of impedance Z_f, and a_0, a_1, b_0, b_1 as well as Z_p and C are the set of parameters that determine the frequency response of the circuital model. It is worth mentioning that this ATLM is valid only if its variables L_{pa} and W_{pa} are kept identical to the ones used in the full-wave analysis.

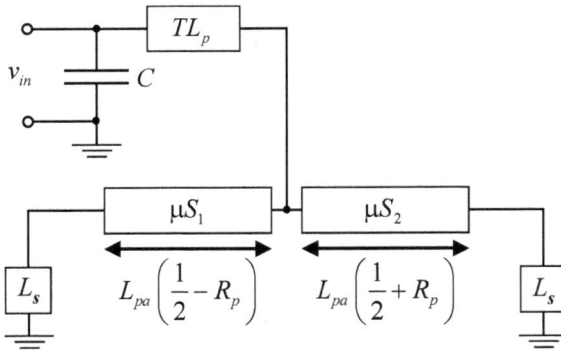

Figure 2. Adaptive transmission line model – ATLM

Once the full-wave simulation $\Gamma_a(f)$ is known, the antenna input impedance $Z_a(f)$ can be easily evaluated. The same is valid for the circuital model analysis in which the reflection coefficient is $\Gamma_c(f)$ and input impedance is $Z_c(f)$. It is important to point out that $\Gamma_a(f)$ data can be exported from the full-wave simulator to the circuit simulator in *Touchstone* format, so $Z_a(f)$ can be utilized by the circuit simulator. The ATLM parameters set is calculated in order to have $\Gamma_c(f)=\Gamma_a(f)$ over the simulation domain $[f_1, f_2]$. The process of finding the values of this parameters set is called ATLM synthesis and it is done with aid of a Gradient optimization tool, usually available in circuit simulators such as Agilent ADS® [15], as follows.

Consider the generalized load reflection coefficient [16] that is written as

$$\Gamma_L = \frac{Z_L - Z_g^*}{Z_L + Z_g},$$

(9)

in which Z_L is the load impedance and Z_g is the generator impedance, with the superscript * denoting the complex conjugate operator. Since for the ATLM the input voltage v_{in} comes from a generator, it follows that $Z_L = Z_c(f)$. By using a Gradient optimization tool with the goal $\Gamma_L = 0$ yields

$$Z_c(f) = Z_g^*,$$

(10)

after the optimization process.

As we want to ensure that $\Gamma_c(f) = \Gamma_a(f)$, i.e., $Z_c(f) = Z_a(f)$, yields

$$Z_g = Z_a^*(f),$$

(11)

which is the generator impedance utilized during the ATLM synthesis. On the other hand, for the circuital simulation afterwards, $Z_g = Z_0$, where Z_0 is the characteristic impedance of the antenna feed network.

Besides, to find a meaningful solution from a physical standpoint, the following two constraints are ensured during the ATLM synthesis

$$\text{Re}\{Z_f\} > 0 \text{ and } \text{Im}\{Z_f\} < 0.$$

(12)

The complete probe-fed microstrip antenna design algorithm is depicted through the flowchart in Figure 3, which can be summarized as follows: perform a full-wave antenna simulation for a given patch length and probe position at a certain frequency range (simulation domain), which results in accurate impedance locus data; synthesize the ATLM based on the most updated full-wave simulation data available; optimize the probe position in order to match the antenna to its feed network through circuital simulation and evaluate the resonant frequency; perform patch length scaling; update the full-wave model with the new values of patch length and probe position; and repeat the whole process in an iterative manner until the goals are satisfied.

Generally, it is difficult to get the input impedance of the circuital model perfectly matched to the one obtained from full-wave simulation over the entire simulation domain $[f_1, f_2]$ (i.e., $Z_c(f) \equiv Z_a(f)$), so it is convenient to set the following goal in the Gradient optimizer,

$$|\Gamma_L| \leq \begin{cases} -30\,\mathrm{dB}, f \in \left[(f_0 - \frac{f_2 - f_1}{4}), (f_0 + \frac{f_2 - f_1}{4})\right] \\ -20\,\mathrm{dB}, f \notin \left[(f_0 - \frac{f_2 - f_1}{4}), (f_0 + \frac{f_2 - f_1}{4})\right] \end{cases}. \qquad (13)$$

The previous goal contributes to reduce the number of iterations required by the Gradient optimization tool to determine the set of parameters. It was found that, in general, the required time for the synthesis of the ATLM is at most 5% of the time spent for one full-wave simulation.

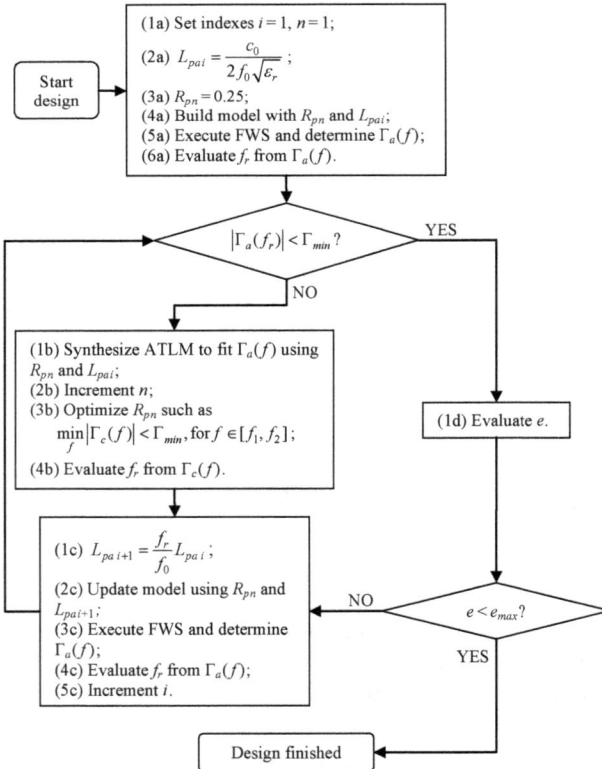

Figure 3. Probe-fed microstrip antenna design algorithm; FWS – Full-Wave Simulation

Regarding the probe position optimization, algorithm step 3b, it can be performed manually by means of a tuning process, a usual feature found in circuit simulators. Thus, R_p is tuned in order to minimize the magnitude of the input reflection coefficient of the circuital model. If desired, the optimization process can be performed employing an optimization tool, e.g., Gradient, Random, also available in circuit simulators. Usually, each circuital analysis takes no longer than 1 second using a simulator such ADS®. But, if one desires to create its own code for the ATLM circuital analysis and probe position optimization, a simple rithm can be implemented to seek the R_p that minimizes $|\Gamma_c(f)|$, and the computational time will be greatly reduced as well.

2.2. Applications

To illustrate the use of the technique proposed before, let us first consider the design of a cylindrical microstrip antenna (Figure 1(b)) with a quasi-rectangular metallic patch mounted on a cylindrical dielectric substrate with a thickness h_s=0.762mm, relative permittivity ε_r= 2.5 and loss tangent tan δ = 0.0022, which covers a copper cylinder (ground layer) with a 60.0-mm radius and 300.0-mm height. The patch centre is equidistant from the top and bottom of the copper cylinder. This radiator was designed to operate at f_0 = 3.5 GHz and the algorithm parameters were chosen as e_{max}=0.1×10^{-2}, Γ_{min}=3.16×10^{-2} (return loss of 30dB), and W_{pa}=1.3L_{pa}. Once it is an electrically thin antenna, the simulation domain was given by f_1 =0.95f_0 and f_2 =1.05f_0.

Following the algorithm (Figure 3), a model was built (step 4a) in the CST® software with L_{pa1} =27.11mm and R_{p1} =0.25, and a first full-wave simulation was performed (step 5a). From the analysis of the obtained reflection coefficient $\Gamma_a(f)$, the determined resonant frequency was f_r=3.384GHz (step 6a) and the reflection coefficient magnitude was -17dB, thus higher than the desired maximum of -30 dB (Figure 4(b)).

Hence, at the first decision point of the algorithm, the reflection coefficient magnitude at resonance is not lower than Γ_{min}, so one must go to the step 1b. Then ATLM was synthesized for L_{pa1} =27.11mm and R_{p1} =0.25 and its parameters set was derived with the aid of the Gradient optimization tool of ADS®. After 55 iterations of the Gradient tool, the following parameters set was found: Z_p=94Ω, C=0.87pF, a_0 =-0.58, a_1 =3.83×10^{-10}s, b_0 =-6.54, and b_1 = 2.21×10^{-9} s. The full-wave impedance locus and the one obtained from circuital simulation of the synthesized ATLM are shown in Figure 4(a), and it can be seen that the locus determined though circuital simulation fits very well the full-wave one.

With the circuital model available, the probe position was optimized through manual tuning of the variable R_p, and since for step 3b it is desired that the reflection coefficient magnitude at the resonance be below Γ_{min}, R_p was tuned such as the ATLM impedance locus crossed the Smith Chart centre (Figure 4(a)), leading to R_{p2} =0.21. The resonant frequency obtained from the circuital simulation with this probe position (step 4b) was f_r=3.392GHz. Following the algorithm, the next step was the scaling of patch length (step 1c) leading to L_{pa2} =26.28mm. After updating the full-wave model with these parameters, a full-wave simulation was executed (step 3c) resulting f_r=3.480GHz with a reflection coefficient magnitude of -54dB (Figure 4(b)). Since $|\Gamma_a|<\Gamma_{min}$, the next step was step 1d where it was found that e=0.57×10^{-2},

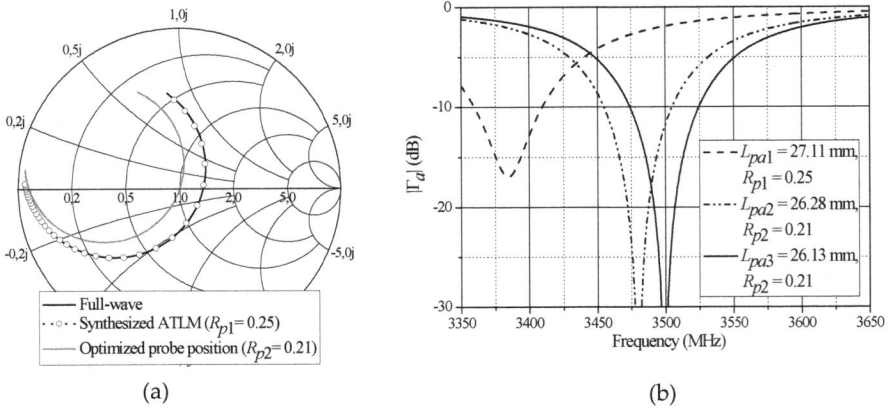

(a)   (b)

Figure 4. Iterations of the algorithm for the probe-fed cylindrical microstrip antenna design: (a) impedance loci of the full-wave and circuital simulations, (b) reflection coefficient magnitude for the full-wave simulations

higher than e_{max}, thus the algorithm went to step 1c, where a second patch length scaling was done leading to L_{pa3} =26.13mm. A last full-wave simulation with R_{p2} =0.21 and L_{pa3} =26.13mm was performed resulting in e=0.03×10^{-2} and return loss of 54dB at resonance, thus satisfying all specifications. This design required only three full-wave simulations in order to guarantee all specifications, what demonstrates the efficiency of the proposed design technique.

Now let us design a probe-fed spherical microstrip antenna, such as the one illustrated in Figure 1(c). A copper sphere (ground layer) of 120.0-mm radius is covered with a dielectric substrate of constant thickness h_s=0.762mm, relative permittivity ε_r=2.5 and loss tangent tanδ=0.0022. A quasi-rectangular patch with length L_{pa} and width W_{pa} is printed on the surface of the dielectric substrate. The design specifications were the same used previously and the steps of the algorithm followed a path similar to the one in the design of the cylindrical radiator. Once again, the algorithm took only three full-wave simulations to perform the design, as observed in Figure 5(a). The ATLM parameter set found was Z_p=91Ω, C=0.63 pF, a_0 = 6.69×10^{-3}, a_1 = 2.32×10^{-10} s, b_0 = -4.10, b_1 = 1.54×10^{-9} s, and the resulting patch parameters were R_{p2} =0.20 and L_{pa3} =26.06mm, which led to a final frequency error e=0.03×10^{-2} and 35-dB return loss at resonance.

As a last example, let us consider the design of a conical microstrip antenna with a quasi-rectangular metallic patch, as shown in Figure 1(d). It is composed of a conical dielectric substrate of constant thickness h_s=0.762mm that covers a 280.0-mm-high cone made of copper (ground layer) with a 40.0° aperture. The dielectric substrate has the same electromagnetic characteristics as the ones employed in the previous examples and the patch centre is located at the midpoint of its generatrix. This radiator was designed to operate at f_0 = 3.5

GHz and the algorithm parameters were chosen as e_{max}=0.1×10^{-2}, Γ_{min}=3.16×10^{-2} (return loss of 30dB), and W_{pa}=1.3L_{pa}. By applying the developed algorithm, the ATLM parameters set found was Z_p=104Ω, C=0.33pF, a_0=-0.26, a_1=3.01×10^{-10}s, b_0=-4.01, b_1=1.53×10^{-9}s, and the determined patch parameters were R_{p2}=0.23 and L_{pa3}=26.18mm, which yielded a final frequency error e = 0.01×10^{-2} and 34-dB return loss at resonance, once again supporting the proposed design technique. Figure 5(b) presents the reflection coefficient magnitudes of the three full-wave simulations required to accomplish the conical microstrip antenna design.

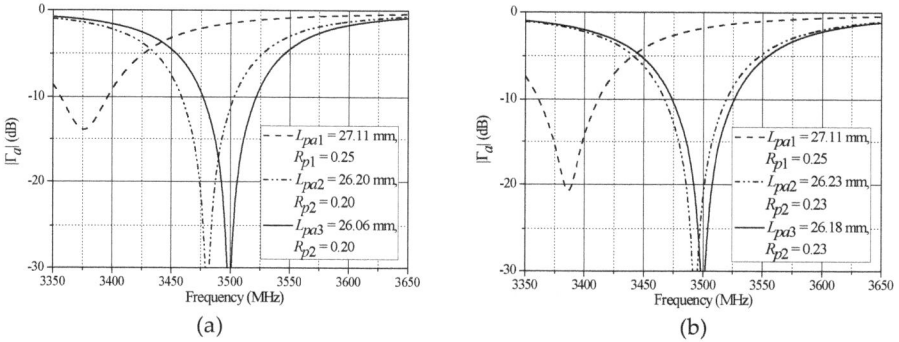

Figure 5. Reflection coefficient magnitudes for each full-wave simulation required for the designs: (a) probe-fed spherical microstrip antenna, (b) probe-fed conical microstrip antenna

3. Radiation pattern synthesis of conformal microstrip arrays

The previous section addressed a computationally efficient algorithm for assisting the design of probe-fed conformal microstrip antennas with quasi-rectangular patches. In order to demonstrate its applicability, three conformal microstrip antennas were synthesized: a cylindrical, a spherical and a conical one. According to what was observed, the algorithm converges very fast, what expedites the antennas' design time.

Another concern in the design of conformal radiators is how to determine the current excitations of a conformal microstrip array to synthesize a desired radiation pattern, in which both the main beam position and the sidelobes levels can be controlled. This section is dedicated to the presentation of a technique employed for the design of conformal microstrip arrays. It is based on the iterative solution of linearly constrained least squares problems [12], so it has closed-form solutions and exhibits fast convergence, and, more important, it takes the radiation pattern of each array element into account in its code, what improves its accuracy. These radiation patterns are determined from the output data obtained through the conformal microstrip array analysis in a full-wave electromagnetic simulator, such as CST® and

HFSS®. Once those data are available, polynomial interpolation is utilized to write simple closed-form expressions that represent adequately the far electric field radiated by each array element, which makes the technique numerically efficient.

The developed design technique was implemented in the Mathematica® platform giving rise to a computer program – called CMAD (Conformal Microstrip Array Design) – capable of performing the design of conformal microstrip arrays. The Mathematica® package, an integrated scientific computer software, was chosen mainly due to its vast collection of built-in functions that permit implementing the respective algorithm in a short number of lines, in addition to its many graphical resources. At the end of the section, to illustrate the CMAD ability to synthesize the radiation pattern of conformal microstrip arrays, the synthesis of the radiation pattern of three conformal microstrip array topologies is considered. First, a microstrip antenna array conformed onto a cylindrical surface is analysed. Afterwards, a spherical microstrip array is studied. Finally, the synthesis of the radiation pattern of a conical microstrip array is presented.

3.1. Algorithm description

The far electric field radiated by a conformal microstrip array composed of N elements and embedded in free space, assuming time-harmonic variations of the form $e^{j\omega t}$, can be written as

$$E = \mathbb{C}\frac{e^{-jk_0 r}}{r} I^t \cdot v(\theta, \phi), \tag{14}$$

where the constant $\mathbb{C}$ is dependent on both the free-space electromagnetic characteristics, μ_0 and ε_0, and the angular frequency ω, $k_0 = \omega\{\mu_0 \varepsilon_0\}^{1/2}$ is the free-space propagation constant,

$$I^t = \begin{bmatrix} I_1 & \cdots & I_N \end{bmatrix}, \tag{15}$$

with $I_n, 1 \le n \le N$, representing the current excitation of the n-th array element and the superscript t indicates the transpose operator,

$$v(\theta,\phi) = \begin{bmatrix} g_1(\theta,\phi) \\ \vdots \\ g_N(\theta,\phi) \end{bmatrix}, \tag{16}$$

in which $g_n(\theta,\phi)$, $1 \le n \le N$, denotes the complex pattern of the n-th array element evaluated in the global coordinate system. Boldface letters represent vectors throughout this chapter.

Based on (14), the radiation pattern of a conformal microstrip array can be promptly calculated using the relation

$$| I^t \cdot v(\theta,\phi) |^2 = w^\dagger \cdot [v(\theta,\phi) \cdot v^\dagger(\theta,\phi)] \cdot w, \tag{17}$$

where the complex weight w is equal to I^*, the superscript $*$ represents the complex conjugate operator and $\dagger$ indicates the Hermitian transpose (complex conjugate transpose operator). Therefore, the radiation pattern evaluation requires the knowledge of both complex weight w and vector $v(\theta,\phi)$.

Once the array elements are chosen and their positions are predefined, to determine the vector v (θ, ϕ) tor $v(\theta,\phi)$ it is necessary to calculate the complex patterns $g_n(\theta,\phi)$, $1 \leq n \leq N$, of the array elements. For conformal microstrip arrays there are some well-known techniques to accomplish this [1], for example, the commonly used electric surface current method [17-19]. However, when this technique is employed to analyse cylindrical or conical microstrip arrays, for instance, it cannot deal with the truncation of the ground layer and the diffraction at the edges of the conducting surfaces that affect the radiation pattern. Moreover, the expressions derived from this method for calculating the radiated far electric field frequently involve Bessel and Legendre functions. Nevertheless, as extensively reported in the literature [20], the evaluation of these functions is not fast and requires good numerical routines. Hence, to overcome these drawbacks and to get more accurate results, in this chapter, the complex patterns $g_n(\theta,\phi)$ are determined from the data obtained through the conformal microstrip array analysis in the CST® package. It is important to point out that other commercial 3D electromagnetic simulators, such as HFSS®, can also be used to assist the evaluation of the complex patterns $g_n(\theta,\phi)$, since they are able to take into account truncation of the ground layer and diffraction at the edges of the conducting surfaces.

From the array full-wave simulation data, polynomial interpolation is applied to generate simple closed-form expressions that represent adequately the far electric field (amplitude and phase) radiated by each array element. In this work, the degree of the interpolation polynomials is established from the analysis of the RMSE (root-mean-square error), which provides a measure of similarity between the interpolated data and the ones given by CST®. For the following examples the interpolation polynomials' degrees are defined aiming at a RMSE less than 0.02.

Considering the previous scenario, to synthesize a radiation pattern in a given plane, it just requires the determination of the current excitations I_n present in the complex weight w. Figure 6 illustrates a typical specification of a radiation pattern containing information about the main beam direction α, the intervals intervals $[\theta_a,\theta_b]$ and $[\theta_c,\theta_d]$ where the sidelobes are located as well as the maximum level R that can be assumed for them.

Based on (17) and following [12], a constrained least squares problem is established in order to locate the main beam at the α direction,

$$\min_{w} w^\dagger \cdot A \cdot w \tag{18}$$

subject to the constraints

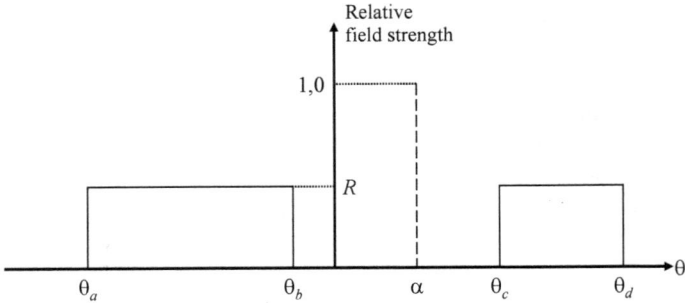

Figure 6. Typical specification of a radiation pattern in a given plane

$$v_s^\dagger \cdot w = 1, \tag{19}$$

$$\mathrm{Re}\{v_d^\dagger \cdot w\} = 0, \tag{20}$$

in which $v_s = v(\alpha, \phi')$, $v_d = \partial v(\theta, \phi)/\partial \theta \, |_{(\theta, \phi) = (\alpha, \phi')}$, ϕ' is the ϕ coordinate of the plane where the pattern is being synthesized, and

$$A = \frac{1}{2} \sum_{\ell=1}^{L} v(\theta_\ell, \phi') \cdot v^\dagger(\theta_\ell, \phi'), \tag{21}$$

with the angles θ_ℓ, $\ell=1,2,\ldots,L$, uniformly sampled in the sidelobes intervals $[\theta_a, \theta_b]$ and $[\theta_c, \theta_d]$. In the next examples the adopted step size between consecutive θ_ℓ is equal to $0.1°$ (for each of the sidelobes intervals).

In order to find a closed-form solution to the problem defined by (18) to (20), we determine its real counterpart [21], that is,

$$\min_{\tilde{w}} \tilde{w}^t \cdot \tilde{A} \cdot \tilde{w} \tag{22}$$

subject to the following linear constraints

$$\tilde{C}^t \cdot \tilde{w} = \tilde{f}, \tag{23}$$

where

$$\tilde{w} = \begin{bmatrix} \mathrm{Re}\{w\} & \mathrm{Im}\{w\} \end{bmatrix}^t, \tag{24}$$

$$\tilde{A} = \begin{bmatrix} \mathrm{Re}\{A\} & -\mathrm{Im}\{A\} \\ \mathrm{Im}\{A\} & \mathrm{Re}\{A\} \end{bmatrix}, \tag{25}$$

$$\tilde{C} = \begin{bmatrix} \tilde{v}_s & \hat{v}_s & \tilde{v}_d \end{bmatrix}, \tag{26}$$

$$\tilde{f} = \begin{bmatrix} 1 & 0 & 0 \end{bmatrix}^t, \tag{27}$$

with

$$\tilde{v}_s = \begin{bmatrix} \mathrm{Re}\{v_s\} & \mathrm{Im}\{v_s\} \end{bmatrix}^t, \tag{28}$$

$$\hat{v}_s = \begin{bmatrix} -\mathrm{Im}\{v_s\} & \mathrm{Re}\{v_s\} \end{bmatrix}^t, \tag{29}$$

$$\tilde{v}_d = \begin{bmatrix} \mathrm{Re}\{v_d\} & \mathrm{Im}\{v_d\} \end{bmatrix}^t. \tag{30}$$

The closed-form solution to the problem (22) and (23) is

$$\tilde{w} = \tilde{A}^{-1} \cdot \tilde{C} \cdot \left(\tilde{C}^t \cdot \tilde{A}^{-1} \cdot \tilde{C} \right)^{-1} \cdot \tilde{f}, \tag{31}$$

from which the complex weight w is promptly evaluated.

After solving the problem (18)-(20) the main beam is located at the α-direction. Nevertheless, it cannot be assured that the sidelobes levels are below the threshold R. In order to get it, the complex weight w is updated by residual complex weights Δw, as follows:

$$w \leftarrow w + \Delta w. \tag{32}$$

A constrained least squares problem, similar to (18)-(20), that ensures the sidelobes levels, is set up for the purpose of calculating the residual complex weights Δw, that is,

$$\min_{\Delta w} \Delta w^\dagger \cdot A \cdot \Delta w \tag{33}$$

subject to the constraints

$$v_s^\dagger \cdot \Delta w = 0, \tag{34}$$

$$\mathrm{Re}\{v_d^\dagger \cdot \Delta w\} = 0, \tag{35}$$

$$v_i^\dagger \cdot \Delta w = f_i, i = 1,2, \dots , m, \tag{36}$$

in which $v_i = v(\theta_i, \phi')$, with θ_i denoting the θ coordinate of the i-th sidelobe, m is the number of sidelobes whose levels are being modified (the maximum m is equal to N–2), and the complex function f_i can be evaluated through

$$f_i = (R - |c_i|)\frac{c_i}{|c_i|}, \tag{37}$$

where

$$c_i = w^\dagger \cdot v_i. \tag{38}$$

It is important to point out that the constraints (34) and (35) retain the main beam located at the α-direction, and the ones in (36) are responsible for conducting the sidelobes levels to the threshold R. A closed-form solution to the problem (33)-(36) is also determined from its real counterpart, analogous to the solution to the problem (18)-(20). The problem (33)-(36) is iteratively solved until the sidelobes levels reach the desired value R. Notice that at each iteration the maximum number of sidelobes whose levels are controlled is equal to $N - 2$, i.e., if the array radiation pattern has more than $N - 2$ sidelobes, we choose the $N - 2$ side-lobes with the highest levels to apply the constraints (36).

The radiation pattern synthesis technique described before was implemented in the Mathe-matica® platform with the aim of developing a CAD – called CMAD – capable of performing the design of conformal microstrip arrays. The inputs required to start the design procedure in the CMAD program are the Text Files (.txt extension) containing the points that describe the complex patterns of each array element – obtained from the conformal microstrip array simulation in CST® package –, the look direction α, the maximum sidelobes level R, and the starting and ending points of the intervals $[\theta_a, \theta_b]$ and $[\theta_c, \theta_d]$ where they are located. As a result, the CMAD returns the current excitations and the synthesized pattern. It is worth mentioning that the use of interpolation polynomials to describe the complex patterns expe-dites the evaluation of both vector $v(\theta, \phi)$ and its derivative; consequently, the CMAD's run time is diminished. In the following three sections, examples of radiation pattern synthesis are provided to demonstrate the capability of the developed CMAD program.

3.2. Cylindrical microstrip array

To illustrate the described pattern synthesis technique, let us first consider the design of a five-element cylindrical microstrip array, such as the one shown in Figure 7(a). For this array, the cylindrical ground layer is made out of copper cylinder with a 60.0-mm radius and a 300.0-mm height. The employed dielectric substrate has a relative permittivity ε_r = 2.5, a loss tangent tan δ = 0.0022 and its thickness is h_s = 0.762 mm. The array patches are identical to the one designed in Section 2.2 to operate at 3.5 GHz. The five elements are fed by 50-ohm coaxial probes positioned 5.49 mm apart from the patches' centres and the interelement spacing was chosen to be $\lambda_0 / 2$ = 42.857 mm (λ_0 is the free-space wavelength at 3.5 GHz).

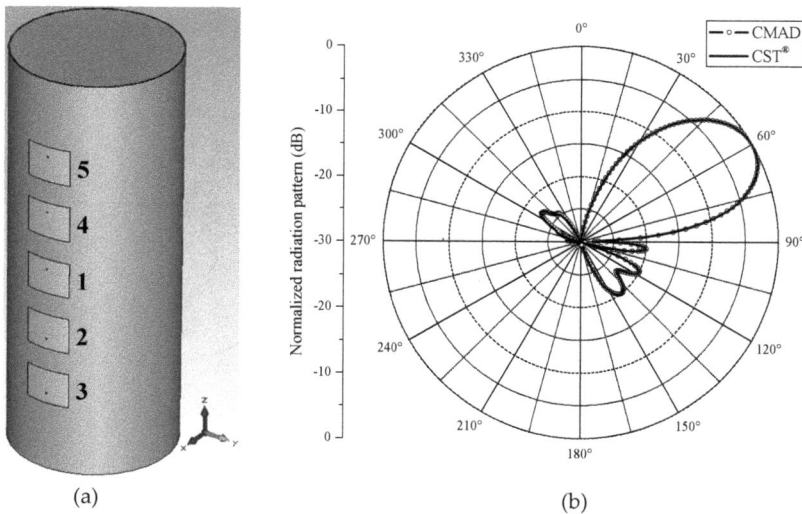

Figure 7. (a) Five-element cylindrical microstrip array, (b) E_θ radiation pattern: xz-plane, α = 60°, R = -20 dB, and f = 3.5 GHz

It is important to point out that the elements close to the ends of the ground cylinder have significantly different radiation patterns than those close to the centre of this cylinder; however, the technique developed in this chapter can handle well this aspect, different from the common practice that assumes the elements' radiation patterns are identical [22]. To clarify this difference among the patterns, Figure 8 shows the radiation patterns of the elements number 1 and 5. In Figure 8(a) they were evaluated in CST® and in Figure 8(b) they were determined from the interpolation polynomials. As observed, there is an excellent agreement between the radiation patterns described by the interpolation polynomials and the ones provided by CST®, even in the back region, where the radiation pattern exhibits low level and oscillatory behaviour. It validates the use of polynomial interpolation functions to represent the far electric field radiated by the conformal array elements.

For this cylindrical array, let us consider that the radiation pattern in the xz-plane must have the main beam located at $\alpha = 60°$ and the maximum sidelobe level allowed is $R = -20$ dB. By using the CMAD program, we get both the array normalized current excitations, depicted in Table 1, and the synthesized radiation pattern, shown in Figure 7(b). In order to validate these results, we provide the normalized current excitations (Table 1) for the array simulation in CST®. The radiation pattern evaluated in CST® is also represented in Figure 7(b). According to what is observed, there is an excellent agreement between the radiation pattern given by the CMAD and the one calculated in CST®, thus validating the developed technique to design cylindrical microstrip arrays.

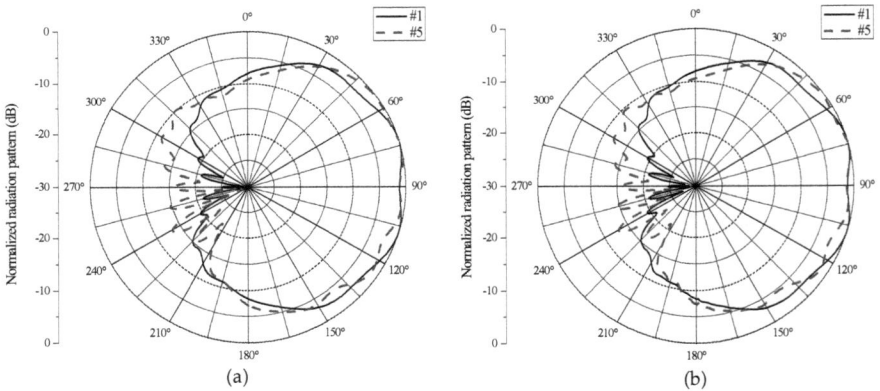

Figure 8. E_θ radiation patterns – elements number 1 and 5: xz-plane and $f = 3.5$ GHz. (a) CST® and (b) interpolation polynomials

Element Number	Normalized Current Excitation
1	$1.0\angle 0.0°$
2	$0.800\angle -82.394°$
3	$0.360\angle 6.211°$
4	$0.781\angle -90.315°$
5	$0.617\angle 172.593°$

Table 1. Cylindrical microstrip array: normalized current excitations

3.3. Spherical microstrip array

Another conformal microstrip array topology used to demonstrate the CMAD's ability to synthesize radiation patterns is the five-element spherical microstrip array, which operates at 3.5 GHz, illustrated in Figure 9(a). For this array, the selected ground layer is a copper sphere with a radius of 120.0 mm. A typical microwave substrate ($\varepsilon_r = 2.5$, $\tan \delta = 0.0022$ and $h_s = 0.762$ mm) covers all the ground sphere and the array patches are the same as the ones designed in Section 2.2. The angular interelement spacing in the θ-direction was chosen to be 20.334°, which corresponds to an arc length of $\lambda_0 / 2$ onto the external microwave substrate surface.

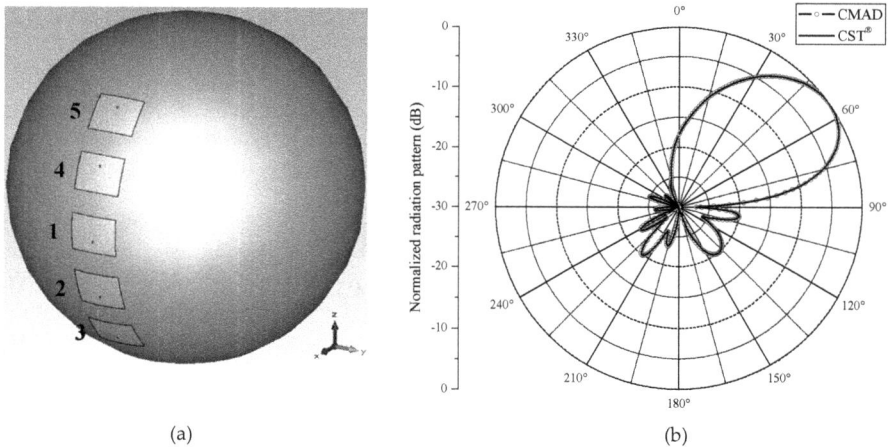

(a)   (b)

Figure 9. (a) Five-element spherical microstrip array, (b) E_θ radiation pattern: xz-plane, $\alpha = 55°$, $R = -20$ dB, and $f = 3.5$ GHz

In this case, the synthesized radiation pattern in the xz-plane must have its main beam located at $\alpha = 55.0°$ direction and the maximum sidelobe level cannot exceed -20 dB. After entering these requirements in the CMAD program, it outputs the normalized current excitations (Table 2) and the synthesized radiation pattern (Figure 9(b)). To verify these results, the normalized current excitations were loaded into the spherical microstrip array simulation conducted in the CST® software. The radiation pattern obtained is also shown in Figure 9(b) for comparisons purposes. As seen, the radiation pattern given by the CMAD program and the one determined in CST® show a very good agreement, thus supporting the proposed radiation pattern synthesis technique. It is important to point out that the interelement spacing could be varied if the array directivity needs to be altered.

Element Number	Normalized Current Excitation
1	0.680∠264.460°
2	0.252∠-6.059°
3	0.160∠156.639°
4	1.0∠0.0°
5	0.728∠-36.758°

Table 2. Spherical microstrip array: normalized current excitations

3.4. Conical microstrip array

Finally, let us consider the radiation pattern synthesis of the four-element conical microstrip array presented in Figure 10(a). For this array, the ground layer is a 280.0-mm-high cone made of copper with a 40.0° aperture. This cone is covered with a dielectric substrate of constant thickness h_s = 0.762 mm, relative permittivity ε_r = 2.5 and loss tangent tan δ = 0.0022. The array elements are identical to the one designed in Section 2.2, so they have a length of 26.18 mm in the generatrix direction, an average width of 34.03 mm in the φ-direction, and the 50-ohm coaxial probes are located 6.02 mm apart from the patches' centres toward the ground cone basis. The interelement spacing in the generatrix direction is of 42.857 mm (= λ_0 / 2) as well as the centre of the element #1 is 110.0 mm apart from the cone apex in this same direction.

The radiation pattern specifications for this synthesis are: main beam direction α = 70° and maximum sidelobe level R = -20 dB, both in the xz-plane. By using the CMAD program, we derive the normalized current excitations, shown in Table 3, and the synthesized radiation pattern in the frequency 3.5 GHz, illustrated in Figure 10(b). Also in Figure 10(b) the array radiation pattern calculated in CST®, considering the normalized current excitations of Table 3, is presented. As observed, the radiation pattern obtained with CMAD matches the one determined in CST®, once again supporting the proposed design approach.

4. Active feed circuit design

As can be seen, the radiation pattern synthesis technique presented in the previous section is suitable for applications that require electronic radiation pattern control, for example. However, it only provides the array current excitations, i.e., to complete the array design it is still necessary to synthesize its feed network. A simple active circuit topology dedicated to feed those arrays can be composed of branches having a variable gain amplifier cascaded to a phase shifter, both controlled by a microcontroller, and a 1 : N power divider, as depicted in

Figure 11. The phase shifters play a role in controlling the phases of the current excitations, as well as the variable gain amplifiers that are responsible for settling their amplitudes. In this section, expressions for calculating the phase shifts ϕ_n and the gains G_n, in terms of the array current excitations and their electrical characteristics, including the self and mutual impedances, are derived. It is worth mentioning that the evaluated expressions take into account the mismatches between the array elements' driving impedances and the characteristic impedance Z_0 of the lines, what improves their accuracy.

At the end of this section, to illustrate the synthesis of the proposed active feed network (Figure 11), the design of the active beamformers of the three conformal microstrip arrays (cylindrical, spherical and conical) that appear along the chapter is described. Furthermore, to validate the phase shifts and gains calculated, the designed feed networks are analysed in the ADS® package.

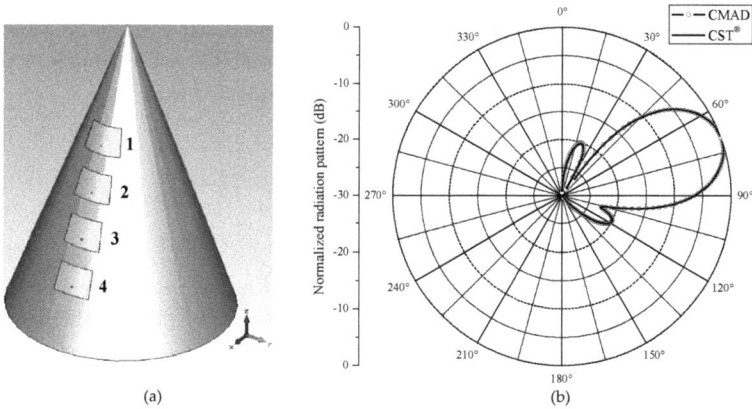

Figure 10. (a) Four-element conical microstrip array, (b) E_θ radiation pattern: xz-plane, $\alpha = 70°$, $R = -20$ dB, and $f = 3.5$ GHz

Element Number	Normalized Current Excitation
1	$0.574\angle-7.835°$
2	$0.875\angle0.149°$
3	$1.0\angle0,0°$
4	$0.625\angle-6.561°$

Table 3. Conical microstrip array: normalized current excitations

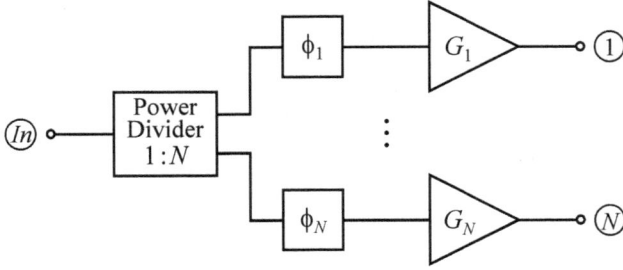

Figure 11. Active feed network

4.1. Design equations

For the analysis conducted here the phase shifters are considered perfectly matched to the input and output lines and produce zero attenuation. Based on these assumptions the scattering matrix $(S_{pf})_n$ of the n-th phase shifter, $1 \leq n \leq N$, assumes the form

$$(S_{pf})_n = \begin{pmatrix} 0 & e^{j\phi_n} \\ e^{j\phi_n} & 0 \end{pmatrix}, \tag{39}$$

with ϕ_n ($0 \leq \phi_n < 2\pi$) representing the phase shift produced by the n-th phase shifter. Notice that the matching requirement can be met to within a reasonable degree of approximation for commercial IC (Integrated Circuit) phase shifters, however those devices frequently exhibit moderate insertion loss. So, to take the insertion loss into account in our analysis model, the gains G_n are either decreased or increased (to compensate the insertion losses).

The variable gain amplifiers are also considered perfectly matched to the input and output lines and they are unilateral devices, i.e., $s_{12n} = 0$. Hence, the scattering matrix $(S_a)_n$ of the n-th variable gain amplifier is given by

$$(S_a)_n = \begin{pmatrix} 0 & 0 \\ s_{21n} & 0 \end{pmatrix}, \tag{40}$$

in which s_{21n} denotes the gain (linear magnitude) of the n-th variable gain amplifier. It is worth mentioning that lots of commercial IC variable gain amplifiers have input and output return loss better than 10 dB and exhibit high directivity, therefore, the preceding assumptions are reasonable. More precise results using the scattering parameters of commercial variable gain amplifiers and phase shifters are presented in [23].

Let us examine the operation of the n-th circuit branch. The input power P_n at the terminals of the n-th array element can be calculated by

$$P_n = \frac{1}{2}\mathrm{Re}\{Zin_n\}\,|\,I_n\,|^2,$$ (41)

where Zin_n is the driving impedance at the terminals of the n-th array element and can be evaluated using

$$Zin_n = \sum_{\kappa=1}^{N} Z_{n\kappa}\frac{I_\kappa}{I_n}$$ (42)

in which $Z_{n\kappa}$ is the n-th array element self-impedance, if $n=\kappa$, and the mutual impedance between the n-th and κ-th array elements, if $n \neq \kappa$. In this chapter the self and mutual impedances will be determined from the array simulation data. However, those impedances could also be obtained from the measurements conducted in the array prototype, what certainly would lead to a more accurate feed network design.

Alternatively, the input power at the terminals of the n-th array element can be expressed in terms of the incident power P_{0n} and the reflection coefficient Γin_n at the terminals as

$$P_n = P_{0n}(1-|\,\Gamma in_n\,|^2),$$ (43)

with

$$\Gamma in_n = \frac{Zin_n - Z_0}{Zin_n + Z_0}.$$ (44)

Combining (41) and (43) results in an expression to evaluate the incident power at the terminals of the n-th array element

$$P_{0n} = \frac{\mathrm{Re}\{Zin_n\}\,|\,I_n\,|^2}{2(1-|\,\Gamma in_n\,|^2)},$$ (45)

which is equal to the n-th variable gain output power, disregarding the losses in the lines.

Based on (45), an equation to determine the gain of the n-th variable gain amplifier is derived:

$$G_n = \frac{P_{0n}}{P_{0m}} = \frac{\text{Re}\{Zin_n\}}{\text{Re}\{Zin_m\}} \frac{1 - |\Gamma in_m|^2}{1 - |\Gamma in_n|^2} \frac{|I_n|^2}{|I_m|^2}. \tag{46}$$

Notice that to evaluate (46) it is necessary to choose one of the circuit branches as a reference, i.e., the gain of the m-th variable gain amplifier is set equal to 1.0.

It is important to highlight that this formulation has relevant importance for arrays whose mutual coupling among elements is strong [23], since it takes this effect into account. For arrays whose mutual coupling among elements is weak and the array elements self-impedances are close to Z_0, (46) is approximated by

$$G_n \cong \frac{|I_n|^2}{|I_m|^2}. \tag{47}$$

Now, to determine the phase shifts ϕ_n, let us consider the current I_n at the terminals of the n-th array element, that is,

$$I_n = I_{0n} e^{j\phi_n}(1 - \Gamma in_n), \tag{48}$$

in which $I_{0n} e^{j\phi_n}$ is the incident current wave at the terminals of the n-th array element.

Once the currents I_n are provided by the algorithm described in the last section, to calculate ϕ_n the phases of the left and right sides of (48) are enforced to be equal. Then,

$$\phi_n = \delta_{nm} - \arg\{1 - \Gamma in_n\} + \arg\{1 - \Gamma in_m\}, \tag{49}$$

with

$$\delta_{nm} = \arg\{I_n\} - \arg\{I_m\}. \tag{50}$$

Also for the determination of the phase shift ϕ_n, the m-th circuit branch was taken as a reference, i.e., its phase shifter does not introduce any phase shift ($\phi_m = 0°$) in the signal.

For arrays whose mutual coupling among elements is weak and the array elements self-impedances are close to Z_0, the phase shift ϕ_n (49) reduces to

$$\phi_n \cong \delta_{nm}. \tag{51}$$

The expressions for evaluating the gains G_n (46) and phase shifts ϕ_n (49) were incorporated into the developed computer program CMAD to generate a new module devoted to design active feed networks, such as the one illustrated in Figure 11. The inputs required to start the circuit design are the array current excitations and the *Touchstone* File (.sNp extension) containing the array scattering parameters – obtained from the conformal microstrip array simulation in a full-wave electromagnetic simulator, for example. In the next section, to demonstrate the capability of this new CMAD feature, the feed networks of the three conformal microstrip arrays previously synthesized will be designed.

4.2. Examples

The normalized current excitations found in Tables 1 to 3 and the scattering parameters of the three conformal microstrip arrays synthesized in this chapter (evaluated in CST®) were provided to the CMAD. As results, it returned the gains and phase shifts of the active feed networks that implement the radiation patterns shown in Figures 7(b), 9(b) and 10(b). These values are listed in Table 4.

To verify the validity of the results found in Table 4, the designed active feed networks were analysed in the ADS® package. As an example, Figure 12 shows the simulated feed network for the conical microstrip array. In this circuit, the array is represented through a 4-port microwave network, whose scattering parameters are the same as the ones used by the CMAD, it is fed by a 30-dBm power source with a 50-ohm impedance, and there are four current probes to measure the currents at the terminals of the 4-port microwave network, which correspond to the array current excitations. Table 5 summarizes the current probes readings for the three analysed feed networks. The comparison between the currents given in Table 5 and the ones presented in Tables 1 to 3 shows that these currents are in agreement, thereby validating the design equations derived before.

Branch Number	Cylindrical Array		Spherical Array		Conical Array	
	Gain (dB)	Phase Shift (deg)	Gain (dB)	Phase Shift (deg)	Gain (dB)	Phase Shift (deg)
1	7.4	351.1	11.1	114.6	0.0	0.0
2	5.7	267.6	3.8	205.0	4.2	7.0
3	0.0	0.0	0.0	0.0	5.4	4.1
4	4.9	260.6	14.0	210.3	1.5	356.6
5	2.4	164.0	10.9	173.0	–	–

Table 4. Gains and phase shifts of the designed active feed networks

Branch Number	Cylindrical Array	Spherical Array	Conical Array
1	0.160∠-10.37°	0.249∠113.2°	0.106∠-9.417°
2	0.128∠-92.76°	0.092∠-157.3°	0.161∠-1.432°
3	0.058∠-4.156°	0.059∠5.375°	0.184∠-1.581°
4	0.125∠-100.7°	0.366∠-151.3°	0.115∠-8.141°
5	0.099∠162.2°	0.267∠172.0°	–

Table 5. Current probes readings (in ampere)

5. Conclusion

In summary, a computationally efficient algorithm capable of assisting the design of probe-fed conformal microstrip antennas with quasi-rectangular patches was discussed. Some examples were provided to illustrate its use and advantages. As seen, it can result in significant reductions in design time, since the required number of full-wave electromagnetic simulations, which are computationally intensive – especially for conformal radiators –, is diminished. For instance, the proposed designs could be performed with only three full-wave simulations. Also in this chapter, an accurate design technique to synthesize radiation patterns of conformal microstrip arrays was introduced. The adopted technique takes the radiation pattern of each array element into account in its code through the use of interpolation polynomials, different from the common practice that assumes the elements' radiation patterns are identical. Hence, the developed technique can provide more accurate results. Besides, it is able to control the sidelobes levels, so that optimized array directivity can be achieved. This design technique was coded in the Mathematica® platform giving rise to a computer program, called CMAD, that evaluates the array current excitations responsible for synthesizing a given radiation pattern. To show the potential of the CMAD program, the design of cylindrical, spherical and conical microstrip arrays were exemplified. Finally, an active feed network suitable for applications that require electronic radiation pattern control, like tracking systems, was addressed. The expressions derived for the synthesis of this circuit take into account the mutual coupling among the array elements; therefore they are also suited for array configurations in which the mutual coupling among the elements is strong. These design equations were incorporated into the CMAD code adding to it one more project tool. In order to validate this new CMAD feature, the feed networks of the three conformal microstrip arrays described along the chapter were designed. The obtained results were validated through the feed networks' simulations in the ADS® software.

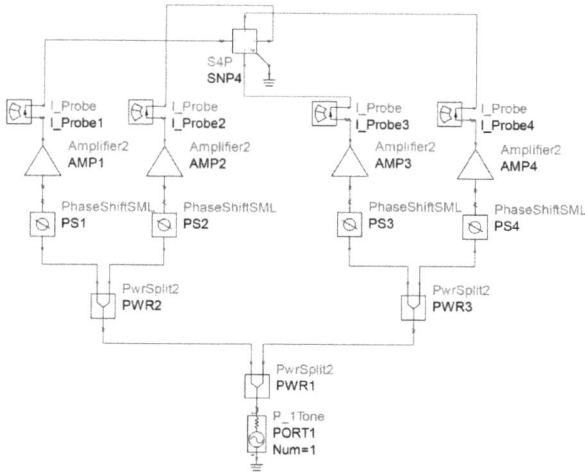

Figure 12. Simulated feed network for the conical microstrip array

Acknowledgements

The authors would like to acknowledge the support given to this work, developed under the project "Adaptive Antennas and RF Modules for Wireless Broadband Networks Applied to Public Safety", with the support of the Ministry of Communications' FUNTTEL (Brazilian Fund for the Technological Development of Telecommunications), under Grant No. 01.09.0634.00 with the Financier of Studies and Projects - FINEP / MCTI.

Author details

Daniel B. Ferreira[1], Cristiano B. de Paula[1] and Daniel C. Nascimento[2]

1 CPqD - Telecommunications R&D Foundation, Brazil

2 ITA - Technological Institute of Aeronautics, Brazil

References

[1] Wong KL. Design of Nonplanar Microstrip Antennas and Transmission Lines. New York: John Wiley & Sons, Inc.; 1999.

[2] Josefsson L, Persson P. Conformal Array Antenna Theory and Design. Hoboken: John Wiley & Sons, Inc.; 2006.

[3] Vasylchenko A, Schols Y, De Raedt W, Vandenbosch GAE. Quality Assessment of Computational Techniques and Software Tools for Planar-Antenna Analysis. IEEE Antennas Propagat. Magazine 2009;51(1) 23–38.

[4] de Paula CB, Ferreira DB, Bianchi I. Algorithm for the Design of Linearly Polarised Microstrip Antennas. In: MOMAG 2012, 5-8 Aug. 2012, Joao Pessoa, Brazil. (in Portuguese)

[5] Pues H, de Capelle AV. Accurate Transmission-Line Model for the Rectangular Microstrip Antenna. IEE Proc.-H 1984;131(6) 334–340.

[6] Bhattacharyya AK, Shafai L, Gary R. Microstrip Antenna – A Generalized Transmission Line. Progress in Electromagnetics Research 1991; 4 45-84.

[7] Kishk AA. Analysis of Spherical Annular Microstrip Antennas. IEEE Trans. Antennas Propagat. 1993;41(3) 338–343.

[8] Richards WF, Lo YT, Harrison DD. An Improved Theory for Microstrip Antennas and Applications. IEEE Trans. Antennas Propagat. 1981;29(1) 38–46.

[9] Rahmat-Samii Y, Michielssen E. Electromagnetic Optimization by Genetic Algorithms. New York: Wiley; 1999.

[10] Maciel DCM. Circularly Polarised Microstrip Arrays for Beam Steering. MSc thesis. ITA – Technological Institute of Aeronautics; 2005. (in Portuguese)

[11] Hansen RC. Phased Array Antennas. New York: John Wiley & Sons; 1998.

[12] Tseng CY, Griffiths LJ. A Simple Algorithm to Achieve Desired Patterns for Arbitrary Arrays. IEEE Trans. on Signal Processing 1992;40(11) 2737–2746.

[13] Mathematica. Wolfram Research. http://www.wolfram.com/products/mathematica/ (accessed 15 June 2012).

[14] Nascimento DC, Lacava JCS. Design of Low-Cost Probe-Fed Microstrip Antennas. In: Nasimuddin N. (ed) Microstrip Antennas. Rijeka: InTech; 2011. p1-26.

[15] ADS. Agilent Technologies. http://www.agilent.com/find/eesof-ads/ (accessed 15 June 2012).

[16] Collin RE. Foundations for Microwave Engineering. 2nd ed. New York: McGraw-Hill; 1992.

[17] Ashkenazy J, Shtrikman S, Treves D. Electric Surface Current Model for the Analysis of Microstrip Antennas on Cylindrical Bodies. IEEE Trans. Antennas Propagat. 1985;33(3) 295-300.

[18] da Silva CM, Lumini F, Lacava JCS, Richards FP. Analysis of Cylindrical Arrays of Microstrip Rectangular Patches. Electron. Lett. 1991;27(9) 778-780.

[19] Tam WY, Luk KM. Far Field Analysis of Spherical-Circular Microstrip Antennas by Electric Surface Current Models. IEE Proc. H Microw., Antennas Propagat. 1991;138(1) 98-102.

[20] Sipus Z, Burum N, Skokic S, Kildal PS. Analysis of Spherical Arrays of Microstrip Antennas Using Moment Method in Spectral Domain. IEE Proc. Microw., Antennas and Propagat. 2006;153(6) 533-543.

[21] Tseng CY. Minimum Variance Beamforming with Phase-Independent Derivative Constraints. IEEE Trans. Antennas Propagat. 1992;40(3) 285–294.

[22] Allard RJ, Werner DH, Werner PL. Radiation Pattern Synthesis for Arrays of Conformal Antennas Mounted on Arbitrarily-Shaped Three-Dimensional Platforms Using Genetic Algorithms. IEEE Trans. Antennas Propagat. 2003;51(5) 1054-1062.

[23] Ferreira DB, de Paula CB, Nascimento DC. Design of an Active Feed Network for Antenna Arrays. In: MOMAG 2012, 5-8 Aug. 2012, Joao Pessoa, Brazil. (in Portuguese)

Chapter 2

Full-Wave Spectral Analysis of Resonant Characteristics and Radiation Patterns of High Tc Superconducting Circular and Annular Ring Microstrip Antennas

Ouarda Barkat

Additional information is available at the end of the chapter

1. Introduction

During the last decades, superconducting antenna was one of the first microwave compo‐
nents to be demonstrated as an application of high-temperature superconducting material [1-3].
High Tc superconducting microstrip antennas (HTSMA) are becoming popular and getting
increased attention in both theoretical research and engineering applications due to their
excellent advantages. Various patch configurations implemented on different types of
substrates have been tested and investigated. In the design of microstrip antennas, anisotrop‐
ic substances have been increasingly popular. Especially the effects of uniaxial type anisotro‐
py have been investigated due to availability of this type of substances. These structures
characterized by their low profile, small size, light weight, low cost and ease of fabrication,
which makes them very suitable for microwave and millimeter-wave device applications [4-6].
HTSMA structures have shown significant superiority over corresponding devices fabricated
with normal conductors such as gold, silver, or copper. Major property of superconductor is
very low surface resistance; this property facilitates the development of microstrip antennas
with better performance than conventional antennas. Compared to other patch geometries, the
circular and annular ring microstrip patches printed on a uniaxial anisotropic substrate, have
been more extensively studied for a long time by a number of investigators [7-11]. Circular
microstrip patch can be used either as radiating antennas or as oscillators and filters in
microwave integrated circuits (MIC's). The inherent advantage of an annular ring antenna is:
first, the size of an annular ring for a given mode of operation is smaller than that of the circular
disc resonating at the same frequency. Second, The presence of edges at the inner and outer
radii causes more fringing than in the case of a circular microstrip antenna, in which fringing
occurs only at the outer edge, and this implies more radiation from these edges, that leads to

high radiation efficiency. These structures are quite a complicated structure to analyze mathematically. Different models are available to model a microstrip antenna as the transmission-line model and the cavity model in simple computer aided design formulas. However, the accuracy of these approximate models is limited, and only suitable for analysing simple regularly shaped antenna or thin substrates. The full-wave spectral domain technique is extensively used in microstrip antennas analysis and design. In this method, Galerkin's method, together with Parsval's relation in Hankel transform domain is then applied to compute the resonant frequency and bandwidth. The integral equation is formulated with Hankel transforms which gives rise to a diagonal form of the Green's function in spectral domain.

The numerical results for the resonant frequency, bandwidth and radiation pattern of microstrip antennas with respect to anisotropy ratio of the substrate, are presented. The Influence of a uniaxial substrate on the radiation of structure has been studied. To include the effect of the superconductivity of the microstrip patch in the full wave spectral analysis, the surface complex impedance has been considered. The effect of the temperature and thickness of HTS thin film on the resonant frequency and bandwidth have been presented. Computations show that, the radiation pattern of the antenna do not vary significantly with the permittivity variation perpendicular to the optical axis. Moreover, it is found to be strongly dependent with the permittivity variation along the optical axis. The computed data are found to be in good agreement with results obtained using other methods. Also, the TM and TE waves are naturally separated in the Green's function. The stationary phase method is used for computing the far-zone radiation patterns.

2. Theory

The antenna configurations of proposed structures are shown in Figure 1. The superconducting patches are assumed to be located on grounded dielectric slabs of infinite extent, and the ground planes are assumed to be perfect electric conductors. The substrates of thickness d are considered to be a uniaxial medium with permittivity tensor:

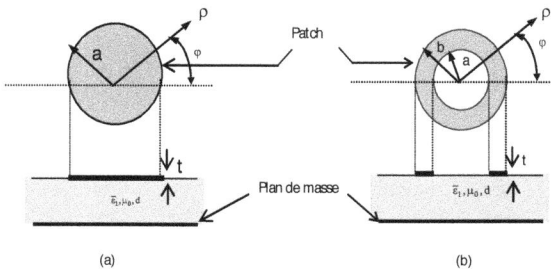

Figure 1. Cross section of a superconducting microstrip patch on uniaxial anisotropic media. (a) circulaire ; (b) annulaire ring

$$\bar{\varepsilon}_j = \varepsilon_0 \begin{vmatrix} \varepsilon_{jx} & 0 & 0 \\ 0 & \varepsilon_{jy} & 0 \\ 0 & 0 & \varepsilon_{jz} \end{vmatrix} \tag{1}$$

Where $\varepsilon_{jx} = \varepsilon_{jy} \neq \varepsilon_{jz}$ (j=1, 2), and the permeability will be taken as μ_0.

Starting from Maxwell's equations in the Hankel transform domain, we can show that the transverse fields inside the uniaxial anisotropic region can be written as [12]:

$$E_\rho(\rho,\phi,z_j) = \sum_{n=-\infty}^{\infty} e^{in\phi} \int_0^\infty dk_\rho k_\rho \ [\frac{\omega\mu_0}{2k_\rho}(\tilde{J}_{n-1}(k_\rho\rho)+\tilde{J}_{n+1}(k_\rho\rho)) \ \cdot \tilde{H}_{jz}\left(k_\rho,z_j\right) + \frac{i\varepsilon_{jz}}{2k_\rho\varepsilon_{jx}}(-\tilde{J}_{n+1}(k_\rho\rho)+\tilde{J}_{n-1}(k_\rho\rho))\frac{\partial \ \tilde{E}_{jz}\left(k_\rho,z_j\right)}{\partial z}] \tag{2}$$

$$E_\phi(\rho,\phi,z_j) = \sum_{n=-\infty}^{\infty} e^{in\phi} \int_0^\infty dk_\rho k_\rho \ [i\frac{\omega\mu_0}{2 \ k_\rho}(\tilde{J}_{n-1}(k_\rho\rho)-\tilde{J}_{n+1}(k_\rho\rho)) \cdot \tilde{H}_{jz}\left(k_\rho,z_j\right) - \frac{\varepsilon_{jz}}{2\varepsilon_{jx}k_\rho}(\tilde{J}_{n+1}(k_\rho\rho)+\tilde{J}_{n-1}(k_\rho\rho))\frac{\partial \ \tilde{E}_{jz}\left(k_\rho,z_j\right)}{\partial \ z}] \tag{3}$$

$$H_\phi(\rho,\phi,z_j) = \sum_{n=-\infty}^{\infty} e^{in\phi} \int_0^\infty dk_\rho k_\rho \ [(\frac{\omega \ \varepsilon_{jz}\varepsilon_0}{2k_\rho}\left(\tilde{J}_{n-1}(k_\rho\rho)-\tilde{J}_{n+1}(k_\rho\rho)\right)\cdot\tilde{E}_{jz}\left(k_\rho,z_j\right) \ + \frac{1}{2k_\rho}(\tilde{J}_{n-1}(k_\rho\rho)+\tilde{J}_{n+1}(k_\rho\rho))\cdot\frac{\partial \ \tilde{H}_{jz}\left(k_\rho,z_j\right)}{\partial \ z}] \tag{4}$$

$$H_\rho(\rho,\phi,z_j) = \sum_{n=-\infty}^{\infty} e^{in\phi} \int_0^\infty dk_\rho k_\rho \ [\frac{\omega\varepsilon_{jz}\varepsilon_0}{2k_\rho}(\tilde{J}_{n-1}(k_\rho\rho)+\tilde{J}_{n+1}(k_\rho\rho))\cdot\tilde{E}_{jz}\left(k_\rho,z_j\right) \ + \frac{i}{2k_\rho}(\tilde{J}_{n+1}(k_\rho\rho)-\tilde{J}_{n-1}(k_\rho\rho))\frac{\partial \ \tilde{H}_{jz}\left(k_\rho,z_j\right)}{\partial \ z}] \tag{5}$$

We can put these equations in the following form:

$$E(\rho,\phi,z_j) = \begin{bmatrix} E_\rho(\rho,\phi,z_j) \\ E_\phi(\rho,\phi,z_j) \end{bmatrix} = \sum_{n=-\infty}^{+\infty} e^{in\phi} \int_0^{+\infty} dk_\rho \cdot k_\rho.\bar{H}_n(k_\rho\rho). \begin{bmatrix} (i\varepsilon_{jz}/\varepsilon_{jx}).\partial \ \tilde{E}_{jz}(k_\rho,z_j)/\partial \ z \\ \omega \ \mu_\rho\tilde{H}_{jz}\left(k \ ,z_j\right) \end{bmatrix} \tag{6}$$

$$H(\rho,\phi,z_j) = \begin{bmatrix} H_\phi(\rho,\phi,z_j) \\ -H_\rho(\rho,\phi,z_j) \end{bmatrix} = \sum_{n=-\infty}^{+\infty} e^{in\phi} \int_0^{+\infty} dk_\rho.k_\rho.\bar{H}_n(k_\rho\rho). \begin{bmatrix} \omega \ \varepsilon_{jz}\varepsilon_0 \ \tilde{E}_{jz}(k \ ,z_j) \\ i\partial \ \tilde{H}_{jz}(k_\rho,z_j)/\partial \ z \end{bmatrix} \tag{7}$$

That is

$$H(\rho,\phi,z_j) = \sum_{n=-\infty}^{+\infty} e^{in\phi} \int_0^{+\infty} dk_\rho.k_\rho.\bar{H}_n(k_\rho\rho).\tilde{e}_n(k_\rho,z) \tag{8}$$

$$E(\rho,\phi,z_j) = \sum_{n=-\infty}^{+\infty} e^{in\phi} \int_0^{+\infty} dk_\rho.k_\rho.\bar{H}_n(k_\rho\rho).\tilde{h}_n(k_\rho,z) \tag{9}$$

The kernel of the vector Hankel transform is given by [12]:

$$\overline{H}_n(k_\rho\rho) = \begin{vmatrix} \dot{J}_n(k_\rho\rho) & -in\,J_n(k_\rho\rho)/k_\rho \\ in\,J_n(k_\rho\rho)/k_\rho\rho & \dot{J}_n(k_\rho\rho) \end{vmatrix}$$

(10)

And

$$\tilde{e}_n(k_\rho,z) = \overline{A}_{nj}\left(k_\rho\right)e^{-i\,\overline{k}_{jz}z} + \overline{B}_{nj}\left(k_\rho\right)e^{i\,\overline{k}_{jz}z}$$

(11)

$$\tilde{h}_n(k_\rho,z) = \overline{g}_j(k_\rho)\,.\{\overline{A}_{nj}\left(k_\rho\right)e^{-i\,\overline{k}_{jz}z} - \overline{B}_{nj}\left(k_\rho\right)e^{i\,\overline{k}_{jz}z}$$

(12)

$\overline{A}$ and $\overline{B}$ are two unknown vectors and $\overline{g}(k_\rho)$ is determined by:

$$\overline{g}_j(k_\rho) = \begin{vmatrix} \omega\varepsilon_{jx}\varepsilon_0 / k_{jz}^h & 0 \\ 0 & k_{jz}^e / \omega\mu_0 \end{vmatrix}$$

(13)

Where

k_0 : is the free space wavenumber,

$k_{jz}^e = (\varepsilon_{jx}k_0^2 - (\varepsilon_{jx}k_\rho^2/\varepsilon_{jz}))^{1/2}$: is TM propagation constants in the uniaxial substrate.

$k_{jz}^h = (\varepsilon_{jx}k_0^2 - k_\rho^2)^{\frac{1}{2}}$: is TE propagation constants in the uniaxial substrate.

In the spectral domain, the relationship between the patch current and the electric field on the microstrip is given by [10, 11]:

$$\tilde{E}\left(k_\rho\right) = \overline{G}\left(k_\rho\right).\tilde{K}\left(k_\rho\right)$$

(14)

$\widetilde{K}(k_\rho)$ is the current on the microstrip which related to the vector Hankel transform of $K(\varrho)$. The unknown currents are expanded, in terms of a complete orthogonal set of basis functions, issued from the magnetic wall cavity model. It is possible to find a complete set of vector basis functions to approximate the current distribution, by noting that the superposition of the currents due to TM and TE modes of a magnetic-wall cavity form a complete set. The current distribution of the nth mode of microstrip patch can be written as [11, 12]:

$$K_n\left(\rho\right) = \sum_{p=1}^{P} a_{np}\Psi_{np}\left(\rho\right) + \sum_{q=1}^{Q} b_{nq}\Phi_{nq}\left(\rho\right)$$

(15)

Here a_{np} and b_{nq} are unknown coefficients.

For superconducting annular ring patch, $(\alpha_{nq}, \beta_{np})$ are the roots of dual equations $\varphi_n(\alpha_{nq}b/a)=0$ and $\psi'_n(\beta_{np}b/a)=0$.

The Hankel transforms of ψ_{np} and φ_{nq} functions are described as [12]:

$$\tilde{\Psi}_{np}(k_\rho) = \begin{cases} \left[\begin{array}{c} \dfrac{\beta_{np}/a)}{((\beta_{np}/a)^2 - k^2_\rho)} Y'_{np}(k_\rho) \\ \dfrac{na}{\beta_{np}k_\rho} Y_{np}(k_\rho) \\ 0 \end{array} \right] \rho \quad & a < \quad < b \\ & \rho \quad \rho \quad > b, \quad < a \end{cases} \tag{16}$$

$$\tilde{\Phi}_{nq}(k_\rho) = \begin{cases} \left[\begin{array}{c} 0 \\ \dfrac{k_\rho a}{(k^2_\rho - (\alpha_{nq}/a)^2)} Z_{nq}(k_\rho) \end{array} \right] \rho \quad & a < \quad < b \\ 0 & \rho > b, \quad \rho < a \end{cases} \tag{17}$$

Where

$$Y_{pp}(k) = \psi_n(\beta_{np}b/a) J_n(k\ b) - \psi_n(\beta_{np}) J_n(k\ a) \tag{18}$$

$$\tilde{\Phi}_{pq}(k) = (b/a)\ '_n(\ b\not{a}) J_n(k\ b) - \ '_n(\ _{nq}) J_n(k\ a) \tag{19}$$

Using the same procedure, the basis functions for superconducting circular patch, are given by the expressions [11]:

$$\tilde{\Psi}_{np}(k_\rho) = \beta_{np}a J_n(\beta_{np}a)\left[\dfrac{j_n(k_\rho a)}{\beta^2_{np} - k^2_\rho} \quad \dfrac{in}{k_\rho\beta^2_{np}a} J_n(k_\rho a) \right]^T \tag{20}$$

$$\tilde{\varphi}_{nq}(k_\rho) = \left[0 \quad \dfrac{k_\rho a j_n(\alpha_{nq}a) J_n(k_\rho a)}{k^2_\rho - \alpha^2_{nq}} \right]^T \tag{21}$$

Where

$\dot{J}_n(\beta_{np}a)=0$ and $J_n(\alpha_{nq}a)=0$.

$J_n(.)$ and $N_n(.)$ are the Bessel functions of the first, and second kind of order n.

$\overline{G}(k_\rho)$ is the spectral dyadic Green's function, and after some simple algebraic manipulation, we determine the closed form of the spectral Green dyadic at z=d for a grounded uniaxial substrate.

$$\overline{G} = \begin{bmatrix} G^{TM} & 0 \\ 0 & G^{TE} \end{bmatrix}$$

(22)

Where

$$G^{TM} = \frac{k_0}{i\omega\,\varepsilon_0\,\,\varepsilon_x.k_0}.\frac{k_z^e.\sin(k_z^e d)}{\cos(k_z^e d) + ik_z^e.\sin(k_z^e d)}$$

(23)

$$G^{TE} = \frac{k_0^2}{i\omega\,\varepsilon_0}.\frac{\sin(k_z^h d)}{k_z^h\cos(k_z^h d) + ik_0\sin(k_z^h d)}$$

(24)

In order to incorporate the finite thickness, the dyadic Green's function is modified by considering a complex boundary condition. The surface impedance of a high-temperature superconductors (HTSs) material for a plane electromagnetic wave incident normally to its surface is defined as the ratio of $|E|$ to $|H|$ on the surface of the sample [13]. It is described by the equation:

$$Z_s = R_s + i\,X_s$$

(25)

Where R_s and X_s are the surface resistance and the surface reactance.

If the thickness t of the strip of finite conductivity σ is greater than three or four penetration depths, the surface impedance is adequately represented by the real part of the wave impedance [13].

$$Z_s = \sqrt{\omega\mu_0/(2.\sigma)}$$

(26)

If t is less than three penetration depths, a better boundary condition is given by [13]:

$$Z_s = 1/t\sigma$$

(27)

Where the conductivity $\sigma = \sigma_c$ is real for conventional conductors. These approximations have been verified for practical metallization thicknesses by comparison with rigorous mode matching result.

For superconductors, a complex conductivity of the form $\sigma = \sigma_n (T/T_c)^4 - i(1 - (T/T_c)^4)/\omega\mu_0\lambda_0^2$, where σ_n is often associated with the normal state conductivity at T_c and λ_0 is the effective field penetration depth.

Now, we have the necessary Green's function, it is relatively straightforward to formulate the moment method solution for the antenna characteristics. The boundary condition at the surface of the microstrip patch is given by:

$$\bar{E}_{scat} + \bar{E}_{inc} - \bar{Z}_s \cdot \bar{K}_n = 0 \tag{28}$$

Here $\bar{E}_{inc}$ and $\bar{E}_{scat}$ are tangential components of incident and scattered electrics fields. Electric field is enforced to satisfy the impedance boundary condition on the microstrip patch, and the current vanishes off the microstrip patch, to give the following set of vector dual integral equations [3]. For superconducting annular ring microstrip antenna, we have:

$$e_n(\rho) = \int_0^\infty dk_\rho k_\rho \bar{H}_n(k_\rho\rho) \cdot (\bar{G}(k_\rho) - \bar{Z}_S)\bar{K}_n(k_\rho) = 0 \qquad a < \rho < b \tag{29}$$

$$K_n(\rho) = \int_0^\infty dk_\rho k_\rho \bar{H}_n(k_\rho\rho) \cdot \bar{K}_n(k_\rho) = 0 \qquad \rho > b, \ \rho < a \tag{30}$$

And for superconducting circular microstrip antenna, we have:

$$e_n(\rho) = \int_0^\infty dk_\rho k_\rho \bar{H}_n(k_\rho\rho) \cdot (\bar{G}(k_\rho) - \bar{Z}_S)\bar{K}_n(k_\rho) = 0 \qquad \rho < a \tag{31}$$

$$K_n(\rho) = \int_0^\infty dk_\rho k_\rho \bar{H}_n(k_\rho\rho) \cdot \bar{K}_n(k_\rho) = 0 \qquad \rho > a \tag{32}$$

Where

$$\bar{Z}_s = \begin{bmatrix} Z_s & 0 \\ 0 & Z_s \end{bmatrix}$$

Galerkin's method is employed to solve the coupled vector integral equations of (29)-(32). Substituting the Hankel transform current expansion of (15) into (29) and (32). Next, multiplying the resulting equation by $\rho\psi_{np}^+(\rho)$ (p=1, 2,....... P) and $\rho\varphi_{nq}^+(\rho)$ (q=1, 2,....... Q) and using Parseval's theorem for VHT, we obtain a system of Q+ P linear algebraic equations for each mode n which may be written in matrix form. Following well known procedures, we obtain the following system of linear algebraic equations:

$$
\begin{bmatrix}
\left(\overline{Z}^{,\Psi\Psi}\right)_{P\times P} & \left(\overline{Z}^{,\Psi\Phi}\right)_{P\times Q} \\
\left(\overline{Z}^{,\Phi\Psi}\right)_{Q\times P} & \left(\overline{Z}^{,\Phi\Phi}\right)_{Q\times Q}
\end{bmatrix}
\cdot
\begin{bmatrix}
(A')_{P\times 1} \\
(B')_{Q\times 1}
\end{bmatrix}
= 0
\tag{33}
$$

Each element of the submatrices $\overline{Z}^{CD}$ is given by:

$$
\overline{Z}_{ij}^{,CD} = \int_0^\infty dk_\rho k_\rho C_{ni}^+\left(k_\rho\right)\cdot\left(\overline{G}\left(k_\rho\right) - \overline{Z}_S\right)\cdot D_{nj}\left(k_\rho\right)
\tag{34}
$$

Where C and D represent either ψ or φ, for every value of the integer n.

The integration path for the integrals of (32) is, in general, located in the first quadrant of the complex plane k_ρ. This integration path must remain above the pole and the branch point of $\overline{G}$. Although other choices of branch cut are possible, the choice made in this paper is very convenient, this treated in section 3. Once the impedance and the resistance matrices have been calculated, the resulting system of equations is then solved, for the unknown current modes on the microstrip patch. Nontrivial solutions can exist, if the determinant of Eq. (35) vanishes, that is:

$$
\det[\overline{Z}'(f)] = 0
\tag{35}
$$

In general, the roots of this equation are complex numbers indicating, that the structure has complex resonant frequencies $(f = f_r + i f_i)$. The bandwidth of a structure operating around its resonant frequency, can be approximately related to its resonant frequency, though the well-known formula $(BW = 2f_i / f_r)$. Once the problem is solved for the resonant frequency, far field radiation, co-polar and cross- polar fields in spherical coordinates are given from.

$$
\begin{bmatrix}
E_\theta(\overline{r}) \\
E_\varphi(\overline{r})
\end{bmatrix}
= \sum_{n=-\infty}^{+\infty} e^{in\varphi}\cdot(-i)^n\cdot e^{ikr}\cdot\overline{T}(\theta)\cdot\overline{V}(k_\rho)\cdot\left(\overline{G}\left(k_\rho\right) - \overline{Z}_S\right)\cdot\overline{K}_n(k_\rho)
\tag{36}
$$

Where

$$
\overline{V}(k_\rho) =
\begin{bmatrix}
1 & 0 & -\dfrac{k_\rho}{k_z} \\
0 & 1 & 0
\end{bmatrix}
\text{ and } (\theta) =
\begin{bmatrix}
\cos\theta & 0 & -\sin\theta \\
0 & 1 & 0
\end{bmatrix}
$$

The losses in the antenna comprise dielectric loss P_d, the conductor loss P_c, and the radiation loss P_r, are given by [4-6]:

$$
P_r = (1/4\eta)\iint \left| E_\theta E_\theta^* + E_\varphi E_\varphi^* \right| r^2 \sin\theta \, d\theta \, d\varphi
\tag{37}
$$

$$P_c = Z_s \iint \left| H_\varphi^2 + H_\rho^2 \right| dS \qquad (38)$$

$$P_d = (\omega \varepsilon tg\delta / 2) \iint \left| E_z \right|^2 dV \qquad (39)$$

The efficiency of an antenna can be expressed by:

$$efficiency = P_r / (P_r + P_c + P_d) \qquad (40)$$

3. Convergence and comparison of numerical results

Computer programs have been written to evaluate the elements of the impedance, resistance matrices, and then solve the matrix equation (35). To enhance the accuracy of the numerical calculation, the integrals of the matrix elements (33) are evaluated numerically along a straight path above the real axis with a height of about $1k_0$ (Figure 2). In this case, the effects of the surface waves are included in the calculation and knowledge of the pole locations is not required, while the length of the integration path is decided upon by the convergence of the numerical results. The time required to compute the integral depends on the length of the integration path. It is found that length of the integration path required reaching numerical convergence at $100k_0$, also Muller's method that involves three initial guesses, is used for root seeking of (35).

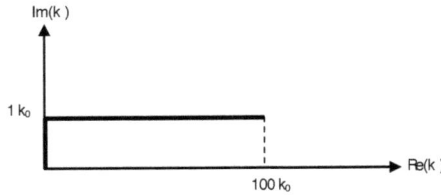

Figure 2. Integration path used for computing the integrals in the complex k_ρ plane

To check the correctness of our computer programs, our results are compared with results of other authors. The comparisons are shown in Table 1 for imperfectly conducting microstrip annular ring antennas. The resonant wave number times the inner radius of the ring is $k_r a$ ($k_r a = 2\pi f_r a \sqrt{\varepsilon_x \varepsilon_0 \mu_0}$), as functions of different sizes of the ratio of the substrate thickness d normalized by the inner radius a is fixed of (0.71cm), and an outer radius of b=2a, the relative permittivity was $\varepsilon_{1z} = \varepsilon_{1x} = 2.65$. Annular ring microstrip antenna is excited in the TM_{11} and TM_{12} modes. We found that, for the TM_{11} mode, the real part of $(k_r a)$ increases as d/a increases. At the same time, for this mode, the imaginary part of $(k_r a)$, which includes the losses by radiation of the structure, is approximately zero. This means that, the TM_{11} mode has narrow bandwidth and weak-radiation. In addition, we observe that, for the TM_{12} mode, the real part of $(k_r a)$ decreases as (d/a) increases, the TM_{12} has relatively wide bandwidth and high-radiation. Therefore, the TM_{11} mode is good for resonator applications and the TM_{12} mode for antennas. It is for that one does the applications of the annular ring microstrip antenna in the TM_{12} mode, better than in the TM_{11} mode. Thus, for this reason, the considered mode in this work is the TM_{12} mode. It is clear from Table 1 that our results agree very well with results obtained by other authors [14, 12].

| d/a | Mode TM_{12} | | | | | | Mode TM_{11} | | | |
| | Results of [14] | | Results of [12] | | Our results | | Results of [14] | | Our results | |
	Re (k,a)	Im (k,a)	Re (k,a)	Im (k,a)	Re (k,a)	Im (k,a)	Re (k,a)	Im (k,a)	Re (k,a)	Im (k,a)
0.005	3.26	0.002	3.24	0.002	3.257	0.002	0.67	$1,6.10^{-4}$	0.676	$1,6.\,10^{-4}$
0.01	3.24	0.003	3.23	0.002	3.248	0.003	0.68	$1,7.\,10^{-4}$	0.682	$1,8.\,10^{-4}$
0.05	3.13	0.008	3.10	0.006	3.085	0.006	0.70	$5,4.\,10^{-4}$	0.695	$5,5.\,10^{-4}$
0.1	3.01	0.014	2.96	0.0103	2.968	0.016	0.71	0.0012	0.705	0.0012

Table 1. Calculated (K,a) of annular ring microstrip antennas.

In table 2, we have calculated the resonant frequencies for the modes (TM_{11}, TM_{21}, TM_{31}, and TM_{12}) for perfect conducting circular patch with a radius 7.9375mm, is printed on a substrate of thickness 1.5875mm. These values are compared with theoretical and experimental data, which have been suggested in [10]. Note that the agreement between our computed results, and the theoretical results of [10], is very good.

In our results, we need to consider only the P functions of ψ_{np}, and the Q functions of φ_{nq}. The required basis functions for reaching convergent solutions of complex resonant frequencies, using cavity model basis functions are obtained with (P=5, Q=0).

| Mode | Results of [10] | | Our results | |
	Resonant Frequency (GHz)	Quality factors (Q)	Resonant Frequency (GHz)	Quality factors (Q)
TM_{11}	6.1703	19.105	6.2101	19.001
TM_{12}	17.056	10.324	17.120	10.303
TM_{21}	10.401	19.504	10.438	19.366
TM_{01}	12.275	8.9864	12.296	8.993

Table 2. Comparison of resonant frequencies of the first four modes of a perfect conducting circular patch printed on a dielectric substrate (a=7.9375mm, $\varepsilon_x = \varepsilon_z = 2.65$, d=1.5875 mm).

4. Resonant frequency of superconducting patch antenna

Shown in Figures 3-4, is the dependence of resonant frequency on the thickness t of super-conducting patch of the antennas. It is observed that, when the film thickness (t) increases, the resonant frequency increases quickly until, the thickness t reaches the value penetration depth (λ_0). After this value, the increase in the frequency of resonance becomes less significant.

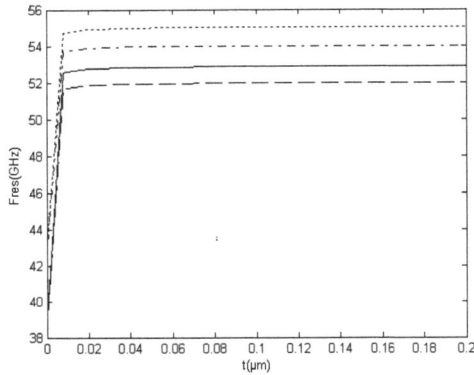

Figure 3. Real part of resonant Frequency against thickness of superconducting annular ring patch antenna. (d=254µm, a=815µm, b=2a, T/ T_c=0.5, λ_0=1500Å, σ_n=210S/mm). (——) ε_x= 9.4, ε_z= 11.6;(– – –) ε_x= 11.6, ε_z= 11.6; (—•—•) ε_x= 13, ε_z=10.3; (......) ε_x=10.3, ε_z= 10.3.

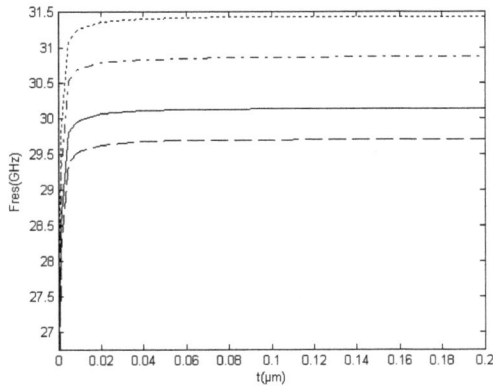

Figure 4. Real part of resonant Frequency against thickness of superconducting circular patch antenna. (d=254µm, a=815µm, T_c =89K, T/ T_c=0.5, λ_0=1500Å, σ_n=210S/mm).(——) ε_x= 9.4, ε_z= 11.6;(– – –) ε_x= 11.6, ε_z= 11.6; (—•—•) ε_x= 13, ε_z=10.3; (......) ε_x=10.3, ε_z= 10.3.

Figures 5-6 demonstrated relations between the real part of frequency resonance, and the normalized temperature (T/Tc), where the critical temperature used here for our data (89K). The variations of the real part of frequency due to the uniaxial anisotropy decrease gradually with the increase in the temperature. This reduction becomes more significant for the values of temperature close to the critical temperature. These behaviours agree very well with those reported by Mr. A. Richard for the case of rectangular microstrip antennas [2].

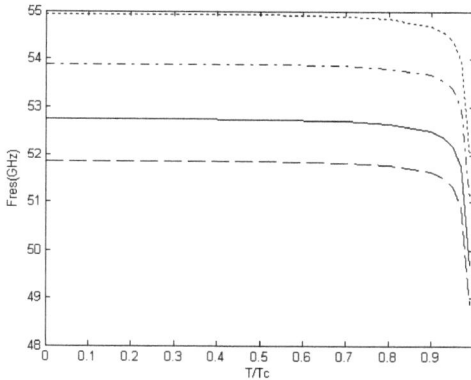

Figure 5. Real part of resonant frequency of superconducting annular ring patch antenna against T/T_c (d=254µm, a=815µm, b=2a, t=0.02µm, λ_0=1500Å, σ_n=210S/mm). (——) ε_x= 9.4, ε_z= 11.6;(– – –) ε_x= 11.6, ε_z= 11.6; (—•—•) ε_x= 13, ε_z=10.3; (......) ε_x=10.3, ε_z= 10.3.

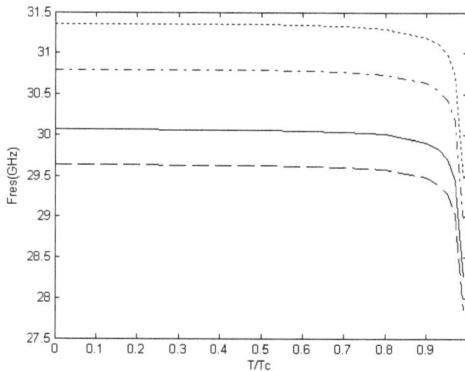

Figure 6. Real part of resonant frequency of superconducting circular patch antenna against T/T_c (d=254µm, a=815µm, T_c =89K, h= 0.02µm, λ_0=1500Å, σ_n=210S/mm). (——) ε_x= 9.4, ε_z= 11.6;(– – –) ε_x= 11.6, ε_z= 11.6; (—•—•) ε_x= 13, ε_z=10.3; (......) ε_x=10.3, ε_z= 10.3.

5. Radiations patterns and efficiency of superconducting patch antenna

The calculated radiations patterns (electric field components, $E\theta$; E_φ), of the microstrip antenna on a finite ground plane in the E plane, and in the H plane are plotted in Figs. 7 -10, printed on
an uniaxial anisotropy substrate thickness ($d=254\mu m$). The mode excited for superconducting
annular ring patch antenna is the TM_{12} and for superconducting circular patch antenna is the
TM_{11}. It is seen that the permittivity ε_z has a stronger effect on the radiation than the permittivity ε_x. The radiation pattern of an antenna becomes more directional as its ε_z increases. Another
useful parameter describing the performance of an antenna is the gain. Although the gain of the
antenna is related to the directivity, the gain of an antenna becomes high as its ε_z increases.

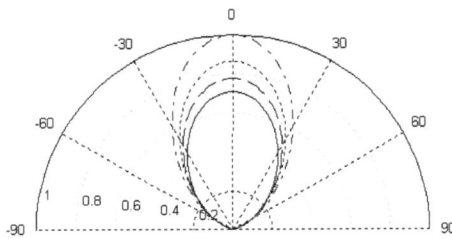

Figure 7. Radiation pattern versus angle θ of superconducting annular ring patch antenna at $\varphi=0°$ plane ($a=815\mu m$, $b=2a$, $t=0.02\mu m$, $d=254\mu m$, $\lambda0=1500\text{Å}$, $\sigma_n=210\text{S/mm}$). (——)$\varepsilon_x= 9.4$, $\varepsilon_z= 11.6$; ($---$)$\varepsilon_x= 11.6$, $\varepsilon_z= 11.6$;(— • — •) $\varepsilon_x= 13$, $\varepsilon_z=10.3$;(......) $\varepsilon_x=10.3$, $\varepsilon_z= 10.3$.

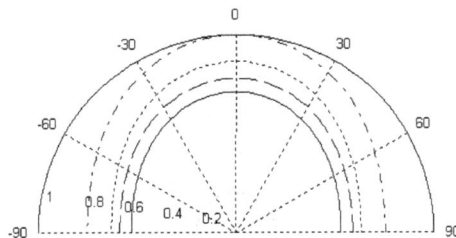

Figure 8. Radiation pattern versus angle θ of superconducting annular ring patch antenna at $\varphi=90°$ plane ($a=815\mu m$, $b=2a$, $t=0.02\mu m$, $d=254\mu m$, $\lambda0=1500\text{Å}$, $\sigma_n=210\text{S/mm}$).(——)$\varepsilon_x= 9.4$, $\varepsilon_z= 11.6$;($---$)$\varepsilon_x= 11.6$, $\varepsilon_z= 11.6$; (— • — •) $\varepsilon_x= 13$, $\varepsilon_z=10.3$; (......) $\varepsilon_x=10.3$, $\varepsilon_z= 10.3$.

In calculation of losses, we have found that, the values of dielectric loss (P_d), the conductor loss
(P_c), and the radiation loss (P_r) will depend on frequency. We use results precedents to calculate
the variation of radiation efficiency as a function of resonant frequency, for various isotropic
dielectric substrates. Our results are shown in Figs. 11 and 12. It is seen that the efficiency
increases with decreasing frequencies. The same behaviour is found by R. C. Hansen [1].

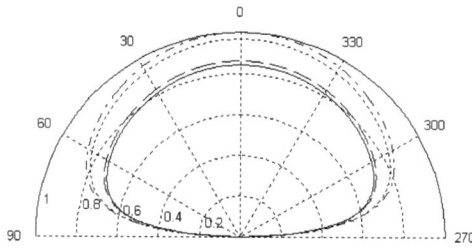

Figure 9. Radiation pattern versus angle θ of superconducting circular patch antenna at $\varphi=0°$ plane (a=815µm, t=0.02µm, T/ T_C=0.5, d=254µm, λ_0=1500Å, $_n$=210S/mm). (——) ε_x= 9.4, ε_z= 11.6; (– – –) ε_x= 11.6, ε_z= 11.6; (— • — •) ε_x= 13, ε_z=10.3 ; (......) ε_x=10.3, ε_z= 10.3.

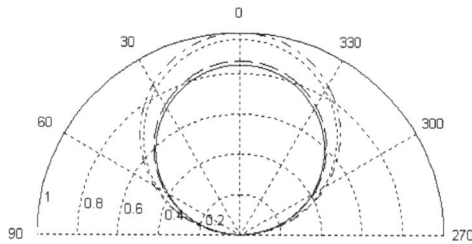

Figure 10. Radiation pattern versus angle θ of superconducting circular patch antenna at $\varphi=\pi/2$ plane (a=815µm, h=0.02µm, T/ T_C=0.5, d=254µm, λ_0=1500Å, $_n$=210S/mm). (——) ε_x= 9.4, ε_z= 11.6; (– – –) ε_x= 11.6, ε_z= 11.6; (— • — •) ε_x= 13, ε_z=10.3 ; (......) ε_x=10.3, ε_z= 10.3.

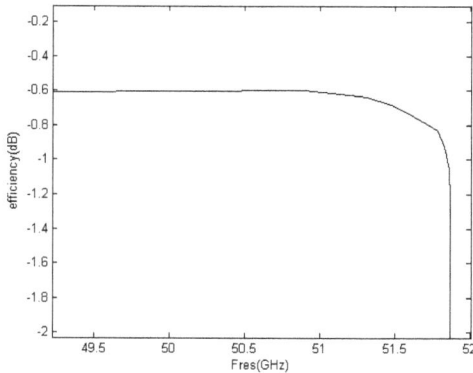

Figure 11. Superconducting annular ring patch antenna efficiency for the mode TM_{12} (d=254µm, a=815µm, b=2a, t=0.02µm, λ_0=1500Å, σ_n=210S/mm, ε_x= ε_z=11.6, δ=0.0004).

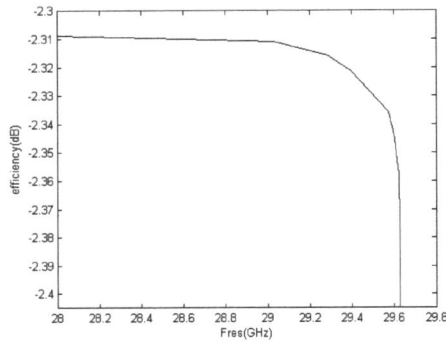

Figure 12. Superconducting circular patch antenna efficiency for the mode TM_{11} ((d=254μm, a=815μm,h=0.02μm, T/T_C=0.5, λ_0=1500Å, σ_n=210S/mm, ε_x= ε_z=11.6, δ=0.0024).

6. Conclusion

This work presents a fullwave analysis for the superconducting microstrip antenna on uniaxial anisotropic media. The complex resonant frequency problem of structure is formulated in terms of an integral equation. Galerkin procedure is used in the resolution of the electric field integral equation, also the TM, TE waves are naturally separated in the Green's function. In order to introduce the effect of a superconductor microstrip patch, the surface complex impedance has been considered. Results show that the superconductor patch thickness and the temperature have significant effect on the resonant frequency of the antenna. The effects of a uniaxial substrate on the resonant frequency and radiation pattern of structures are considered in detail. It was found that the use of such substrates significantly affects the characterization of the microstrip antennas, and the permittivity (ε_z) along the optical axis has a stronger effect on the radiation of antenna. Thus, microstrip superconducting could give high efficiency with high gain in millimeter wavelengths. A comparative study between our results and those available in the literature shows a very good agreement.

Author details

Ouarda Barkat*

Electronics Department, University of Constantine, Constantine, Algeria

References

[1] Hansen, C R. Electrically Small, Superdirective, and Superconducting Antennas. John Wiley& Sons, Inc, Hoboken, New Jersey, (2006).

[2] Richard, M A. Bhasin B K & Clapsy C P. Superconducting microstrip antennas: an experimental comparison of two feeding methods. *IEEE Transactions on Antennas Propagation, AP*-(1993). , 41, 967-974.

[3] Barkat, O, & Benghalia, A. Radiation and resonant frequency of superconducting annular ring microstrip antenna on uniaxial anisotropic media. *Springer, Journal of Infrared, Millimeter, and Terahertz Waves*, (2009). , 30(10)

[4] Garg, R, & Bhartia, P. Bahl, & Ittipiboon A I. *Microstrip Antenna Design Handbook.* Artech House, Boston, London, (2001).

[5] Lee F H & Chen W*Advances in microstrip and printed antennas.* New York, John Wiley & Sons, (1997).

[6] James, R J, & Hall, P S. *Handbook of Microstrip Antennas.* Peter Peregrinus Ltd, London, (1989).

[7] Yang, M G, Xing, X, Daigle, A, Obi, O, Liu, M, Lou, J, Stoute, S, Naishadham, K, & Sun, N X. Planar Annular Ring Antennas With Multilayer Self-Biased NiCo-Ferrite Films Loading. *IEEE Transactions on antennas and propagation.*, (2010). , 58(3)

[8] Richard, M A, Bhasin, K B, & Claspy, P C. Performance of a K-band superconducting annular ring antenna. *microwave and optical technology letters.*, (1992). , 5(6)

[9] Gürel, S Ç, & Yazgan, E. Modified cavity model to determine resonant frequency of tunable microstrip ring antennas. *Taylor and Francis, Electromagnetic,191999*, 443-455.

[10] Losada, V. BOIX R R & Horno M. Resonant modes of circular microstrip patches in multilayered substrates, IEEE Transactions on Microwave Theory and techniques (1999). , 47, 488-497.

[11] Chew W C & Kong J AResonance of non axial symmetric modes in circular micro strip disk. *J. Math. Phys.* (1980). , 21(10)

[12] [12]Ali, S M, Chew, W C, & Kong, J A. Vector Hankel transform analysis of annular ring microstrip antenna. *IEEE Transactions on Antennas and Propagation*, July (1982). , AP-30(4)

[13] Cai Z & Bornemann JGeneralized Spectral Domain Analysis for Multilayered Complex Media and High Tc Superconductor Application. *IEEE Transactions on microwave Theory and Techniques*, (1992). , 40(12)

[14] Chew, W C. A broad band annular ring microstrip antenna. *IEEE Transactions on Antennas and Propagation.* (1982). , AP-30

Chapter 3

Bandwidth Optimization of Aperture-Coupled Stacked Patch Antenna

Marek Bugaj and Marian Wnuk

Additional information is available at the end of the chapter

1. Introduction

The microstrip antennas have been one of the most innovative fields of antenna techniques for the last three decades. Microstrip antennas have several advantages such as lightweight, small-volume, and that they can be made conformal to the host surface. In addition, these antennas are manufactured using printed-circuit technology, so that mass production can be achieved at a low cost. Microstrip antennas, which are used for defense and commercial applications, are replacing many conventional antennas [1]. However, the types of applications of microstrip antennas are restricted by the narrow bandwidth (BW). Accordingly, increasing the BW of the microstrip antennas has been a primary goal of research for many years. This is reflected in the large number of papers on the subject published in journals and conference proceedings. In fact, several broadband microstrip antennas configurations have been reported in the last few decades. They have additional advantages: simplicity of production, small weight, narrow section, easiness of integration of radiators with feeding system. However, this construction also has the following disadvantages: narrow band, limited power capacity, not sufficient efficiency of radiation. Still ongoing search for solutions to obtain a broad range of work in the microstrip antenna. The aim of this study is to demonstrate that it is possible to build a microstrip antenna with wide range of work.

The basic configuration of microstrip antenna consists of metallic strip printed on thin earthed dielectric base. The feeding is accomplished through concentric cable, it runs perpendicularly through a substrate or a strip line runs on a substrate in the plane of aerial[15]. The aperture coupling feed was proposed by Pozar and it has many advantages over other types of feeds. These include shielding of the antenna from spurious feed radiating, the use of suitable substrates for feed structure and the antenna, and the use of thick substrates for increasing the antenna bandwidth [5].

The methods of analysis and projection of microstrip antennas have developed simultaneously with the development of aerials. Nowadays several methods of analyzing the antennas on dielectric surface are used, however, the most commonly used ones are the full wave model based on Green's function and the method of moments where analysis relies on solution of integral equation, concerning electric field, with regard to unknown currents flowing through elements of the antenna and its feeding system [7].

Millimeter wave printed antennas can take on many forms, including microstrip patch elements and a variety of proximity coupled printed radiators. The microstripline-fed printed slot and the aperture coupled patch are examples of the latter type and may be useful in certain planar array applications [2].

The paper presents the model of the antenna on which the influence of parameters on antenna bandwidth simulation was conducted (the influence of changes permittivity and thickness of every layer). One of the most important parameters which have been calculated is the bandwidth. Its value depends on antenna parameters (thickness and permittivity of every layer). The paper shows that as a result of optimization which has been demonstrated we can create a planar antenna with wide range of work. The analysis process of multilayer microstrip antennas is complex and time consuming[4,11].

One of the main problems associated with the use of planar antennas in radio links is their relatively narrow frequency band of operation, it has been indicated at the outset of this chapter. This problem has been described in literature, in books such as monographs [6] or [9, ch. 3 and 6] and in paper such as [8]. The next literature is worth noting the work of Borowiec and Słobodzian, who described [10] a method of increasing bandwidth of planar antennas. Unfortunately, the bandwidth of the antenna work, which was obtained only at 15-18 %. Increasing the frequency band planar antenna work can be done through the use of:

- Tuning frequency planar radiators,

- Dual frequency transmitters,

- Multilayer structures.

2. Analyzed antenna

Aperture coupling antenna feeding method was first suggested in 1985 by David.M. Pozar [9]. Using this method of feeding antenna we have hope to obtain a wide operating bandwidth B, which has tried to prove, by various authors [9]. The basic condition for obtaining satisfactory results is to optimize the parameters of multilayer antennas.

An exploded view of this type antenna is shown in Figure 1. The antenna consists of four layers with two radiators. Radiators are microstrip patches; they are on layers of h_2 and h_4. The microstrip patch is feed by microstrip line trough slot in the common ground plane. The width microstrip line is W and is printed on a substrate described by h_1 and ε_1. Coupling of the slot to the dominant mode of the patch and the microstrip line occurs because the slot interrupts the longitudinal current flow in them[3].

For an analyzed planar structure activity of the antenna is more important than surrounding space, so the analyzed area is not to large. But it is large enough so that the results would reflect analyzed area precisely and strictly, taking into account limited computing power of a computer. One of the most important parameters, which will be calculated, is bandwidth whose value is dependent from parameters of antenna. The described aerial is structure multi resonance.

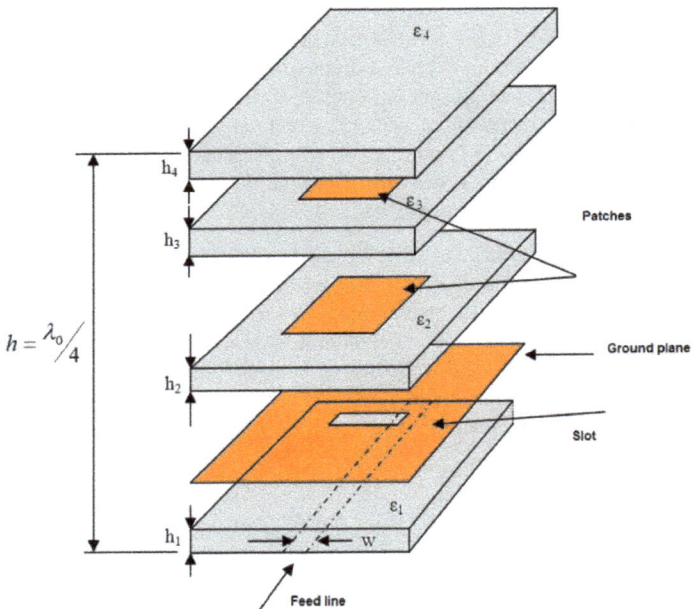

Figure 1. An aperture coupled microstrip patch antenna.

3. Optimization

Choice of laminates and other materials for the implementation of the antenna and antenna system is one of the most important steps in the process of designing microstrip antennas. Planar microstrip antenna can be built with the theoretically infinite number of layers. This solution, unfortunately, leads to a reduction in antenna efficiency and leads to excitation of surface waves. In order to provide the required bandwidth efficiency of the antenna work and we need to limit the number of layers to this, to obtain both wide bandwidth operation and high efficiency.

In this case the optimization is folded, since all possible combinations of parameters will be examined. This optimization is time-consuming and complex process. In Figure 2 an algorithm of the optimization was presented. Bandwidth planar antenna can be expressed using the equation (1):

$$B = \frac{VSWR - 1}{\frac{c\sqrt{\varepsilon_e}}{4 f_0 h}} \tag{1}$$

where:

Q - quality factor

f_0 - center frequency of band

ε_e - permittivity

h - thickness of antenna

Optimization parameters:

- Layers permittivity: $\varepsilon_1,\ \varepsilon_2,\ \varepsilon_3,\ \varepsilon_4$

- Layers thickness: $h_1,\ h_2,\ h_3,\ h_4$ (Fig.1)

In order to optimize the iterative method was used. It is time consuming but allowed us to investigate the effect of dielectric parameters of the antenna operating band (fig.3). In the optimization process the following steps of changes of input parameters were accepted:

- electric penetrability of layers ($\varepsilon_1,\ \varepsilon_2,\ \varepsilon_3,\ \varepsilon_4$) – 0.05

- thickness of layers ($h_1,\ h_2,\ h_3,\ h_4$) – 0.05 [mm]

4. Limitations of the optimization

Permeability and thickness of each layer are limited typical values offered by various manufacturers such as ROGERS, ARLON, etc. Different companies offer dielectrics with very similar structural parameters. Therefore, the values of h and ε_r values are limited discrete values occurring in the family.

As a result of computational series this way conducted for every of layers three-dimensional graphs giving the full image and the inspection of all possible combinations to the thickness

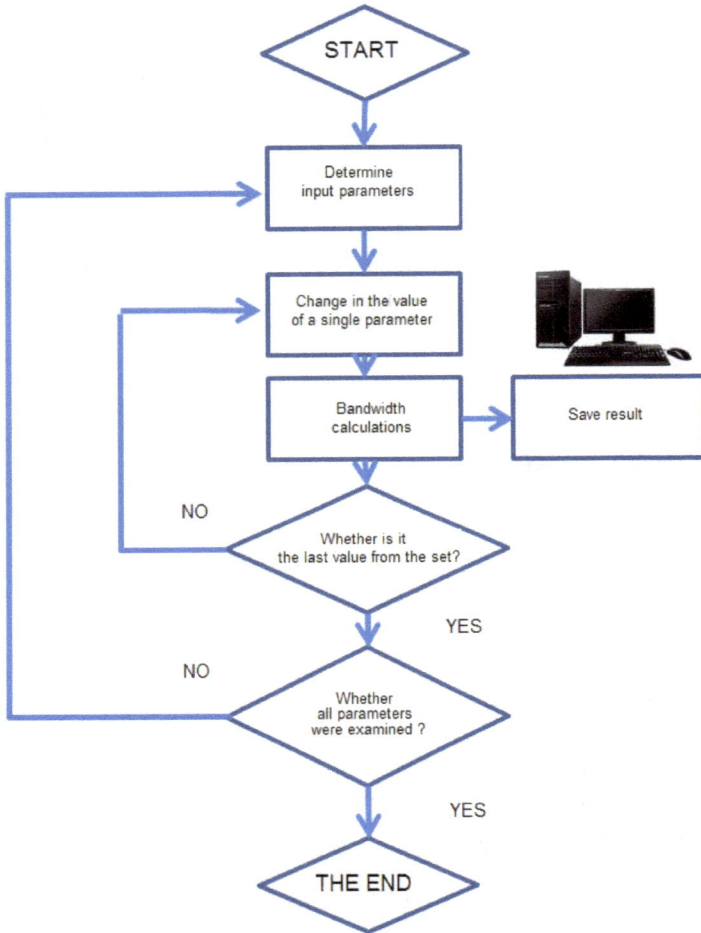

Figure 2. Algorithm of the optimization

and permittivity (Figure 3). On each of graphs optimum areas marked on the color red. Layer h_3 has the optimum area smallest of all examined layers.

5. The results of the optimization process

The object of the study is primarily broadband properties of the antenna input impedance and directional properties, which determine the further use of the model. Particular attention is paid to the bandwidth of the work, which was the basic parameter determining the final construction of the structure. Using the current state of knowledge should lead to the use of dielectrics with relatively lowest values of electrical permittivity. Optimization was carried out by examining the every of the layers in structure. The results to optimize were presented on the graphs below. The optimization process was as follows: for each layer has been studied sufficiently large set of combinations of thickness and permeability, and for each of them determined the bandwidth of operation. As a result of a series of calculations for each of the layers was formed three-dimensional graphs which give a complete picture and review all possible combinations of thickness and permeability. On each of the graphs obtained areas with the best bandwidth - in the diagrams shown in red. As can be seen in the following analysis for each layer there is at least some optimum combinations. A limitation of choosing the right laminate is the availability of mass-produced laminates of sufficient thickness and permeability. Laminates as well as other electronic components such as resistors and capacitors are produced in the so-called series with typical values, with defined thicknesses and permittivity. This is some major difficulty for designers.

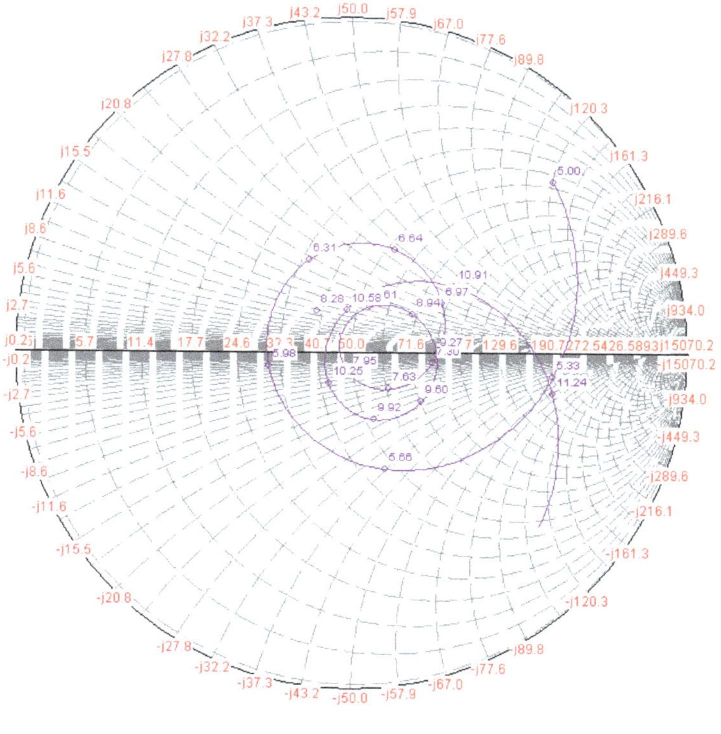

Figure 3. Results of optimization for the parameters of the four layers of the aperture-coupled microstrip antenna

The results also show how little room for maneuver there is for choosing the right laminate structure. For some structures may unfortunately find that the physical realization is impossible because in our area of interest is not in any laminate. Then we would look for a solution by changing the dimensions of the radiating elements so as to be able to use existing laminates.

As a result of optimization selected the following dielectrics:

h_1=1,57 mm, ε_1 =2.2 - Rogers-RT/duroid 5880

h_2 =3,048 mm, ε_2 =2.6 -Rogers-RT/ULTRALAM 2000

h_3 =3 mm, ε_3 =1.07- polymethacrylamid hard foam.

h_4 =0,25mm, ε_4 =3.50 - Rogers/RO3035

6. Influence of dielectric parameters on antenna bandwidth

The thickness of layers in a multilayer structure is one of the more important elements affecting directly on the bandwidth of the work the whole structure of radiation. Analysis the impact of the thickness the layers on the bandwidth in the structure consisted of the cyclic change in thickness (in steps of 0.05 mm) layer and setting the bandwidth (VSWR <2) when the other elements in the construction of the antenna in the same state. Maintaining other elements of the antenna can be sure that the effect is obtained only from changes in thickness. The analysis was performed for all layers occurring in the structure. For obtained in the optimization process the parameters of laminates determined parameters of the antenna.

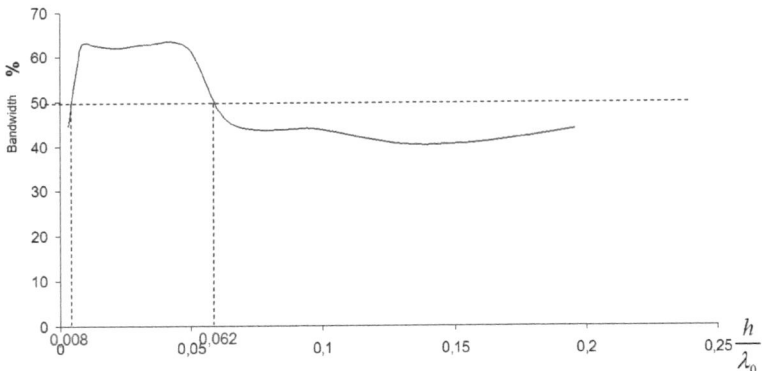

Figure 4. Influence thickness of h_4 layer on bandwidth

The first was analyzed h_4 layer. It is located at the top of the antenna. The results are illustrated in Figure 4. The results can be concluded that the bandwidth of the work at a level above 50% can be achieved by changes in the thickness of this layer in the range from 0.008 to 0.062λ_0.

When we increase the thickness of the layer above the limit then the range bandwidth is narrows in the area of the upper frequencies.

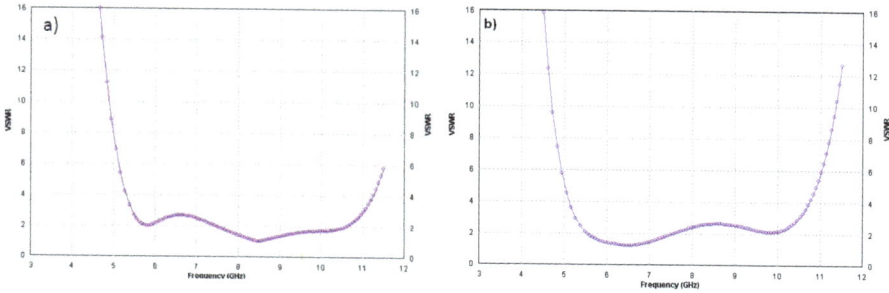

Figure 5. VSWR for values outside of the optimal range (a) thickness >0.062λ_0 (b) thickness <0.008λ_0

Using the same calculation process, we determined the influence of thickness h3 to h1 on the bandwidth of the antenna.

7. Study the impact of the total thickness of the antenna operating bandwidth

Using the results obtained during the optimization and additional numerical calculations can be concluded that the multilayer structure achieves the highest bandwidth for the measured quantity of work, as the ratio of thickness of the antenna to the wavelength and containing in the range of 0.21-0.27 λ_0. When the thickness of the structure has a value outside this range is rapidly declining bandwidth operation. For all layers of optimum thickness ratio of the total structure to the radiated wave length (center frequency 8.27 GHz antenna operation) at which it achieves the greatest band of operation is in all cases very close to the value of 0.25λ_0 (fig.6)

Mistake how we can make when selecting layer h_1 and h_2 is the smallest of all the layers, both in terms of thickness and permeability. Minor deviations from the optimum value (due to the bandwidth of operation) the effect of limiting the bandwidth of operation well below 50%. For working bandwidth> 50% of the thickness and permittivity of each layer should contain, respectively, in the intervals:

- layer h1: thickness from 0.041 to 0.062 λ_0 and permittivity ε_{r1} from 1.78 to 2.62;

- layer h2: thickness from 0.084 to 0.121 λ_0 and permittivity ε_{r2} from 2.28 to 2.77;

- layer h3: thickness from 0.077 to 0.176 λ_0 and permittivity ε_{r3} from 1 to 1.36;

- layer h4: thickness from 0.008 to 0.062 λ_0 and permittivity ε_{r4} from 2.05 to 4.62;

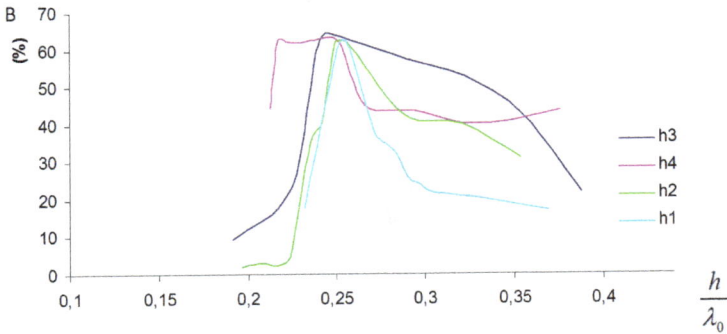

Figure 6. Bandwidth as a function of thickness of the multilayer structure due to the change in thickness of each layer.

8. Calculation of parameters for optimum model

For optimized model antenna the electrical parameters were determined: VSWR, impedance, characteristic of radiation. Figure 6 shows the VSWR of optimal single patch antenna. However, the input impedance is shown in Figures 7 (reactance and resistance). Resistance in the range of work fluctuates around the value of 50Ω. the reactance value fluctuates around the value of the 0.

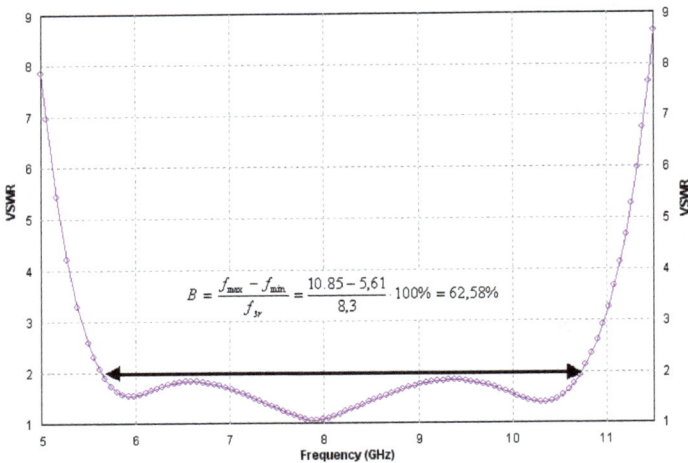

$$B = \frac{f_{max} - f_{min}}{f_{sr}} = \frac{10.85 - 5.61}{8.3} \cdot 100\% = 62{,}58\%$$

Figure 7. VSWR of optimal single patch antenna.

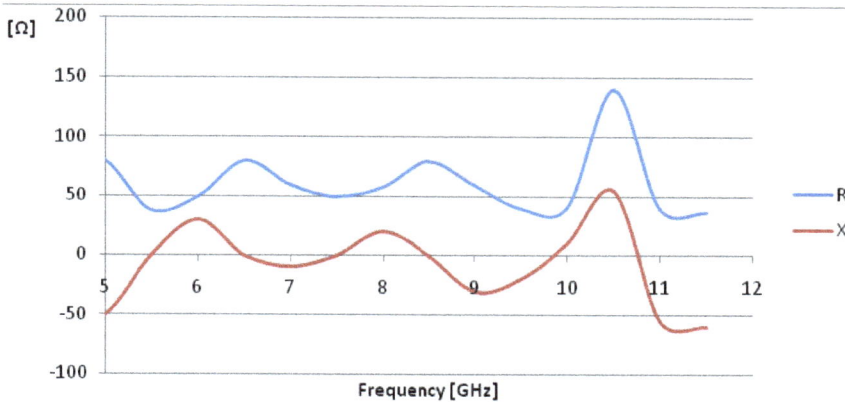

Figure 8. Reactance and resistance of optimal single patch antenna.

The resultant input impedance allows the full use of the antenna in the whole range of work. As a result of optimizing the working range was achieved 62.5% (VSWR<2).

The process of optimization and testing of all parameters of the antenna was carried out using IE3D software (Zeland Software) based on the method of moments. In order to verify the correctness of their analysis before the construction of physical model of antenna we one more time made calculations the VSWR and radiation characteristics of the final model using a different calculation method (fig.10 and 11). To perform calculations in order to verify the selected software CST Microwave Studio based on FDTD method. Smith chart of wideband multilayer antenna is show in figure 9.

Input impedance of the antenna in operating band varies in the range of 30 to 70 [Ω]. This input impedance makes full use of the antenna in the whole operating band. Such changes in the impedance is a side effect for maximum bandwidth. It should be noted that despite these changes the antenna VSWR <2.

The radiation pattern of antenna has stable shape in the whole operating band. In the whole frequency range the gain of antenna (calculated) is around 5 [dBi]. The width of the radiation characteristics (- 3dB) is around 100 °.

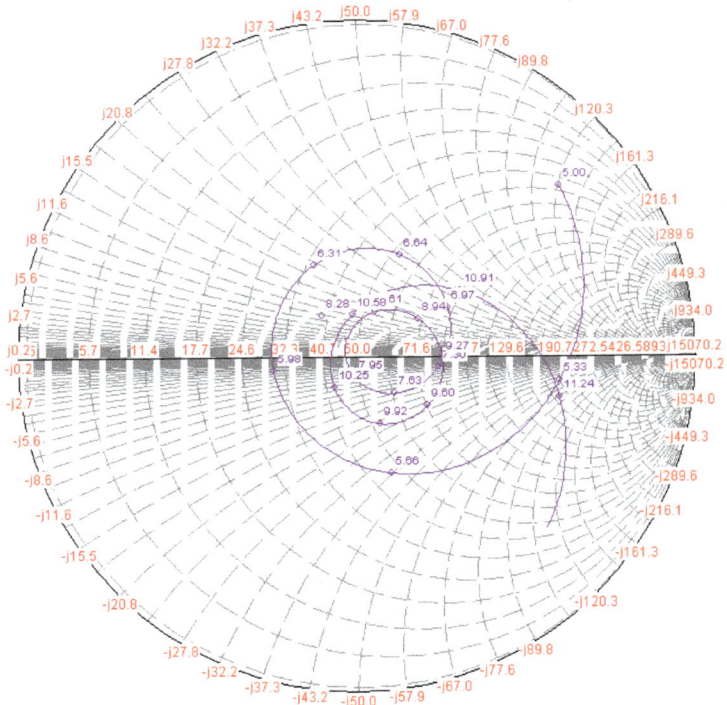

Figure 9. Smith chart of wideband multilayer antenna.

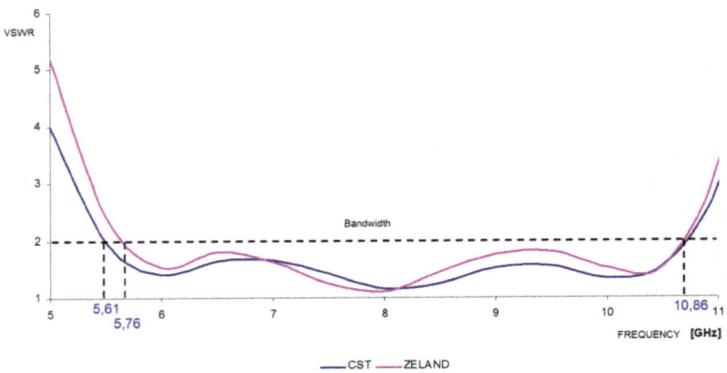

Figure 10. Figure 10. Verification of model with using CST.

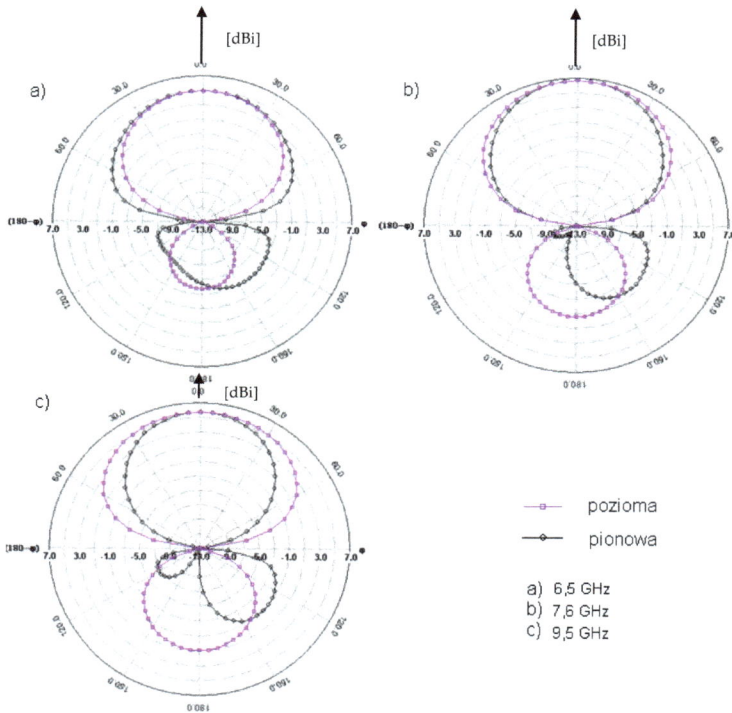

Figure 11. Radiations pattern of multilayer antenna-simulations

Selection of components of the antenna (thickness and permeability layers) gives the value of VSWR<2. The antenna's radiating patch has very similar level and distribution of currents. The main radiating element is side edges of the patches are the same as in single-layer structures. The distribution of currents guarantees consistent with the direction of the feed line.

9. Physical model of the antenna

In the figure 13 shows the antenna layers. First from left is put layer h1 with coupling slot. Holes in layers of the antenna were carried out intentionally with a view to precise connecting with oneself layers, they aren't bringing changes of parameters of the aerial in.

In Figure 13 a ready model of the aerial was described, the structure of aerials is coated with foamed PCV. They aren't bringing changes of parameters of the aerial in (checked in measurements).

Max E-Current = 5.9981 (A/m)

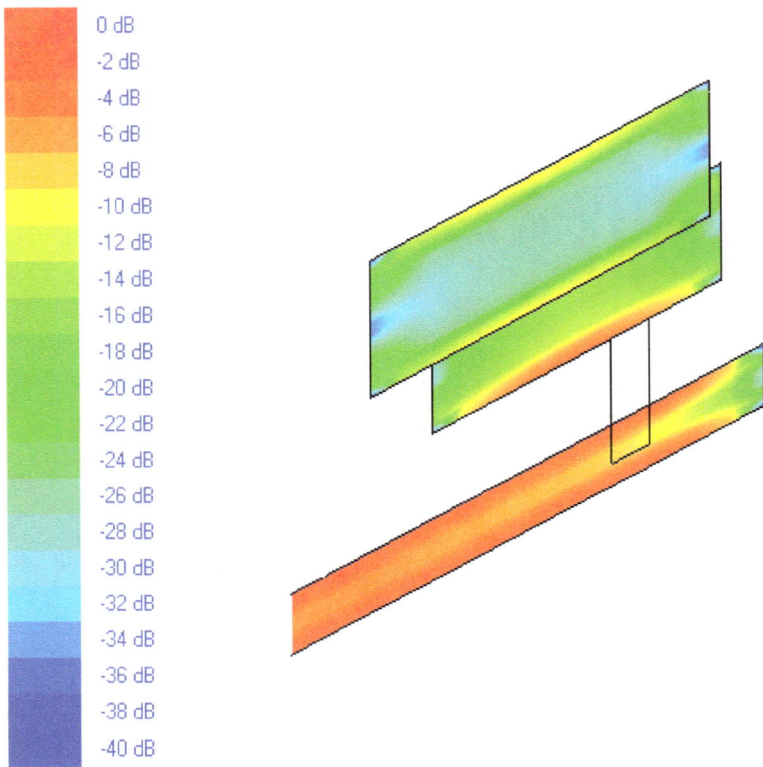

Figure 12. Current distribution in antanna elements

Figure 13. Layers of antenna and model of antenna.

Obtained in a multi-layer band work is not possible in single layer antennas [2]. In the whole range of work input impedance of the antenna is stable. Antenna input impedance in the band of work is changing in the range of 36 to 67 [Ω]. Such changes in the impedance are a side effect for a maximum bandwidth of work.

The result for the physical model confirms the correctness of the analysis and calculations obtained using the calculation method based on the method of moments. Three of the four laminates used in its construction include teflon, and one of the layers is made of polimetha-crylamid foam. This caused some difficulties in the technological process of joining them together and generates small inaccuracies in the contact of subsequent layers. This gave slight differences between calculations and measurements.

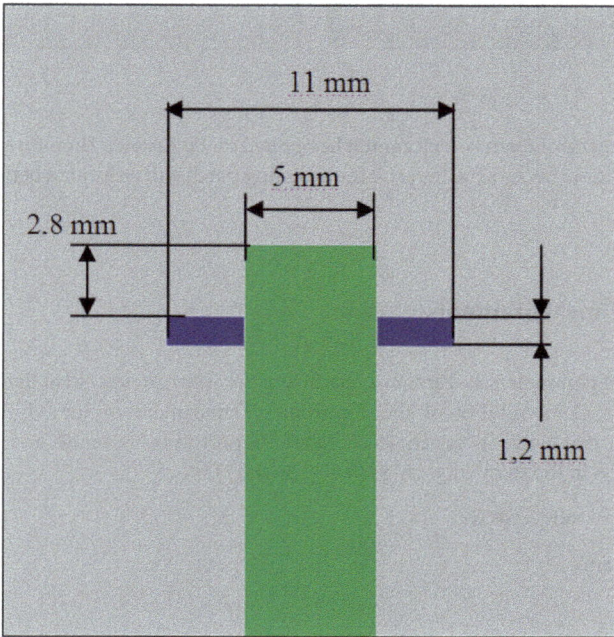

Figure 14. Dimension of h_1 layer

The resulting optimization model of the antenna is shown in Figures 9, 10 and 11. The antenna is a multi-layered structure having in its construction of two radiating elements in the form of "patches" located on successive layers of the dish. Patch located in the h2 layer is fed by a line through a gap made in the plane of the masses. Above it there is another patch that by using the resonance (of another already in the structure) greatly improves the bandwidth of operation. The power supply line located on the layer defined by the h1 and ε1.

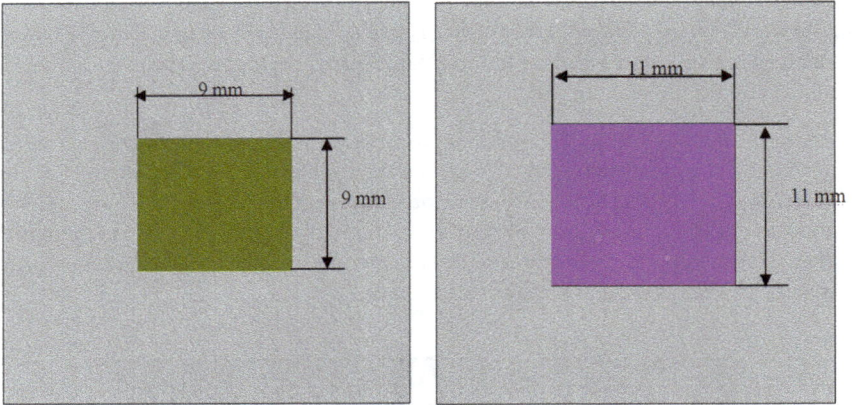

Figure 15. Dimension of h_2 and h_4 layers

The h2 layer is made of foam, which cannot be applied metallization. Therefore, the patch was placed on the bottom layer of h4. In result the radiating patch will be located between the layers h3 and h4.

10. Result of measurements

Measurement of parameters and antenna characteristics were carried out at the Laboratory for Electromagnetic Compatibility of the Electronics Department of the Military Technical Academy equipped with an anechoic chamber. A prototype is made and the following measured results in the frequency range (5,5-11,5 GHz) for :

• Standing wave ratio VSWR

• Input impedance Z

• Resistance R

• Reactance X

• Radiation patterns

Figure 16 shows the model of antenna VSWR during measurements. In the next Figure 17 shows the results of measurements of VSWR, which reaches a value of less than 2 in frequency range from 5.8 GHz to 10.15 GHz. during the measurement parameters and antenna characteristics of dielectric parameters was also studied based on the methodology described in the articles[12,13,14].

Figure 16. Model of antenna during VSWR measurmeants.

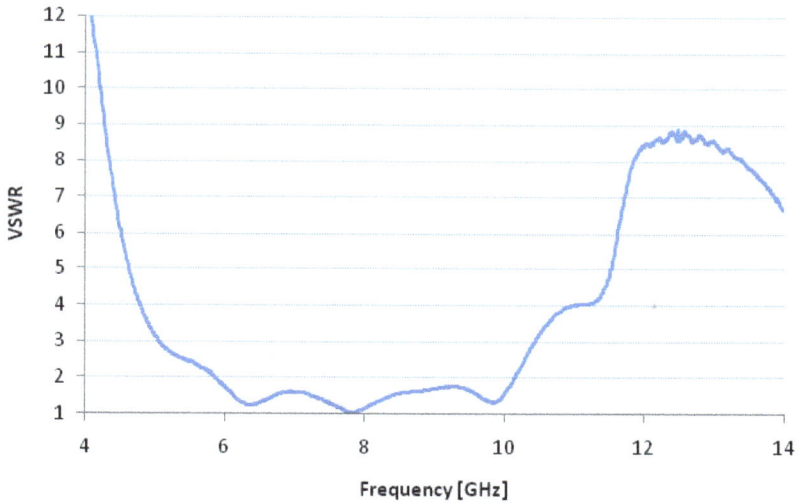

Figure 17. Measured VSWR of optimal single patch.

Obtained in the measurement bandwidth of the work is:

$$B = \frac{2(f_{max} - f_{min})}{f_{max} + f_{min}} \cdot 100\% = \frac{2(10150 - 5800)}{(10150 + 5800)} \cdot 100\% = \frac{2 \cdot 4350}{15950} \cdot 100\% \approx 55\% \tag{2}$$

Figure 18 shows the measured values of resistance and reactance of antenna model.

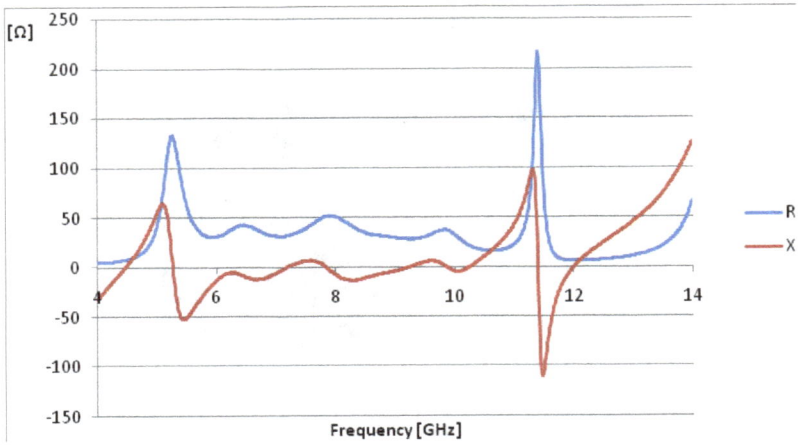

Figure 18. Measured input impedance of optimal single patch antenna.

Figure 18 shows the measured values of resistance and reactance of antenna model. The measured resistance value ranges from 35 to 55 ohms. Reactance varies in the range from -10 to 5 ohms.

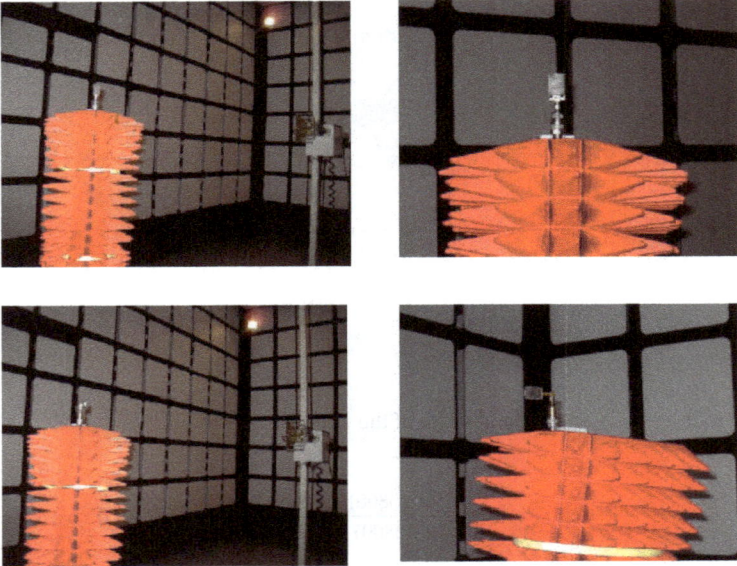

Figure 19. Photos of the antenna during measurments in the anechoic chamber.

The subject carried out by the author toprimarily study broadening the antenna bandwidth properties, which confirm the elimination of the problem of narrowband in the structures of planar antennas and directional properties. Particular attention is paid to the impedance properties defined by examining the standing wave ratio (VSWR) in the bandwidth of the antenna, the ratio was determined relative to a standard value of impedance 50Ω. Radiation characteristics was investigated for frequencies from 5.5 GHz to 11.5 GHz in steps of 1 GHz. The paper presents the characteristics for a center frequency band operation (7.6 GHz) and for two frequencies distant by about 1 GHz bandwidth from the ends of the work, the frequency of 6.5 GHz and 9.5 GHz. For the same frequency were also determined and presented in the work sheet in the computer simulations.

Figure 20 shows the far field radiation characteristics of the antenna at the center frequency (7.6 GHz) and at 6.5 and 9.5 GHz

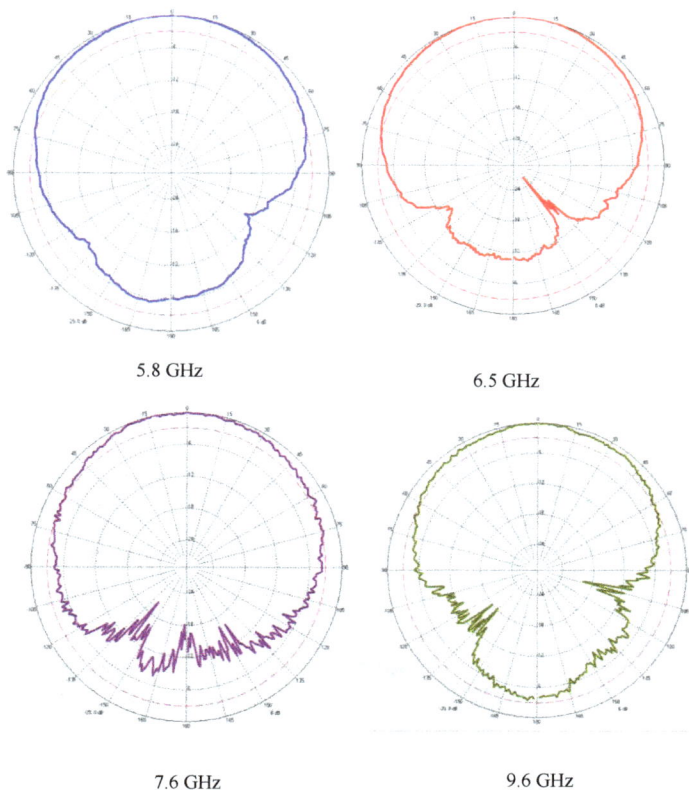

5.8 GHz

6.5 GHz

7.6 GHz

9.6 GHz

Figure 20. Radiation paterrns(H - plane).

Width of the main lobe at the level of -3 [dB] in the whole operating band is approximately 140 °. Radiation pattern has a stable shape in the frequency domain.

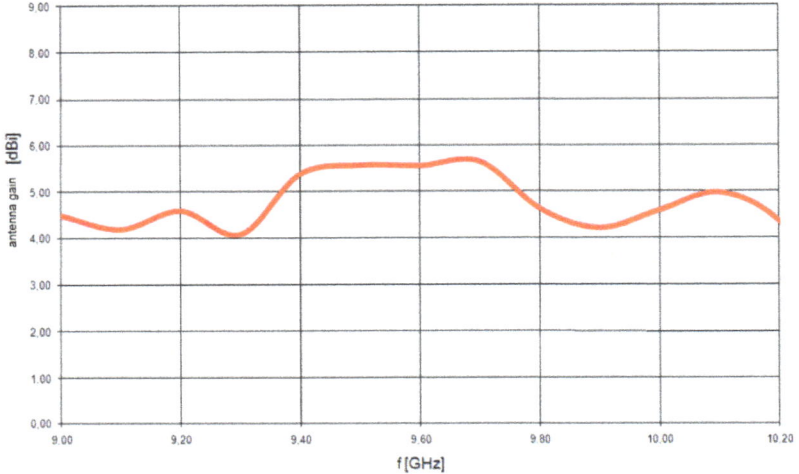

Figure 21. Measured antenna gain versus frequency.

Antenna gain is in the range from 4 dBi to 5.7 dBi. For multilayer structure with a bandwidth greater than 50% this is considered to be very good gain values.

Presented in a multi-layer structure despite the complex structure can be used to construct the antenna array, thereby increasing antenna gain. For this purpose, designed four element array, whose design is shown in Figure 21.

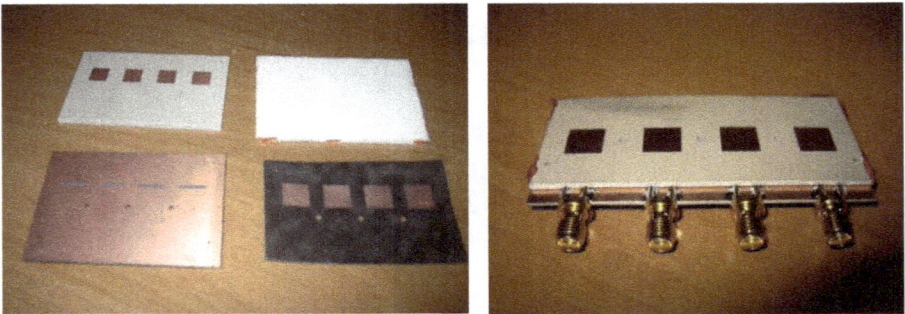

Figure 22. Antenna array.

Parameters and characteristics of the antenna array were also measured in the anechoic chamber (fig.21). For the model was made the following we measured in this same frequency range (5,5-11,5 GHz) following parameters: standing wave ratio VSWR, input impedance Z, resistance R, reactance X, radiation patterns. Measurement results are shown in Figures 23 and 24.

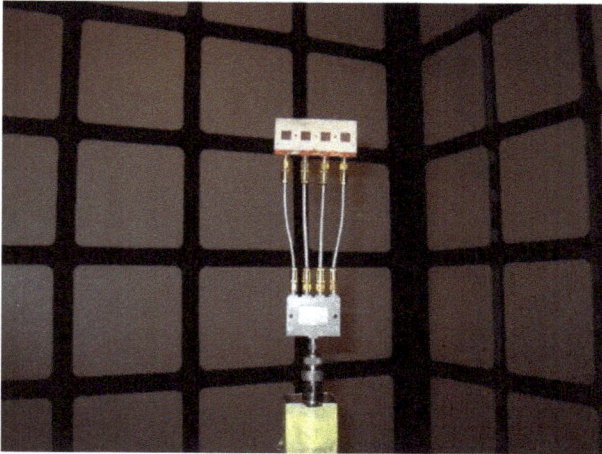

Figure 23. Antenna array during measurments in anechoic chamber.

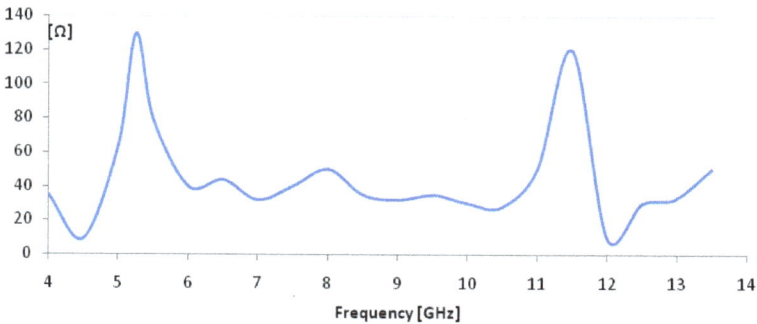

Figure 24. Impedance of antenna array- measurmeants.

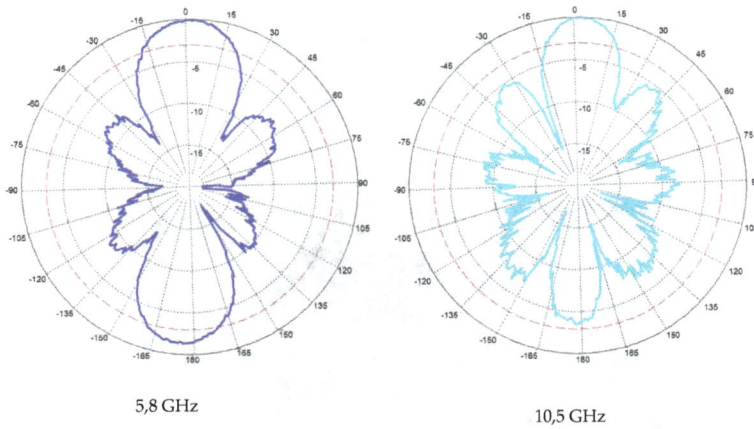

5,8 GHz

10,5 GHz

Figure 25. Radiation paterrns of array.

11. Comparison of simulations and measurements

The most important parameter in view of the article is the course of VSWR as a function of frequency. The result obtained for the physical model confirms the correctness of the analysis and calculations obtained using the calculation method based on the method of moments, as well as the FDTD method.

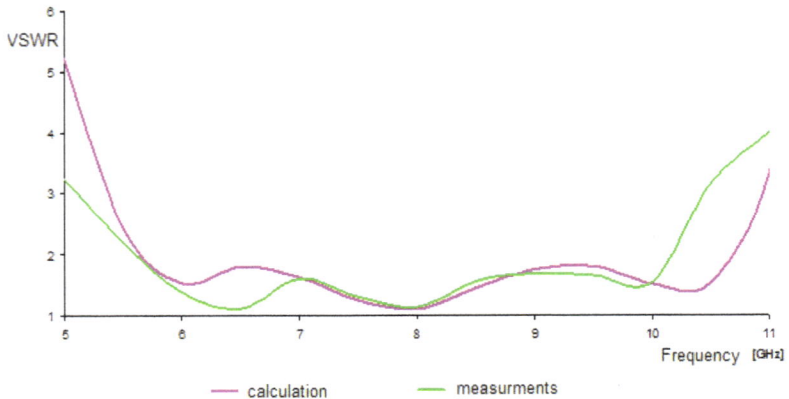

Figure 26. VSWR – simualtions and measurments.

Reduction of bandwidth in the upper-frequency of band (as confirmed by further simulations and calculations exploring the effects of precision interfaces between layers of the antenna on its parameters), the inaccuracy of the interfaces between layers of the dish. Three of the four laminates used in its construction include Teflon, and one of the layers is made of foam polimethacrylamid. This causes some technological difficulties in the process of connecting them together and produce a minimum of contact inaccuracy of subsequent layers. The simulations confirm the situation.

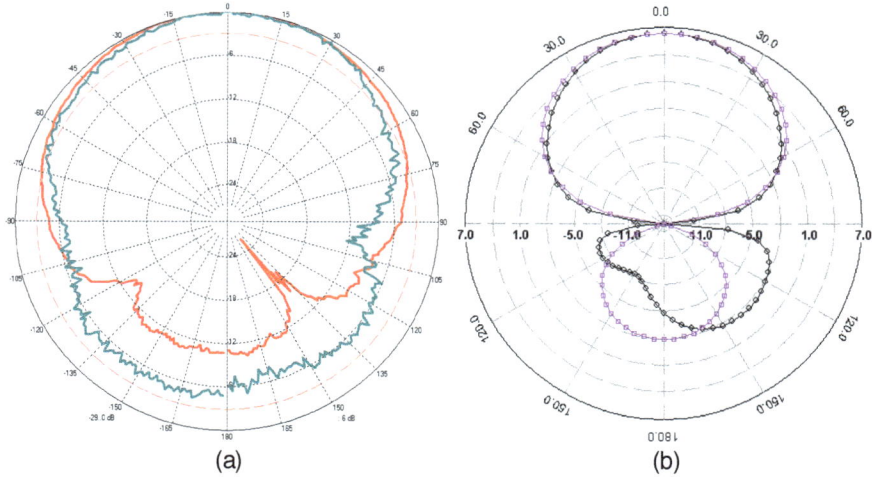

Figure 27. Radiation paterrns of antenna measurments (a) and simualtions (b).

12. Conclusion

This article presents issues related to the theory and technique of multilayer planar antennas fed by the slot. This paper describes multilayer configuration to increase the BW of the antenna. This configuration has many advantages, including wide BW, reduction in spurious feed network radiation, and a symmetric radiation pattern with low cross-polarization. The antenna configuration with a resonant aperture yields wide BW by proper optimization of the coupling between the patch and the resonant slot. The basic characteristics and the effects of various parameters on the overall antenna performance are discussed.

The results of the study are satisfactory. The paper describes a clear advantage of multilayer antennas over monolayer ones, where the bandwidth is significantly narrower. The coupling aperture antenna after optimization can result in a considerable increase of bandwidth. The bandwidth of aerial is the outcome of the way feed as well as the utilization to build the multilayer substrates and they parameters. Calculations show how difficult it is to choose optimal value for laminates when we want to obtain wide band. Permittivity and thickness are equally important for bandwidth antenna. The analysis of the antenna was done by the usage of IE3D - Zeland Software (method of moments - MoM). The method has been applied to the microstrip-fed slot antenna and to the aperture coupled antenna with a good result when compared with measured data.

The model of antenna allows use of all the positive properties of planar antennas with simultaneous work in a wide frequency range, which so far has been the main element to eliminate this type of the antenna of use in many designs. Given the trend for miniaturization of antenna devices and the development of radio technology and integrated systems presented in the paper of construction seems to be very prospective. The design of this antenna is a modern solution to the antenna, which is especially important in the use of the moving objects. The results of experimental studies measuring fully confirm the possibility of constructing planar antenna with wide bandwidth operation. In addition, ease of implementation of antenna arrays using planar antennas opens up new possibilities in the use of this construction. Designed antenna operates in X-band and certainly could be an interesting alternative to the currently used antenna antennas operating in this band.

Author details

Marek Bugaj* and Marian Wnuk

*Address all correspondence to: marek.bugaj@wat.edu.pl

Faculty of Electronics, Military University of Technology, Warsaw, Poland

References

[1] Turker, N, Gunes, F, & Yildirim, T. Artificial Neural Design of Microstrip Antennas", Turk J Elec Engin, (2006). , 14(3)

[2] Kumar, G, & Ray, K. P. Broadband Microstrip Antennas" Artech House, (2003). 1-58053-244-6

[3] Garg, R. pp. Bhartia, I. Bahl "Microstrip antenna design handbook" Artec Hause, INC

[4] Bi, Z, & Wu, K. Ch. Wu, J. Litva "And dispersive boundary condition handicap microstrip component analysis using the FDTD method", IEEE Trans. Antennas and Propagation, (1992). , 40, 774-777.

[5] Kin-Lu WongCompact and bradband miccrostrip antennas", (2002). by John Wiley & Sons, Inc., New York, 0-47122-111-2

[6] Pozar, D. M. Microstrip Antenna Aperture-Coupled to a microstrip line", Electron. Lett., (1985). , 21, 49-50.

[7] Pozar, D. M. A Reciepprocity Mothod of Analisis for Printed Slot and Slot-Coupled Antennas," IEEE Trans.Antennas Propaga., Dec. (1986). , AP-34(12)

[8] Wood, C. Improved Bandwidth of Microstrip Antennas using Parasitic Elements", IEE Proc., Pt.H, 4/(1980). , 127

[9] Pozar David MA Microstrip Antenna Aperture Coupled to a Microstrip line", Electron. Lett.Jan.17 (1985). , 21, 49-50.

[10] Borowiec, R, & Slobodzian, P. Multilayer microstrip structure" Phd. dissertation Institute, of Telecommunications and Acoustics, University of Wroclaw

[11] Bugaj, M, & Wnuk, M. The analysis of microstrip antennas with the utilization the FDTD method" in XII Computional Methods Experimental Measurements, WIT Press 1746-4064page 611-620

[12] Kubacki, R, Nowosielski, L, & Przesmycki- Technique, R. for the electric and magnetic parameter measurement of powdered materials, Computotional methods and experimental measurements XIV (2009). Wessex Institute of Technology, Anglia, str. 241÷250, 0174-3355X,., 48

[13] Kubacki, R, Nowosielski, L, Przesmycki, R, & Frender- The, R. technique of electric and magnetic parameters measurements of powdered materials, Przegląd Elektrotechniczny (Electrical Review), 0033-2097R. 85 NR 12/(2009).

[14] Przesmycki, R, Nowosielski, L, & Kubacki-, R. The improved technique of electric and magnetic parameters measurements of powdered materials, Advances in Engineering Software, November (2011). 0965-9978, 42(11)

[15] Gupta, K, & Benalla, A. Microstrip Antenna Design", Artech House, London (1980).

Multiband Planar Antennas

Chapter 4

Shared-Aperture Multi-Band Dual-Polarized SAR Microstrip Array Design

Shun-Shi Zhong and Zhu Sun

Additional information is available at the end of the chapter

1. Introduction

Since the 1970s, the remote sensing synthetic aperture radar (SAR) systems have been developing increasingly. The SIR-C/X SAR mounted on the American Space Shuttle Endeavour firstly completed the high resolution 3D imaging all over the globe in Feb. 2000 [1]. It employs three dual-polarized sub-arrays operating at L-, C- and X-bands, respectively, as shown in Fig. 1[2]. The multi-band dual-polarized array receives more target information than that of a single-band one-polarization counterpart, and thus enhances the capability of target detection and identification. The common bands of space-borne SARs are L (center at 1.275 GHz), S (3.0 GHz), C (5.3 GHz) and X (9.6 GHz) bands. To minimize the volume and weight of a SAR antenna, a common aperture configuration for dual or multiple bands are expected, which will share the sub-systems behind the array as well. As a result, a lot of shared-aperture dual-band dual-polarization (DBDP) integrated planar antennas have been proposed in the last two decades [2].

The typical configurations of the shared-aperture DBDP planar arrays include the perforated structure, the interlaced layout and the overlapped layout. The perforated structure mainly includes perforated-patch/patch [3,4], ring/patch [5-7] and cross-patch/patch[8]. The interlaced layout includes interlace patch with dipole/slot [9-11] and interlaced slot with slot/dipole [12-13], *etc.* [14-15] present the review and comparison for these arrays. Both the perforated structure and interlaced layout are commonly adopted in space- or air-borne applications because of their low profile performance. On the other hand, the overlapped layout can provide further improvement in the bandwidths of dual bands, but with larger antenna profile. In [16], the thick air layer and an L-probe feed are applied to design a dual-band element of overlapped layout and a wide bandwidth is achieved at both low and high bands. Meanwhile, in the element aspect, some shapes [17-18] and a considerable number of feed methods, such as

Figure 1. Antenna of SIR-C/X SAR

probe-feed [19], aperture-coupled [17-18], proximity-coupled [20], and L-shape feed [21] are proposed to improve the key performances. Besides, balanced-feed [22, 23] and hybrid-feed [24-27] are also proposed to enhance the port isolation or cross-polarization performance. Recently, a DBDP array with an impressive $S_{11} \leq$ -15 dB bandwidth of 8% in the low band is introduced by using the feed skills above mentioned [28].

In this chapter a universal multi-band SAR array synthetic method is introduced to extend the shared-aperture DBDP array to a shared-aperture MBDP SAR array. As a validation, a shared-aperture L/S/X tri-band dual-polarized (TBDP) array prototype was fabricated and measured with the central frequencies of 1.25 GHz, 3.5 GHz, and 10 GHz for L-, S- and X- bands, respectively [29]. The array design and measured results are presented and discussed.

2. Array configuration

2.1. TBDP array configuration

From the system view, a similar beamwidth in the transverse (elevation) direction for each band is desired, which means that the transverse dimension of aperture for each band should be in proportion to its wavelength [3]. As an example, the transverse dimensions of aperture in L-, S- and X- bands for the SIR-C/X-SAR are 2.95 m, 0.75 m, and 0.4 m, respectively, which have the ratio of 7.375:1.875:1. Then for a TBDP array with three independent apertures of L/S/X bands, its corresponding transverse dimension of aperture may be 2.88 m for L-band, 1.08 m for S-band and 0.36 m for X-band with the ratio of 8:3:1, as seen in Fig.2a.

In this chapter a novel design method is proposed, as shown in Fig.2b, where L/S and L/X sub-arrays are set on the two sides, and L-band sub-array is located in the middle. Then the transverse dimensions of aperture is 1.08+1.44+0.36 =2.88m for L-band, 1.08 m for S-band and 0.36 m for X-band, with ratio of 8:3:1 that is close to the central wavelength ratio of L-, S- and X-bands. Thus S-band and X-band elements are physically separated by a large spacing to avoid mutual interference, and a 33% off in aperture size is achieved as compared to its counterpart with independent tri-band aperture of 1.08 + 2.88 + 0.36 = 4.32 m. Actually, this method divides a TBDP shared-aperture array into two DBDP shared-aperture sub-arrays and one single-band DP sub-array so that the foregone works of DBDP shared-aperture arrays can be consulted and eventually the development of a TBDP shared-aperture array will be simplified.

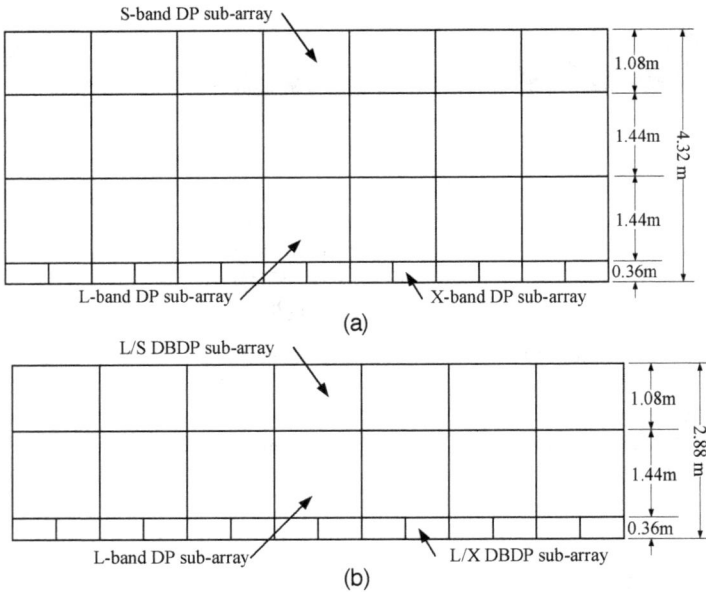

Figure 2. TBDP array configuration

2.2. L/S, L/X & L sub-array configuration

The interlaced configuration of Fig.3 is adopted for the L/S and L/X DBDP shared-aperture sub-array designs, which is more flexible for the odd frequency ratio than the perforated one[10]. Moreover, the "pair-wise anti-phase feeding" technique [30] is used in both S- and X-bands to improve the cross-polarization performance and the polarization isolation of whole array as well.

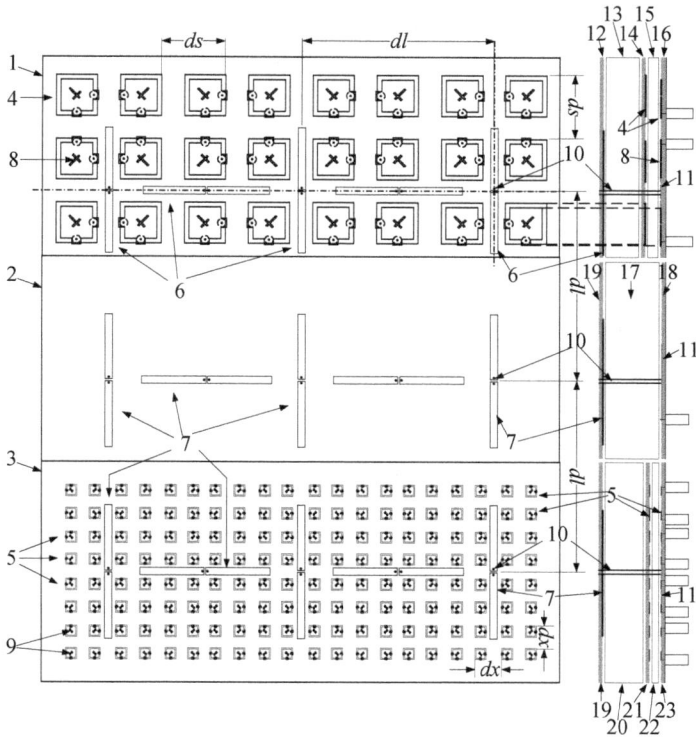

1 L/S DBDP sub-array 2 L band DP sub-array 3 L/X DBDP sub-array 4 S band stacked patch
5 X band stacked patch 6 L band print dipole (L/S) 7 L band print dipole (L/X & L)
8 isolation-slot etached on driven patch (S-band) 9 isolation-slot etached on driven patch (X-band)
10 Two-wire feeder 11 Balun of L-band 12/14/16/18/19/21/21 Substrate (Rogers 6002)
13/15/17/20/22 Foam

Figure 3. TBDP array configuration

The S-band and X-band elements in L/S and L/X DBDP arrays are distributed in square lattices, with the element spacing of $0.65\ \lambda_0$ at X-band and $0.62\ \lambda_0$ at S-band for convenience in the integration and satisfying the requirement of scanning capability ($\pm 25°$). In the L/S, L/X and L sub-arrays, the same L-band element is used to keep uniform radiation characteristic and S parameters. Besides, on the purpose of realizing dual polarization as well as enhancing the polarization isolation, "T"-shaped orthogonal-set microstrip dipoles are applied. Since that the phase centers for H- and V- polarization will deviate from each other as the "T"-shaped configuration is adopted, to ensure the superposition of the phase centers for H- and V-polarization, one more vertical microstrip dipole is added in the array prototype. This makes a little difference in the beamwidths of H- and V- polarization, however, this may not be a problem when the array aperture becomes larger, which is the case for most space-borne SAR antennas, e.g. the lengthwise aperture dimension of the SIR-C/X-SAR for all three bands is 12 m.

Since this is a phased array with two-dimensional scanning capability, each element is terminated to a T/R module and a beam controller is employed to steer the beam rather than a fixed feed network. Thus final output of each element is connected to a SMA connector.

3. Element design

3.1. S- & X-band element

To meet the requirements of bandwidth and cross-polarization as well as to simplify the design and fabrication, stacked patches are adopted for the S-band (in L/S sub-array) and X-band (in L/X sub-array) elements. To realize the dual-polarized radiation, the element should be symmetric in two dimensions, so that circular and square (rectangular) shapes are the most common ones. In this design, the square patch is adopted for its better cross-polarization performance [27] and lower fabrication tolerance, as shown in Fig.4.

The feeding approach greatly refers to the isolation and the cross-polarization level of an element. [17, 21] have presented profound studies on the feeding technique of stacked patches. In [26], the excellent isolation of better than 40dB is measured by using the hybrid excitation and a balanced feed. As the cost, however, these feeding approaches are too complicated as they are applied to the array environment, because that the aperture coupling and the balanced feed are difficult for the vertical connection, while the balanced feed needs a 180 degree feed network. To ensure the fabrication reliability, the dual-probe feed is finally employed, although its isolation is not as good as the aforementioned methods.

To overcome this drawback, an asymmetric slot is etched on the driven patch to improve the port isolation [31]. Fig.5 and Fig.6 are the calculated S parameters and radiation patterns simulated by HFSS 10.0, showing the change of isolation and cross-polarization with the slot loaded. Different slot shapes are used for the S- and X- band driven patches to improve the isolation. In addition, it is found that the "H"-shaped slot in Fig.4b has less impact to S_{11} than the "cross"-shaped slot in Fig.4a, thus one has to rematch the element of latter while the "H"-shaped needs not. As seen from Fig.7, the field disturbance introduced by the slot deteriorates the element cross-polarization level. Fortunately, this would not be a problem for an array as the "pair-wise anti-phase feed" technique is used in the array configuration. The detailed design of the S/X stacked patches has been introduced in [32].

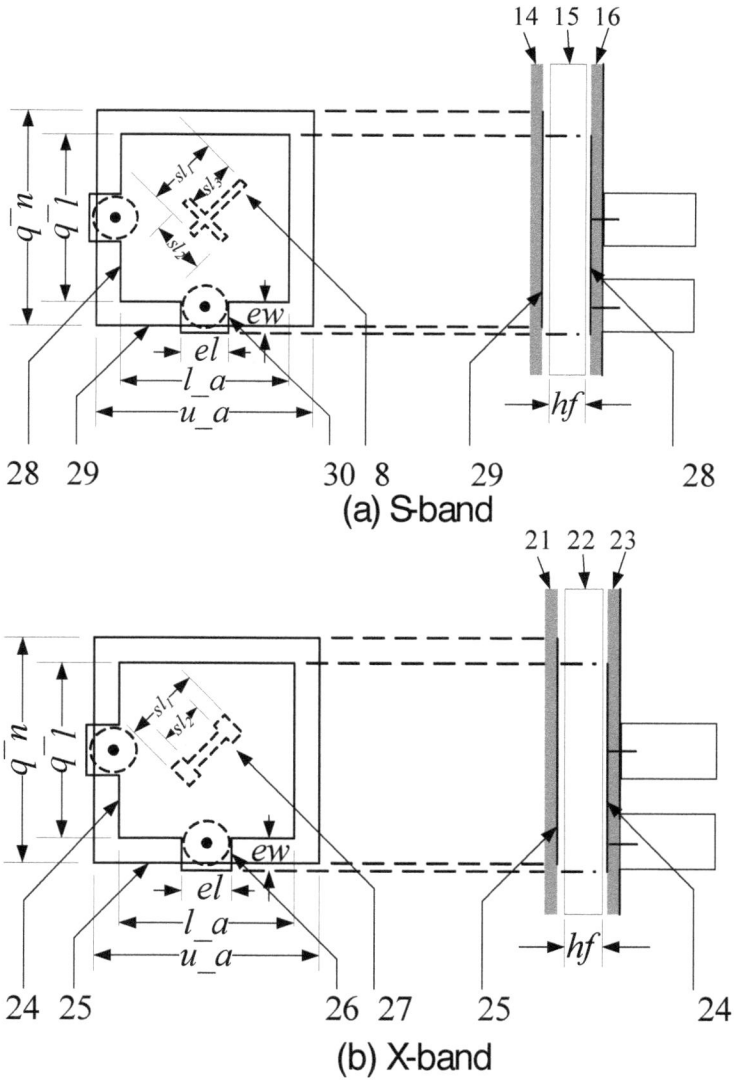

14/16 Substrate (Rogers 6002) 15 Foam 8 Isolation-slot etched on driven patch
28 Driven patch 29 Parasitic patch 30 Probe feed
21/23 Substrate (Rogers 6002) 22 Foam 24 Driven patch
25 Parasitic patch 26 Feed point 27 Isolation-slot etched on driven patch

Figure 4. S-and X-band stacked patches with slot etched

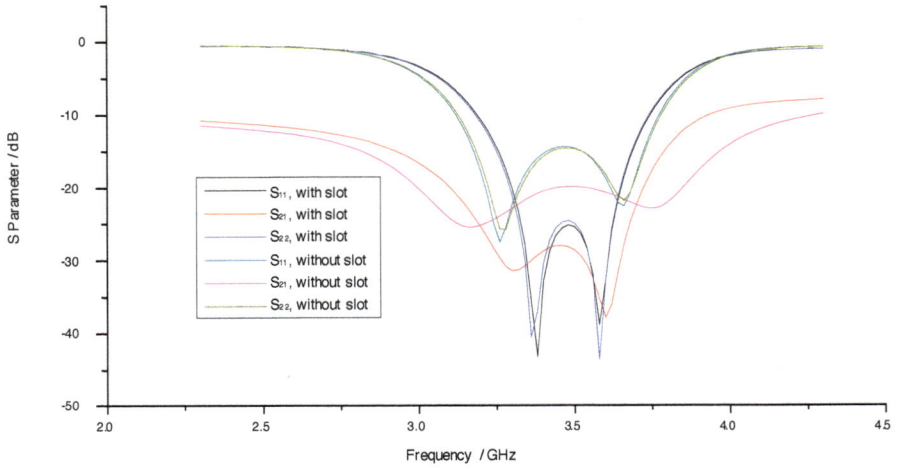

(a) S-band

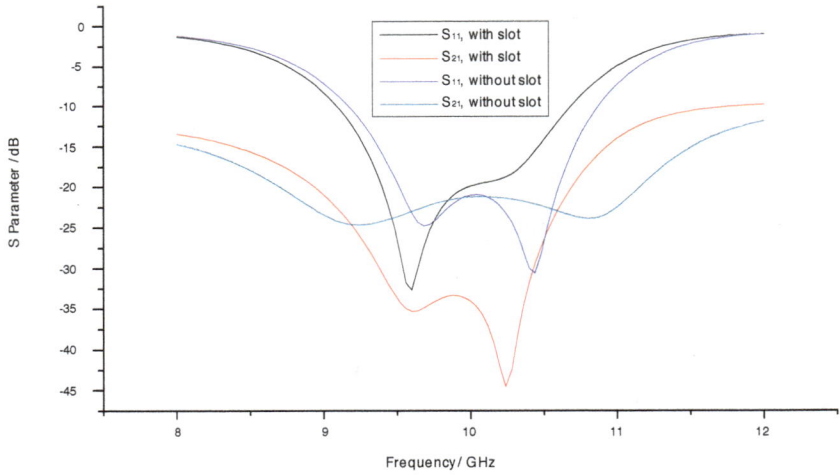

(b) X-band

Figure 5. Isolation improvement of slot-loaded stacked patches

(a) S-band

(b) X-band

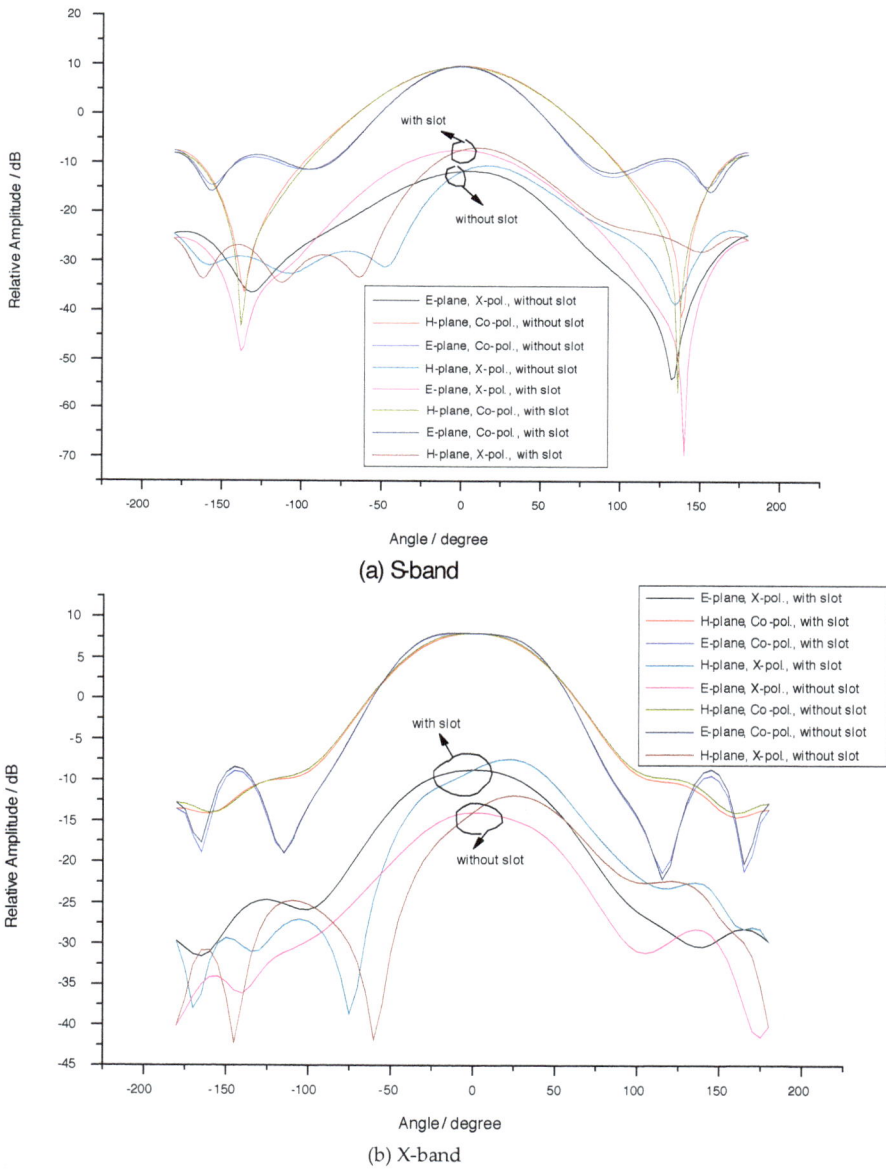

Figure 6. Radiation patterns of slot-loaded elements

3.2. L-band element

For L/S, L/X and L sub-arrays, the structure of L-band microstrip dipole is almost the same. The only difference is the length of L-band dipole in L/S sub-array, which is shortened by 20% than its counterparts in L/X and L sub-arrays. In order to avoid the conflicting in sub-array splitting caused by the odd frequency ratio of about 3:1 for L- to S-band, the capacities are intentionally introduced in its feed network to realize a minimized design (See Fig.4).

The structure of L-band dipole is detailed in Fig.6. The microstrip dipole is printed on the top layer with a height of $\lambda_0/4$ above the ground, while the ground, in this case, serves as a reflecting metal plate. An microstrip feed network is designed to generate 180 degree phase shift, and then it is vertically transferred to a parallel two-wire line, finally this line connects to the microstrip dipole. In addition, an open-ended microstrip stub is adopted at the exciting point to realize the impedance match.

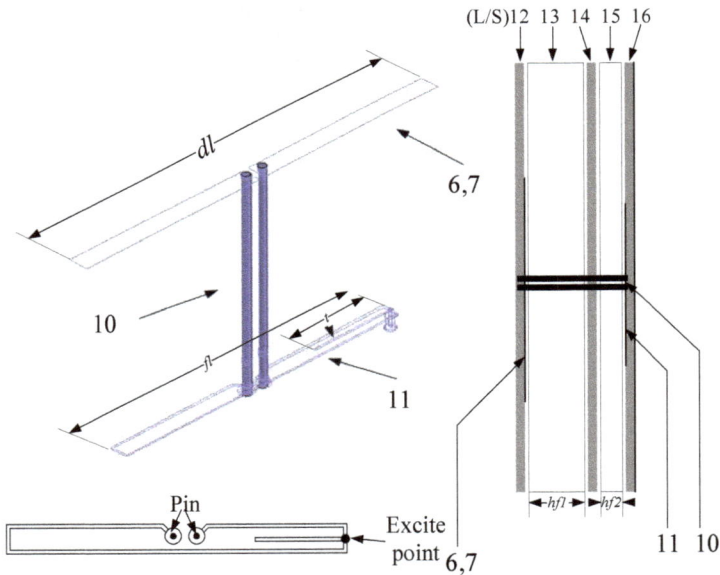

6/7 Printed dipole 10 Two-wire feeder 11 180 degree phase shifter 12/14/16 Substrate 13/15 Foam (Rogers 6002)

Figure 7. L-band microstrip dipole

This element as well as its feed structure is very slim and thus can be flexibly interlaced in the gap of S- and X-band elements. Although the profile of L-band element is somewhat high, its intrinsic broadband and good cross-polarization performance make it suitable for the TBDP SAR antenna mission.

4. Measurement results

A shared-aperture L/S/X TBDP array prototype was fabricated and measured to validate the design. Figs.8 a and b show the back and the front side of the array, respectively, which is incorporated with the L/X, L and L/S sub-arrays (from up to bottom in Fig.8a). The S-parameters were measured using the Agilent 8722ES and the measured radiation patterns were obtained in an anechoic chamber, while all the losses such as the insert losses of coaxial line, connectors and power dividers were compensated in the process.

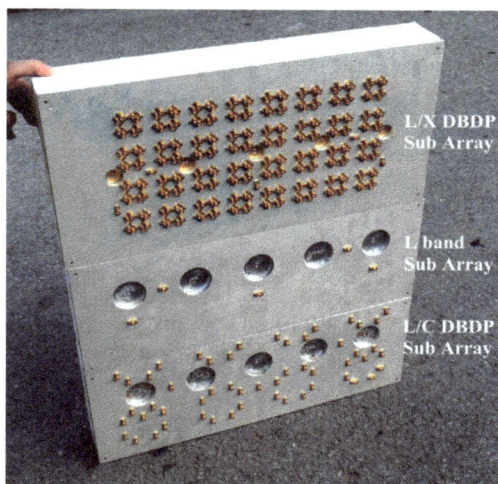

(a)

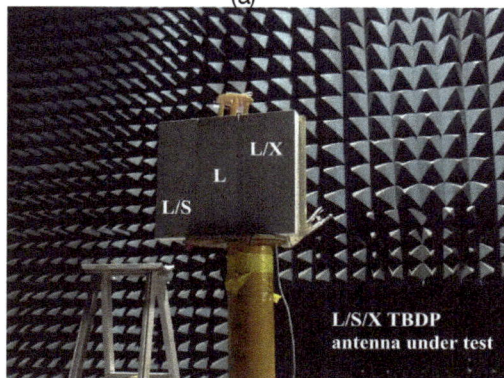

(b)

Figure 8. Photos of TBDP array

4.1. L-band

The measured L-band S_{11} and radiation patterns are shown in Fig.9 and Fig.10. It is seen that though the boundary conditions of L-band elements in L, L/S and L/X sub-arrays are quite different, similar return loss performance is measured except for LS4, which perhaps is caused by the solder false. The measured VSWR≤2 bandwidth is 167MHz (1.163-1.330GHz, 13.4%), and the array isolation is better than 37dB over the whole bandwidth (See Fig.9b). The array isolation hereinafter is defined as $-10\log_{10}|S_{21}|$, which is a positive dB value.

Although the whole L-band array is compounded by the elements in three separate sub-arrays, the measured radiation patterns keep in accordance with the theoretical ones, which means that the compound method raises little impact on the resulting radiation patterns. The cross-polarization level remains lower than -30 dB within the main lobe. The measured gain is 13.2dB (H-port) and 14.6dB (V-port), corresponding to the antenna efficiency of 62% and 61.2%, respectively. These lower figures may be related to its long electric-length of the microstrip and two-wire line feeder.

4.2. S-band

The narrowest VSWR≤2 bandwidth among all elements is defined as the array VSWR≤2 bandwidth. In S-band, an array VSWR≤2 bandwidth of 14.8% (3.25-3.768GHz) is measured according to Fig.11a, while the measured array isolation remains better than 45dB over the bandwidth (Fig.11b).

The measured radiation patterns agree well with the theoretical ones and the cross-polarization level keeps lower than -30dB in the main lobe (See Fig.12). Its measured front-back ratio is 37.8dB. The measured gain of 18.6dB is achieved for the 8×2 elements aperture, which means an antenna efficiency of 92.4%. Moreover, instead of the phase shifter, the coaxial line with custom-made length is applied to realize the scan experiment, as shown in Fig.13. From Fig. 14, a scan capability of ±27°is observed.

4.3. X-band

As shown in Fig.15, an array VSWR≤2 bandwidth of 16.8% (9.098- 10.781GHz) is measured in X-band with the array isolation better than -43dB over the whole bandwidth. From Fig.16, the measured radiation patterns fit well with the theoretical ones with the cross-polarization level better than -35dB in the main lobe. The measured front-back ratio is 42.3dB at the broadside direction, which is more than 4dB better than that of S band. It benefits from the larger electric size of the ground. Measured gains of 22.19dB (16×2 elements, V port) and 21.79dB (16×2 elements, H port) are realized, which mean the efficiencies of 92.7% and 84.5% at V- and H-ports, respectively. The scan radiation patterns of 30° in both ports and both planes were also measured. As shown in Fig. 17a-c,, the grate lobe of a little higher than -10dB appears around 80°-90°.

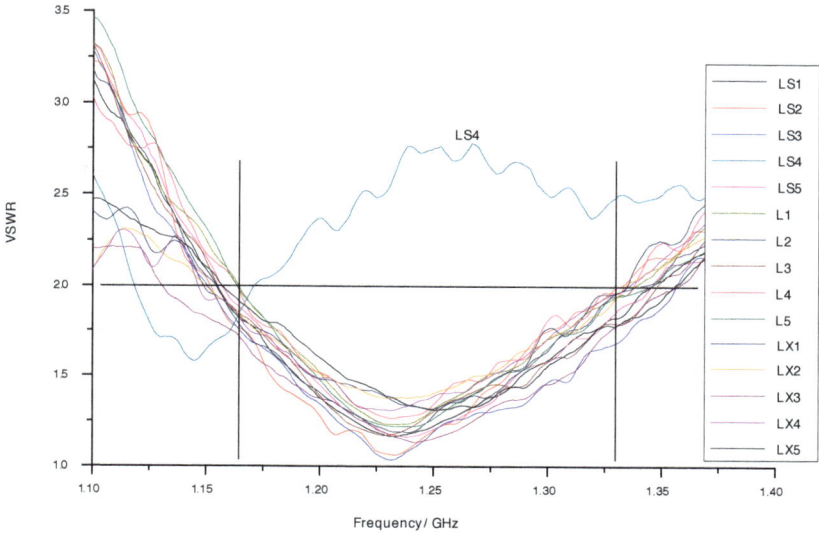

(a) VSWR of L-band elements (in L/S, L and L/X sub-arrays)

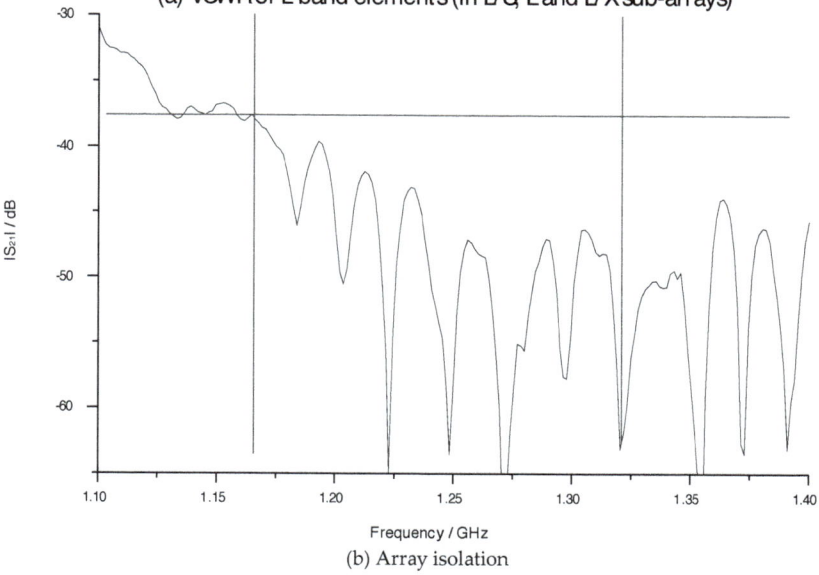

(b) Array isolation

Figure 9. Measured S parameters in L- band

(a) Horizontal-port

(b) Array isolation

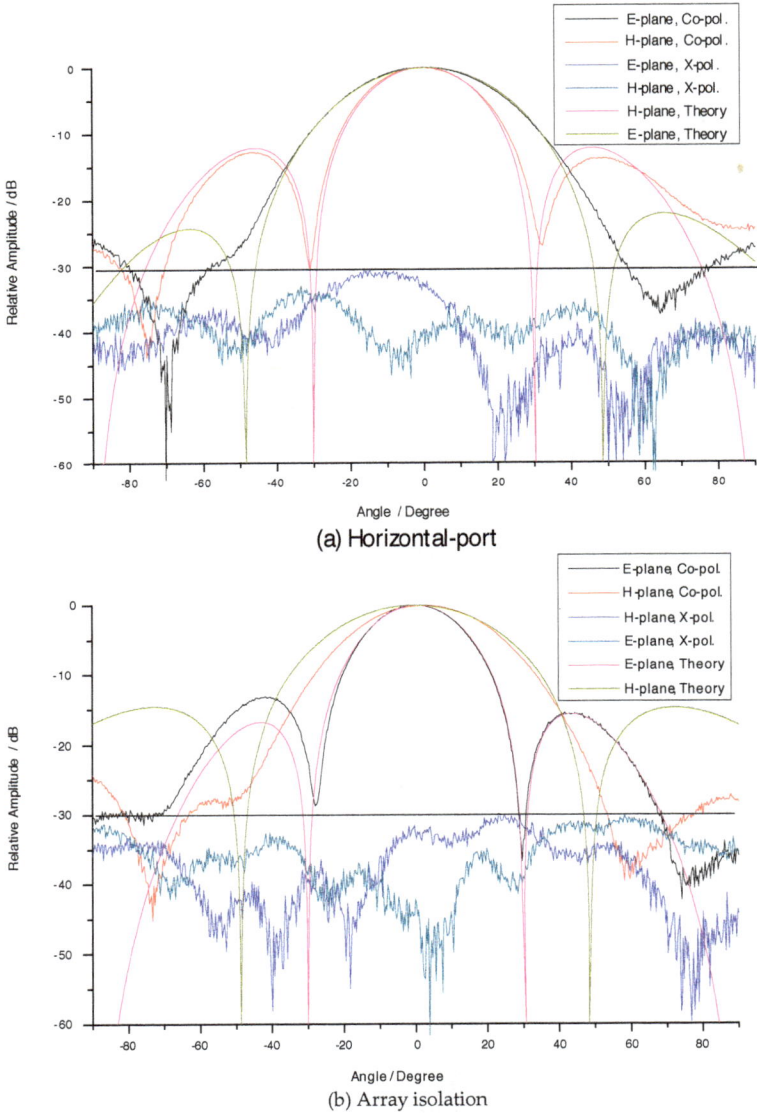

Figure 10. Measured and theoretical radiation patterns in L-band

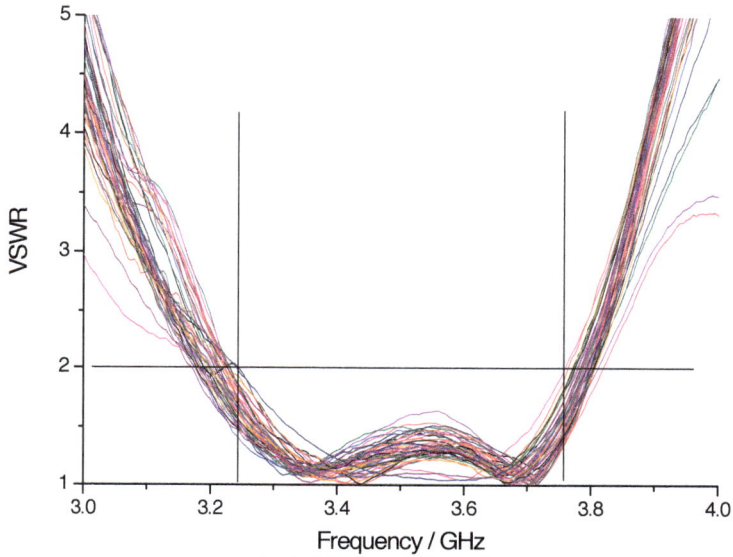

(a) VSWR of S-band elements

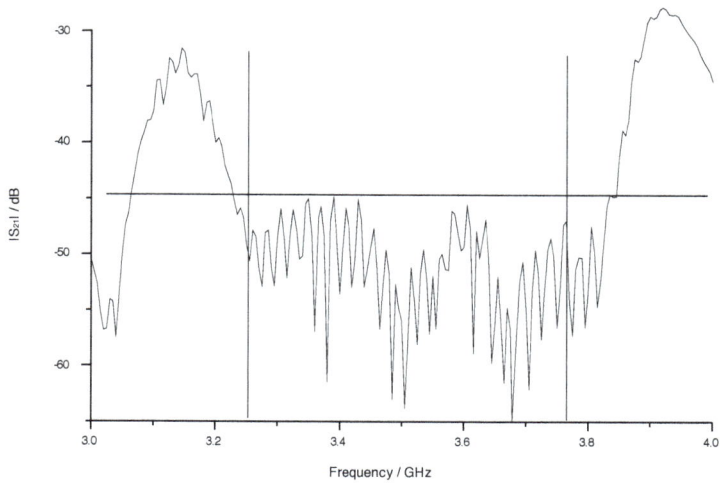

(b) Array isolation

Figure 11. Measured S parameters in S-band

(a) Horizontal-port

(b) Vertical-port

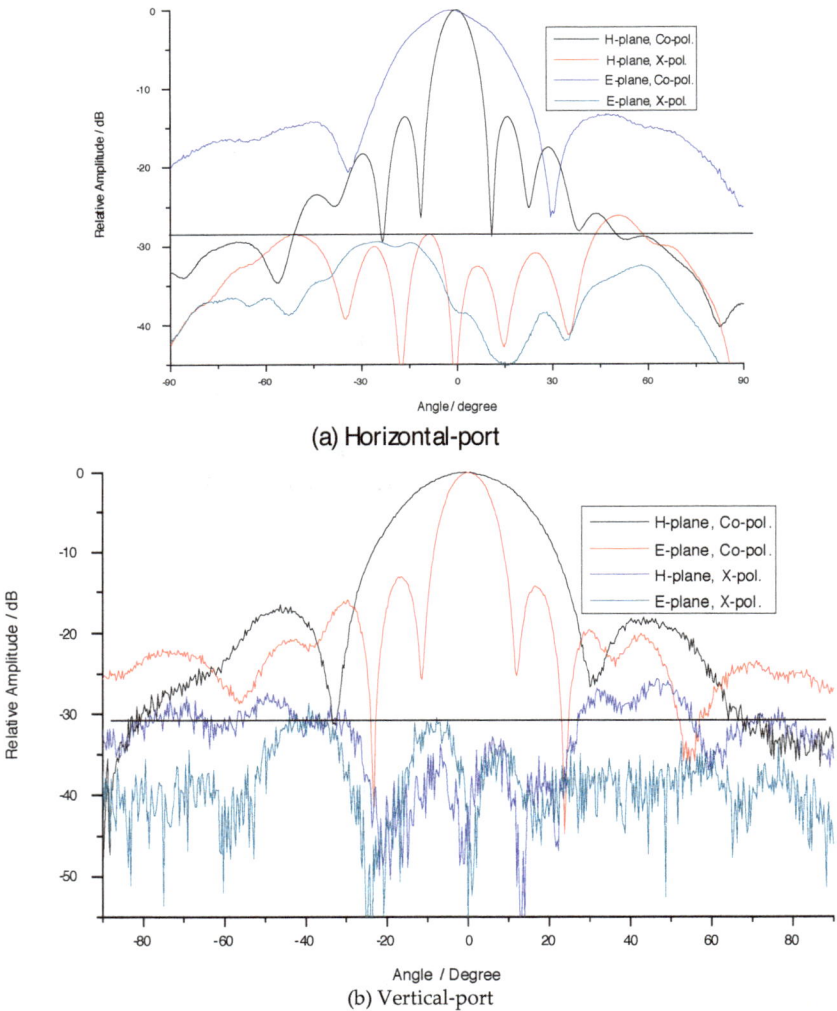

Figure 12. Measured radiation patterns in S-band

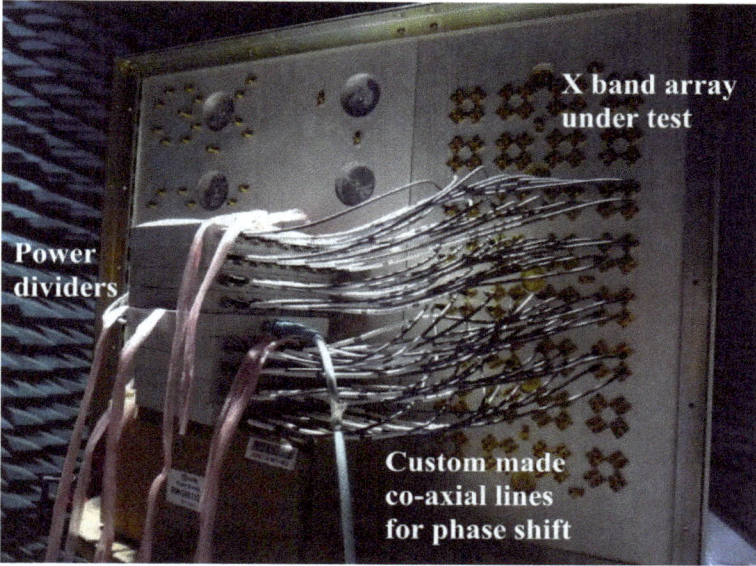

Figure 13. Custom-made coaxial lines instead of phase shifters

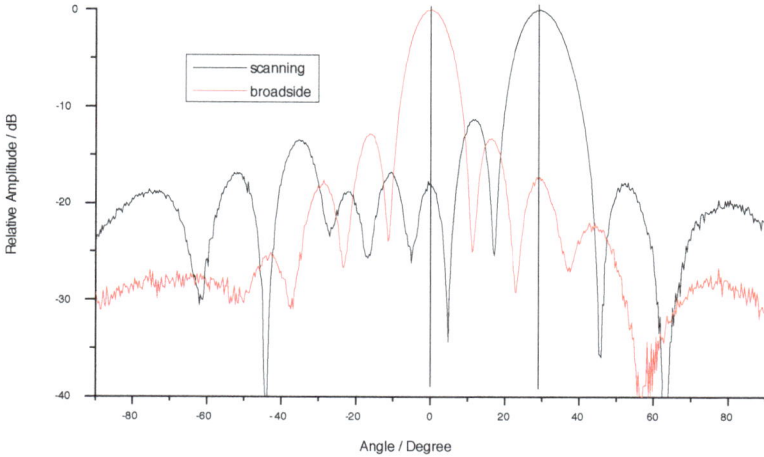

Figure 14. Scanning radiation patterns in S-band

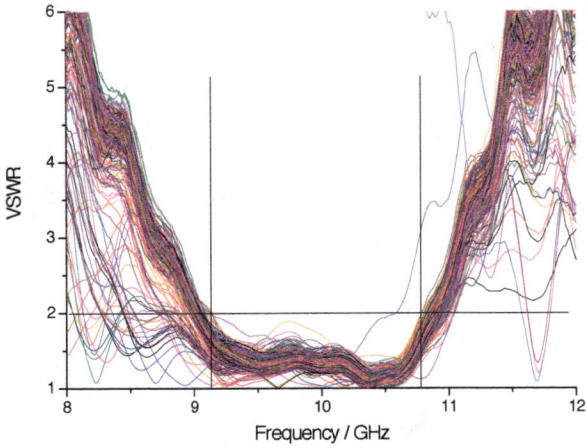

(a) VSWR of X-band elements

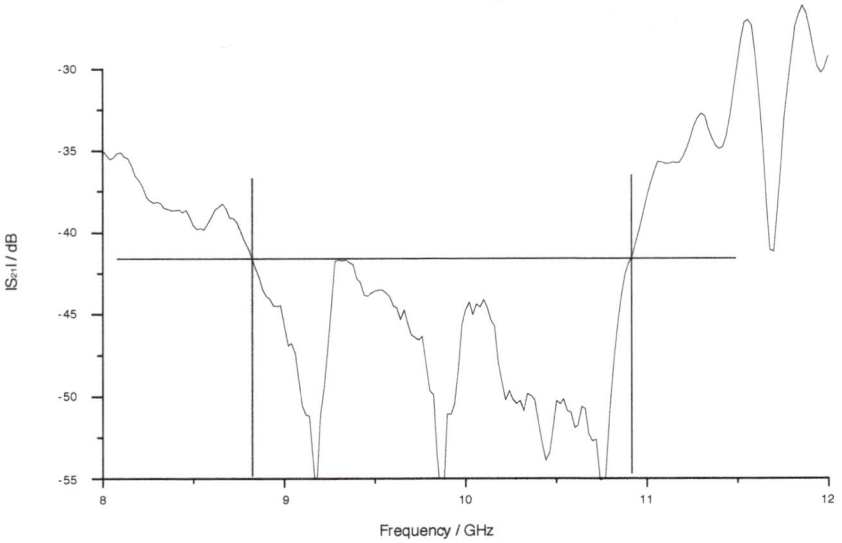

(b) Array isolation

Figure 15. Measured S-Parameters in X-band

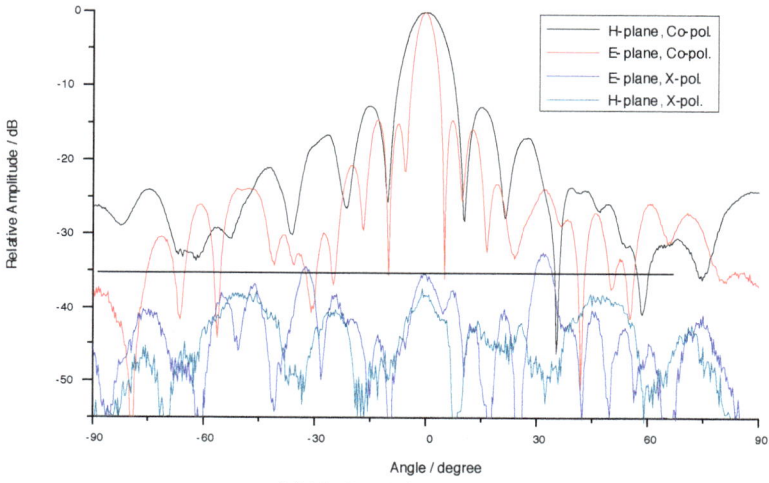

(a) Horizontal-port

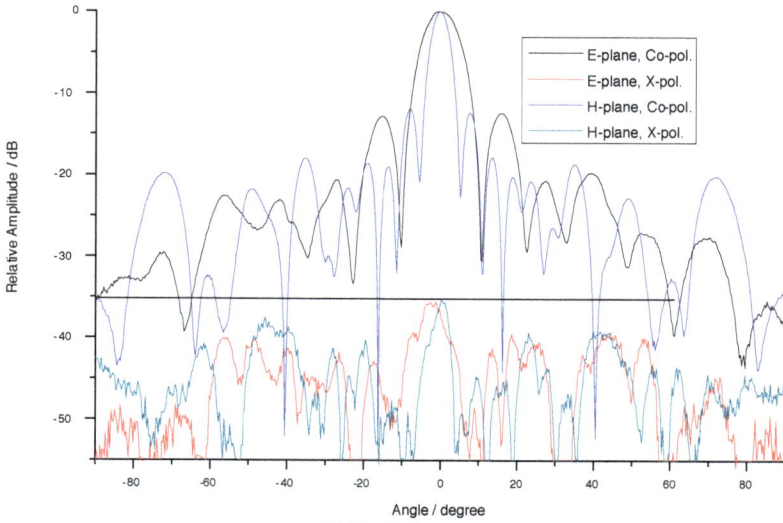

(b) Vertical-port

Figure 16. Measured radiation patterns and cross-polarization in X-band

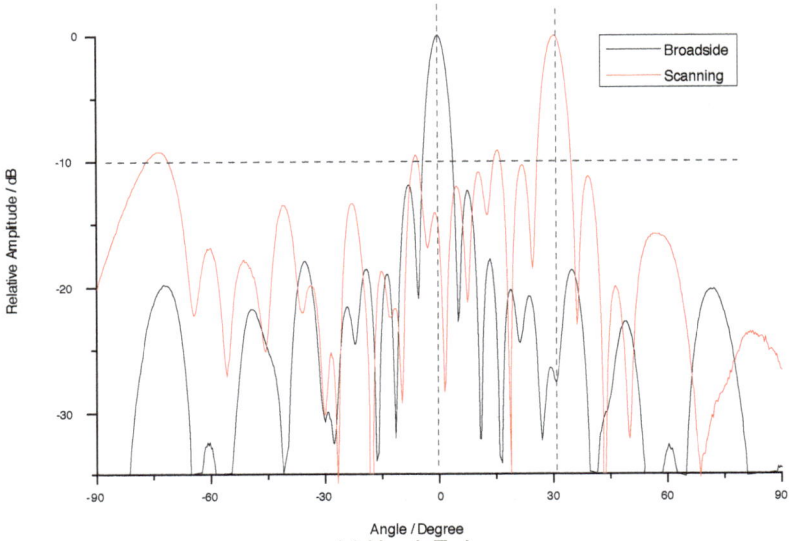

(a) H-pol. E-plane

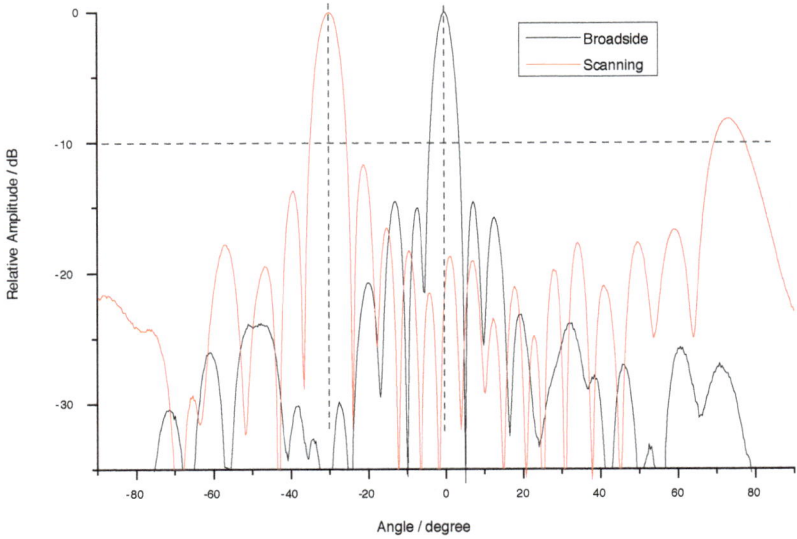

(b) V-pol. H-plane

(c) V-pol. E-plane

(d) H-pol. H-plane

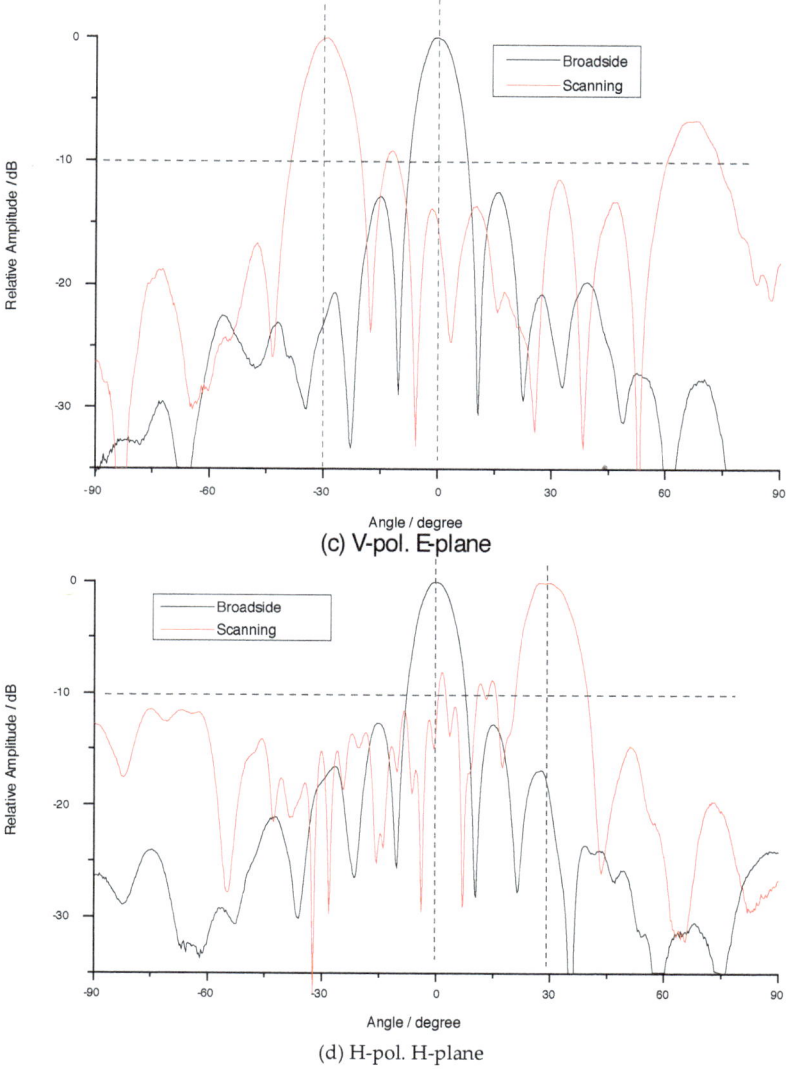

Figure 17. Measured Scan Radiation Patterns in X-band

5. Conclusion

The method of assembling two DBDP shared-aperture sub-arrays and one single-band DP sub-array to form a MBDP shared-aperture array has been introduced. An array prototype has been fabricated and measured to validate the method. According to the measured results, the array prototype achieves satisfactory results: similar bandwidth of 13.4%~16.7% in three bands, the array isolation of better than 37dB for all bands and the cross-polarization level of lower than -30dB within the main lobe region and the scanning capacity of ±27 degree at S- and X- bands. The array prototype has a 33% off in the aperture size as compared with tri-band independent aperture antenna and exhibits robust characteristics throughout the bandwidths. This array design method can be extended to the shared-aperture arrays with more than three bands.

Acknowledgements

This work was supported by the National Nature Science Fund of China under Grants No. 60871030, No.61171031, and the National High-Technology Research and Development (863) Project of China under Grant No. 2007AA12Z125.

Author details

Shun-Shi Zhong and Zhu Sun

Shanghai University, China

References

[1] Jordan, R. L, Huneycutt, B. L, & Werne, M. "The SIR-C/X-SAR synthetic aperture radar system", IEEE Trans. on Geoscience and Remote Sensing, (1995). , 33(4), 829-839.

[2] Zhong, S..-S., "DBDP SAR Microstrip Array Technology", in Nasimuddin ed., Microstrip Antennas, InTech, Croatia, Chapter 17, (2011). , 433-452.

[3] Pozar,D. M.,and Targonski,S. D., "A shared-aperture dual-band dual-polarized microstrip array", IEEE Trans. Antennas Propagat., Feb. (2001). , 49(2), 150-157.

[4] Vetharatnam, G., Kuan, C. B., and Teik, C. H., "Combined feed network for a shared-aperture dual-band dual-polarized array", IEEE Antennas and Wireless Propagat. Letters, (2005). , 4, 297-299.

[5] Mangenot, C, & Lorenzo, J. "Dual band dual polarized radiating subarray for synthetic aperture radar", IEEE Antennas and Propagat. Society Int. Symp., (1999). , 3, 1640-1643.

[6] Hsu , S.-H., Ren, Y.-J., and Chang, K., "A dual-polarized planar- array antenna for S-band and X-band airborne applications", IEEE Trans. Antennas Propagat., Aug. (2009). , 51(4), 70-78.

[7] Soodmand, S., "A novel circular shaped dual-band dual-polarized patch antenna and Introducing a new approach for designing combined feed networks", 2009 Loughborough Antennas & Propagat. Conf., Loughborough, UK, Nov. (2009). , 401-404.

[8] Vallecchi, A, Gentili, G B, & Calamia, M. "Dual-band dual polarization microstrip antenna", IEEE Antennas and Propagat. Society Int. Symp.,, (2003). , 4, 134-137.

[9] Pozar, D. M, Schaubert, D. H, Targonski, S. D, & Zawadski, M. "A Dual-band dual-polarized array for spaceborne SAR", IEEE Antennas and Propagat. Society Int. Symp., Atlanta, GA, (1998). , 4, 2112-2115.

[10] Qu, X, Zhong, S. S, Zhang, Y. M, & Wang, W. "Design of an S/X dual-band dual-polarised microstrip antenna array for SAR applications", IET Microw. Antennas Propag., (2007). , 1(2), 513-517.

[11] Gao G. , Zhang, Y., Li, A., Zhao, J., &Cheng, H., "Shared-aperture Ku/Ka bands microstrip array feeds for parabolic cylindrical reflector", 2010 International Conference on Microwave and Millimeter Wave Technology (ICMMT),, (2010). , 1028-1030.

[12] Uher, R. P.J., & Pozar, D. M., "Dual-frequency and dual-polarization microstrip antennas for SAR applications", IEEE Trans. Antennas Propagat., Sept. (1998). , 46(9), 1289-1296.

[13] He , S. & Xie, J., "Analysis and design of a novel dual-band array antenna with a low profile for 2400/5800-MHz WLAN systems", IEEE Trans. Antennas Propagat., Feb. (2010). , 58(2), 391-396.

[14] Zhong, S.-S., Sun, Z., & Tang, X.-R., " Progress in dual-band dual polarization shared-aperture SAR antennas", Frontiers of Electrical and Electronic Engineering in China,Sept.(2009). (3), 323-329.

[15] Schippers, H, Verpoorte, J, Jorna, P, Hulzinga, A, Thain, A, Peres, G, & Van Gemeren, H. "Development of dual-frequency airborne satcom antenna with optical beamforming",, 2009 IEEE Aerospace conference, Big Sky, MT, March (2009).

[16] Li, P, Luk, K. M, & Lau, K. L. "A Dual-feed dual-band L-probe patch antenna", IEEE Trans. Antennas Propagat., July (2005). , 53(7), 2321-2323.

[17] Parker, G. S, Antar, Y. M. M, Ittipiboon, A, & Petosa, A. "A dual polarised microstrip ring antenna with good isolation", Electronics Letters, May (1998). , 34(11), 1043-1044.

[18] Gao, S.-C., Li, L.-W, Leong, M.-S, & Yeo, T.-S., "Dual-polarized slot-coupled planar antenna with wide bandwidth", IEEE Trans. Antennas Propagat., Mar. (2003). , 51(3), 441-448.

[19] Waterhouse, R. B. "Design of probe-fed stacked patches", IEEE Trans. Antennas Propagat., Dec. (1999). , 47(12), 1780-1784.

[20] Gao ,S. & Sambell, A., "Dual-polarized broad-band microstrip antennas fed by proximity coupling ", IEEE Trans. Antennas Propagat., Jan. (2005). , 53(1), 526-530.

[21] Wong H., Lau, K.-L. , & Luk, K.-M., "Design of dual-polarized L-probe patch antenna arrays with high isolation", IEEE Trans. Antennas Propagat., Jan. (2004). , 52(1), 45-52.

[22] Brachat, P, & Baracco, J. M. "Printed radiating element with two highly decoupled input ports", Electronics Letters, Feb. (1995). , 31(4), 245-246.

[23] Wong, K.-L., Tung, H.-C., & Chiou, T.-W., "Broadband dual-polarized aperture-coupled patch antennas with modified H-shaped coupling slots", IEEE Trans. Antennas Propagat., Feb. (2002). , 50(2), 188-191.

[24] Yamazaki, M, Rahardjo, E. T, & Haneishi, M. "Construction of a slot-coupled planar antenna for dual polarisation", Electronics Letters, Oct. (1994). , 30(22), 1814-1815.

[25] Guo, Y.-X., Luk, K.-M., & Lee, K.-F., "Broadband dual polarization patch element for cellular-phone base stations", IEEE Trans. Antennas Propagat., Feb. (2002). , 50(2), 251-253.

[26] Chiou, T.-W., & Wong, K.-L., "Broad-band Dual-polarized single microstrip patch antenna with high isolation and low cross polarization", IEEE Trans. Antennas Propagat., Mar. (2002). , 50(3), 399-401.

[27] Sun ,Zhu, Zhong, Shun-Shi, & Tang, Xiao-Rong, "C-Band dual-polarized stacked patch antenna with low cross-polarization and high isolation", EuCAP 2009, Berlin, German, March (2009).

[28] Wincza, K, Gruszczynski, S, & Grzegorz, J. "Integrated dual-band dual-polarized antenna element for SAR applications", 2009 IEEE Wireless and Microwave Technology Conference (WAMICON'09), Clearwater, FL, April (2009).

[29] Zhong, S.-S., Sun, Z., Kong, L.- B., Gao, C., Wang ,W., & Jin, M.-P., "Tri-Band Dual-Polarization Shared-Aperture Microstrip Array for SAR Applications", IEEE Trans. Antennas Propagat., Sep. (2012). , 60(9)

[30] Woelders , K., & Granholm, J., "Cross-polarization and sidelobe suppression in dual linear polarization antenna arrays", IEEE Trans. Antennas Propagat., Dec. (1997). , 45(12), 1727-1740.

[31] Uz Zaman, A., Manholm, L.,. &. Derneryd, A., "Dual polarised microstrip patch antenna with high port isolation", Electronics Letters, Sept. (2007). , 43(10), 551-552.

[32] Kong, L.-B., Zhong, S.-S., & Sun, Z., "Broadband microstrip element design of a DBDP shared-aperture SAR array", Microwave and Optical Technology Letters, Jan. (2012). , 54(1), 133-136.

Compact Planar Multiband Antennas for Mobile Applications

Ahmad Rashidy Razali, Amin M Abbosh and Marco A Antoniades

Additional information is available at the end of the chapter

1. Introduction

This chapter focuses on the design of compact planar multiband antennas intended for existing wireless services including LTE, GPS, GSM, PCS, DCS, GPS, UMTS, WLAN and Wi-MAX bands. The present techniques available in the open literature include the modification of the main radiator via bending, folding, meandering and wrapping. Each approach offers different advantages, depending on the required application. The constraint for the lower band generation is the main challenge in radiator miniaturization. The quarter wavelength radiator that is subjected to miniaturization may suffer from limited bandwidth and low radiation efficiency. An alternative approach which relies on modifications to the ground plane is a promising technique, which often has been previously overlooked by antenna designers. The introduction of a ground slot in a finite antenna ground plane can be further extended to include reconfigurable features. Thus, such antennas that are compact and have multiband capability can be promising candidates for many wireless applications.

2. Design guidelines for planar antenna configurations

Antennas for wireless communications are commonly developed in the form of passive planar structures, which consist of a main radiating element, a supporting ground plane, a supporting substrate, and a feeding structure. The design configurations, sizes and type of substrates will depend on the desired frequency of operation and its radiation performances. Nowadays, planar antennas for wireless devices are mainly constructed using printed planar technology. Fundamentally, the small antenna design is based on the configuration of a *monopole*. Following the expansion of research on antenna design, the monopole configuration started to change, forming many alternative designs such as the popular *planar inverted-F antenna* abbreviated as PIFA and co*planar inverted-F antenna* abbreviated as CIFA.

In general, there are two main considerations that govern planar antenna designs; antenna miniaturization techniques and multiband operation. In antenna designs, multiband operation can be achieved by modifications to the main radiator applied using two strategies. The first strategy is to create several radiators for different resonances from a single feeder. The second strategy is to elongate the main radiator's physical length to achieve multiple resonant modes. However, creating several radiating branches may occupy more space, hence making the antenna physically larger than the desired volume. For this reason, while designing a multiband antenna, the designs must also apply miniaturization approaches, such as the ones presented in the following design strategies section.

3. Design strategies

3.1. Modification of the main radiator

The main radiator of a monopole/PIFA/CIFA type antenna plays a major role in determining the resonant frequency of the antenna. For a small antenna, particularly the $\lambda/4$ radiator configuration, miniaturization can be achieved in many ways. The main radiator arm can be modified by changing its configuration in such a way that it occupies optimally the limited area/volume provided for it.

One of the design techniques to achieve multiband operation is to have branches for the main antenna radiator. The branches are of monopole strips or arms to create different current paths for different resonances. This technique allows the excitation of multiple resonant frequencies at their fundamental mode. Multi-stacking or multi-layering is another technique that offers similar operation as the multi-branching technique. Similarly, it is capable of creating different current paths and thus different resonances.

A multi-resonator configuration is also achievable through proper application of some sort of slot or slit, sometimes also known as a notch geometry. In this approach, the radiator is modified in such a way that its original geometry is introduced with fine-tuned defined configurations of slots or slits. The resulting modification separates the main current stream into several other paths, which in turn create different resonators. It is worthwhile to mention that designs utilizing slots are not necessarily limited to straight lines or rectangular geometries. Other shapes have also been proposed such as a U-slotted PIFA, a V-slot loaded patch antenna, a Z-shaped slot antenna, or an open-ended Rampart-slot antenna.

The second strategy for the antenna design has the goal of obtaining a compact design and multi-band operation while using a single radiator without splitting it into branches. A considerable number of work employing this design strategy can be identified in the antenna literature. Among the popular techniques to elongate a single radiator to achieve multiple mode resonances without diminishing the antenna's compact feature include spiralling, looping, folding into a 3-D geometry, and bending.

Miniaturization can also be accomplished by meandering the antenna structure. The working principle is similar to that of spiralling. The meandered geometry preserves the original length of the radiating element but miniaturizes its overall size. The resulting input impedance of the meandered geometry is usually different from its spiral counterpart. Several factors determine the variation of the input impedance such as the gap between the opposite meandered lines and the width of the meandered structure. The meandered geometry can be applied in a variety of antenna types.

All the above-considered techniques for compact multi-band antennas offer reasonable solutions. However, they introduce considerable complexity with regard to the control of matching the input impedance of the antenna. Because of this problem, some antenna designers have focused their attention on providing better control of the input impedance match without being constrained by miniaturization. This strategy is the main motivation behind the approach involving parasitic elements within the antenna geometry. The aim of using parasitic elements is to compensate the impedance mismatch and to achieve better resonant frequency tuning.

3.2. Modification of the ground plane

As a result of the advancement in RF transceivers, the space allocated for its radiating element has become smaller. The reduced space is due to the increase in the number of new circuitry needed to provide better data channelling. The modification of the radiating element in a compact volume has been very challenging. An alternative solution is a better utilization of its ground plane. Even though the geometry of the main radiating element plays a main role in determining the resonant frequency and other performances of an antenna, the importance of the ground plane as the natural complimentary agent to a radiating current must not be neglected. The modification includes size variation, location of radiating element within the ground plane area or inserting slots in the ground plane.

Variations in a finite ground plane size and geometry affect the performance of an antenna. The resonant frequency of a conventional PIFA starts to converge to that of the case of an infinite ground plane when the size of the finite ground plane is increased above a unit wavelength. Also, the bandwidth increases with the length of the ground plane. With respect to the PIFA antenna used in a cellular phone, it was shown that an increased bandwidth, especially with respect to the lowest resonant frequency of operation, is achieved with a longer ground plane.

With respect to the complimentary role of ground plane to the main radiator, any modification to its geometry should also affect the overall antenna performance. Inserting slots is one obvious example. The insertion of ground slots creates some sort of discontinuity which causes the electric current launched by the primary radiator to reroute its path along the conducting surface of the ground. As a result, the electrical length of the ground is increased. With the strong coupling from the radiator, the ground slots cause a considerable impact on the input impedance. This positive impact includes the introduction of new resonances which are advantageous for the multiband design

3.3. Reconfigurable approach

With multiband capability, reconfigurable antennas can utilize more efficiently the radio frequency spectrum, facilitating better access to wireless services in modern radio transceivers. Reconfigurable antennas are generally divided into two main categories: frequency tunable and pattern diversity antennas. Furthermore, the selection of electronic switches is of paramount importance. Depending on the type of antennas, switches such as RF MEMS, varactors and PIN diodes can be used. The choice is governed by electrical specifications, fabrication complexity, bias requirement, switching time, and price.

4. Design examples

4.1. Coplanar IFA with fixed ground slots

Coplanar inverted-F antennas (CIFA) feature a low profile, compact size, and easy integration with an RF front-end. On the other hand, they feature a narrow operational bandwidth. Several techniques to increase the operational bandwidth or achieve multiband operation can be applied as discussed in [1, 2]. However, most of these techniques focus on modifications of the radiating element to either provide several radiating branches or elongating the radiator's dimension to generate multiple resonant modes. This approach faces a problem when the antenna has to be embedded into a small space as demanded by a compact transceiver. An alternative technique to provide multiple resonant frequencies or bandwidth enhancement is through a better utilization of the ground plane [3-10]. In this technique, secondary radiators are formed by ground slots, which introduce new resonant frequencies or enhance the already existing ones. The feasibility of this approach to enhance an impedance bandwidth has been demonstrated for a planar inverted-F antenna (PIFA) [6-8]. It has been shown that with the proper tuning of slot parameters, new resonant frequencies can be generated to provide multiband operation or increase the bandwidth [7, 8]. In all of these designs, a coaxial probe was used to feed the radiating patch. This configuration requires ground slots to be in close proximity of the primary radiator to excite efficiently new resonances. A shortfall of such a configuration is a limited means for tuning the slot dimensions. Also, the restricted slot locations may limit the antenna integration with the RF circuitry [9]. Recently, an alternative approach involving a CIFA and a microstrip feedline coupled to ground slots has been proposed [10]. According to the work described in [10], the use of the microstrip feedline in conjunction with the CIFA eliminates the shortcoming of a coaxial probe-fed patch. The reason is that this feeder can be positioned arbitrarily on the printed circuit board (PCB) and thus offers a more flexible coupling with ground slots to introduce new resonant frequencies. The work presented in [10] considers a single ground slot and was limited to WLAN frequency bands. In the following design, the work is extended to multiple ground slots and includes detailed simulation and experimental investigations.

Two configurations of CIFAs with slots in the ground plane are simulated, fabricated, and measured to validate the proposed multiband design technique. The design is accomplished with the aid of CST Microwave Studio 2009 [11].

4.1.1. Single and double ground slot configurations

The two investigated configurations of CIFAs are shown Figure 1. The design assumes a 0.508 mm thick RO4003 substrate with a relative permittivity $\varepsilon_r = 3.38$ and tan $\delta = 0.0027$. The total length of the meandered CIFA tail is designed to be of quarter wavelength of its operational frequency which is 2.45 GHz. The meandered tail including the shorting strip is set to be contained within an area of $W_p \times L_p = 6 \times 13$ mm^2. A 50 Ω microstrip line which is arranged in the same plane as the main radiator is used to feed this antenna. Note that the length and position of the microstrip feedline is only limited by the size of the finite ground, otherwise it can be arbitrary. The shorting strip of the CIFA is connected to the ground plane via the substrate. The other parameter of the CIFA are the shorting strip to the antenna feeder gap, $g = 1.15$ mm, while the width and gap of the meandered tail are set to 1 mm. On the opposite side of the substrate, the size of the ground plane is set to $W \times H = 40 \times 50$ mm^2.

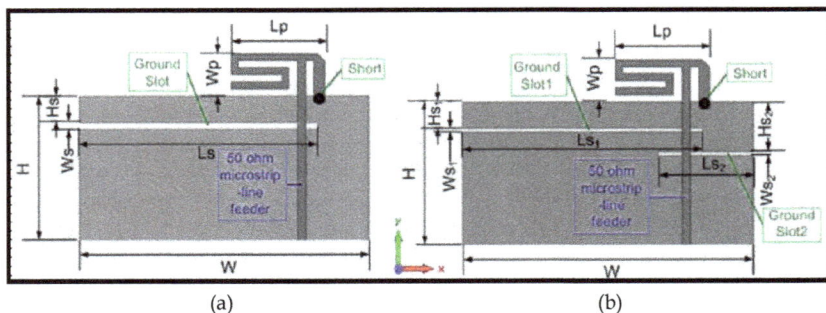

Figure 1. Antenna configurations. (a) Single ground slot CIFA for dual-band operation. (b) Double ground slot CIFA for tri-band operation.

Two configurations of CIFAs are introduced with two different ground slot outlines, as shown in Figure 1. As observed in Figure 1(a), the introduced slots are open-circuited in one arm and short-circuited at the other. The open-circuited arms of the ground slots are responsible for introducing new resonances, while the short-circuited ones act as tuning stubs. The initial lengths of the radiating arms are about one quarter-wavelength. The coupling locations along the microstrip feedline are selected to avoid an adverse loading of the primary radiator. The final dimensions are obtained from a parametric analysis performed using CST Microwave Studio. For the first configuration (Figure 1 (a)), an open-end slot with an optimized distance of $W_s \times L_s = 1 \times 33$ mm^2 and at a distance of $H_s = 4$ mm from the upper ground edge is designed. The second configuration (Figure 1 (b)) features two slots of dimension $W_{s1} \times L_{s1} = 0.5 \times 33$ mm^2 and $W_{s2} \times L_{s2} = 0.5 \times 13.2$ mm^2, respectively, placed at opposite edges. The distance of both slots with respect to the upper ground edge are H_{s1} = 4 mm *and* H_{s2} =7.5 mm. These two configurations indeed indicate that the microstrip feedline feed offers more flexibility to the ground slot coupling than a coaxial probe operating in conjunction with a PIFA.

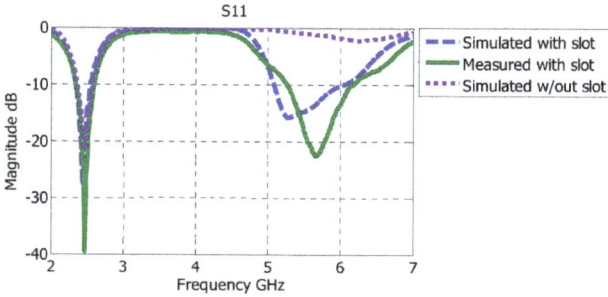

Figure 2. Simulated and measured reflection coefficient, $|S_{11}|$, for the CIFA of Figure 1(a) with and without a single slot.

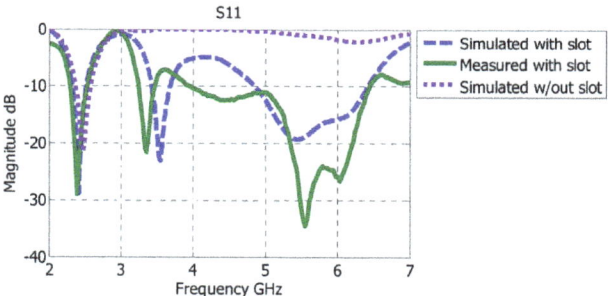

Figure 3. Simulated and measured reflection coefficient, $|S_{11}|$, for the CIFA of Figure 1(b) with and without a double slot.

The performances of both the CIFA configurations are assessed in terms of their reflection coefficients, radiation characteristics, and gain. Figure 2 shows the simulated and measured reflection coefficients of the first slot configuration of the CIFA. The presented results validate the proposed idea that the coupling of a single open-ended slot with a microstrip feedline is capable of generating a new resonant frequency at about 5.5 GHz (900-MHz impedance bandwidth with $|S_{11}|$ below -10 dB). Together with the fundamental resonance of the CIFA radiator at 2.4 GHz, the dual-band CIFA can support the WLAN 2.4/5.5-GHz system. There is good agreement between the simulation and experimental results. The simulated and measured reflection coefficients for the CIFA configuration with a double ground slot are shown in Figure 3. The presented results demonstrate a promising multiband operation of the CIFA. The proposed technique introduces two additional resonant frequencies. The first one is above 3 GHz, and another one is above 5 GHz. The impedance bandwidths are quite wide and accommodate not only the 2.4/5.5-GHz WLAN, but also include 2.5/3.5/5.5 GHz for WiMAX. Again, there is a relatively good agreement between the simulated and measured results for the reflection coefficients. In addition, the proposed coupling to the ground slots does not affect the reflection coefficient performance of the CIFA around the original resonance of 2.4 GHz. This means that the design procedure does not introduce any extra complexities.

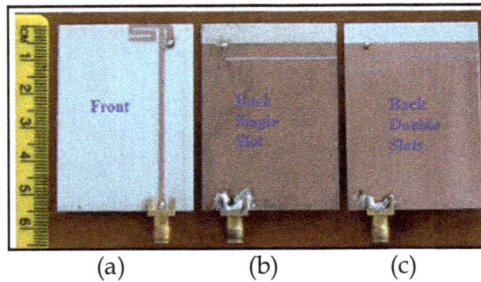

(a)   (b)   (c)

Figure 4. Fabricated prototypes of the meandered-tail CIFA. (a) Top view. (b) Bottom view of the single-ground-slot CIFA for dual-band operation. (c) Bottom view of the double-ground-slot CIFA for tri-band operation.

The measured radiation patterns for these antennas indicate approximately omni-directional characteristics along the principal planes. The measured gains of the dual-band CIFA are greater than 1.6 dBi, while for the tri-band CIFA they are more than 2.2 dBi. The photographs of the two manufactured varieties of CIFAs with ground slots are shown in Figure 4. They confirm a simple manufacturing structure for these multi-band antennas. Their overall size is quite compact and the slotted ground leaves still plenty of space for inclusion of RF and signal processing electronics.

4.1.2. Parallel and perpendicular ground slot configurations

In the following work, an investigation to study the effect of different orientations of open-end ground slots that are coupled to the microstrip feedline is presented. In contrast to the previous work, the CIFA configuration with a straight quarter wavelength radiating arm accompanied by differently oriented ground slots is considered. The substrate used is Rogers RO4003 with a dielectric constant of 3.38, tan δ = 0.0027 and thickness of 0.508 mm. The antennas are fabricated and their prototypes are experimentally tested. In the investigated antennas, a basic CIFA with a straight radiating arm is assumed, as shown in Figure 5. The microstrip feedline was then chosen to achieve coupling to ground slots that were located and oriented in different positions. In Figure 5(a), the open-end ground slot is offset by some distance from the CIFA and its orientation was chosen to be parallel to the CIFA radiating arm. Assuming a 0.508 mm thick RO4003 substrate, the total length of the PIFA arm is a quarter wavelength at 2.45 GHz. The width of the radiating arm is set to 1 mm. The total area of the antenna including the shorting strip is set to be contained within an area of $l_a \times h_a = 24$ mm $\times$ 3.5 mm. A 50 Ω microstrip feedline is arranged in the same plane as the CIFA. The length of this feedline is governed by the location of the input port of the RF front-end to which it has to be connected. A shorting connection of the CIFA is made using a conducting strip through the substrate. The shorting strip to the antenna feeder gap is $g = 1.15$ mm.

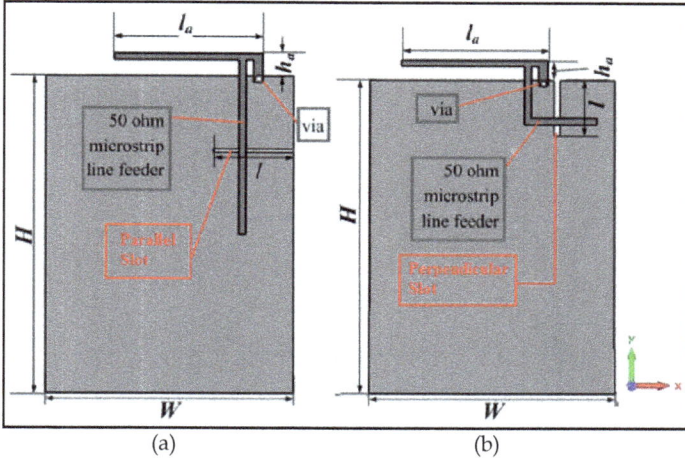

Figure 5. The first (a) and second (b) configurations of CIFAs with ground slots parallel and perpendicular to the main radiating arm, respectively.

On the opposite side of the substrate, the size of the ground plane is set to $W \times H = 40$ mm $\times$ 55 mm. An open-end slot with a length (l) and width of 13 mm $\times$ 0.5 mm is inserted in a parallel position with respect to the radiating arm as shown in Figure 5(a). The second configuration with the ground slot perpendicular to the CIFA is shown in Figure 5(b). In this case, the microstrip feeder is bent halfway exactly at a right angle towards the RHS edge of the ground plane, as shown in Figure 5(b). Concerning the length of the feeder, it can be arbitrary. However it must be properly coupled to the slot. Notice that the overall location of the primary radiator has been shifted by a few millimeters to the left to avoid the microstrip feed point from overpassing the edge. This time, the open-end slot with length (l) and width of 10 mm $\times$ 1 mm is inserted in a position that is perpendicular with respect to the radiating arm. It is noticeable that the length, width and position of the slots for both configurations are different. Their optimum dimensions are derived from parametric analyses performed with the use of CST Microwave Studio.

Figures 6 and 7 show the simulated reflection coefficients for the CIFAs without and with the slots. It is apparent from the results for the first configuration of CIFA shown in Figure 6

that the introduction of the slot has successfully introduced two additional resonant fre-
quency bands apart from the original 200 MHz band ($|S_{11}|$ below -10 dB) at around 2.45
GHz due to the primary CIFA radiating element. From the simulated reflection coefficient,
the new frequency bands can approximately cover a bandwidth of almost 300 MHz at 3.3
GHz and up to 1.14 GHz around 5.5 GHz with reference to the -10 db $|S_{11}|$ coefficient crite-
rion. These bands cover WLAN 2.45/5.25/5.85 GHz, HiperLAN/2 5.6 GHz and WiMAX
2.5/5.5 GHz. The 3.3 GHz resonant frequency band however is close to WiMAX 3.5 GHz.
Figure 7 shows the simulated reflection coefficient for the second configuration of CIFA.
Again, the multiband operation of this antenna is clearly observed. The slot insertion is
again responsible for the introduction of a new frequency band of 720 MHz bandwidth ($|$
$S_{11}|$ below -10 dB) at around 5.4 GHz. Together with the original 2.45 GHz resonant frequency band for the primary radiating element, the second configuration of the antenna covers
WLAN 2.45/5.25 GHz, HiperLAN/2 5.6 GHz and WiMAX 2.5/5.5 GHz.

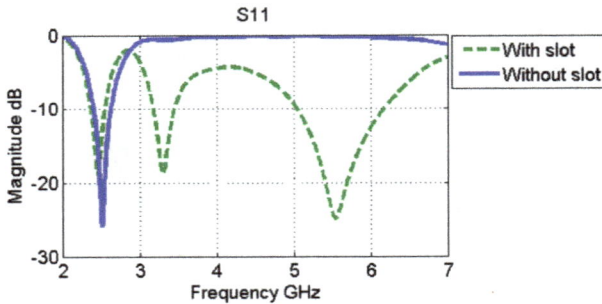

Figure 6. Simulated reflection coefficient, $|S_{11}|$, for the first configuration of the proposed CIFA shown in Figure 5 (a) with and without a parallel slot.

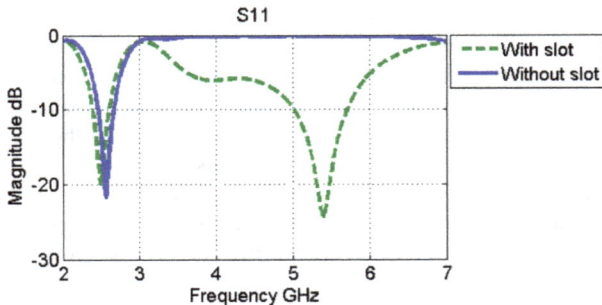

Figure 7. Simulated reflection coefficient, $|S_{11}|$, for the second configuration of the proposed CIFA shown in Figure 5 (b) with and without a perpendicular slot.

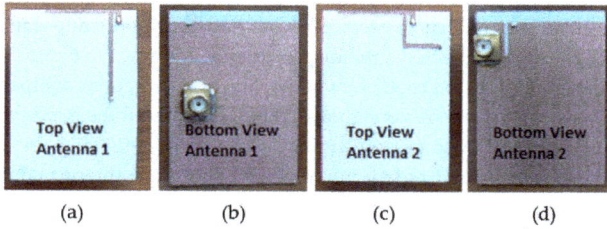

(a)   (b)   (c)   (d)

Figure 8. Fabricated prototypes of the two antenna configurations. (a) Top view of the first configuration of the CIFA. (b) Bottom view of the first configuration of the CIFA. (c) Top view of the second configuration of the CIFA. (d) Bottom view of the second configuration of the CIFA.

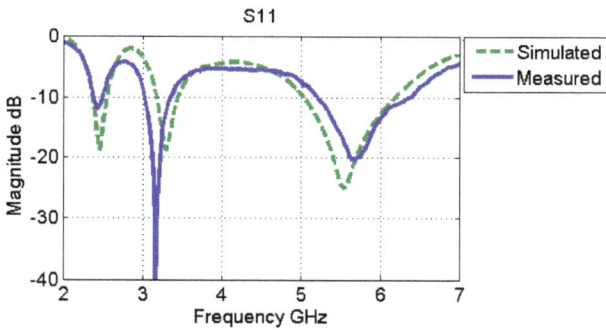

Figure 9. Measured reflection coefficient, $|S_{11}|$, for the first configuration of the CIFA prototype shown in Figures 8 (a) and (b), compared to the simulated result.

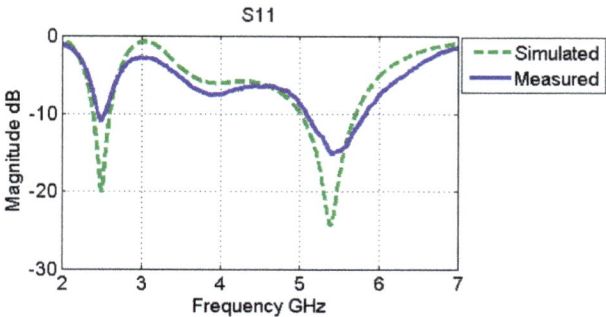

Figure 10. Measured reflection coefficient, $|S_{11}|$, for the second configuration of the CIFA prototype shown in Figures 8 (c) and (d), compared to the simulated result.

To verify the simulated performance, the two antennas were fabricated and experimentally tested with respect to their reflection coefficients, radiation patterns and gain. Photographs (front and back view) of the fabricated antenna prototypes are shown in Figure 8. Figures 9 and 10 present the measured reflection coefficient of the fabricated antennas compared with the simulated results presented earlier. A very good agreement between the experimental and simulation results for the two antenna prototypes is obtained confirming the validity of the presented designs. The measured radiation patterns for both the antenna configurations are shown in Figures 11 and 12. They were obtained in an indoor far-field range using a broadband linearly polarized horn antenna as a receiving antenna. For the first configuration of the CIFA, the co-polar component of the radiation pattern, except for one dip, is nearly omni-directional in the xy plane at 2.45 GHz, 3.3 GHz and 5.5 GHz, as observed in Figure 11. The cross-polar component in that plane is quite weak at 2.45 GHz but increases with frequency and is comparable to the cross-polar component at 5.5 GHz. In the remaining xz and yz planes the co- and cross-polar components are comparable in magnitude across the frequencies of 2.45, 3.3 and 5.5 GHz. This is the desired feature for portable wireless devices in mobile applications. For the second configuration of CIFA, the cross-polar component of the merasured radiation patterns at 2.45 GHz and 5.5 GHz show nearly omni-directional properties in the xy-plane. In this plane the cross-polar component is considerably larger than the cross-polar component. For the remaining xz and yz planes, the co- and cross-polar components are of similar strength. The measured gain of the antennas ranges between -0.45 to 3.95dBi.

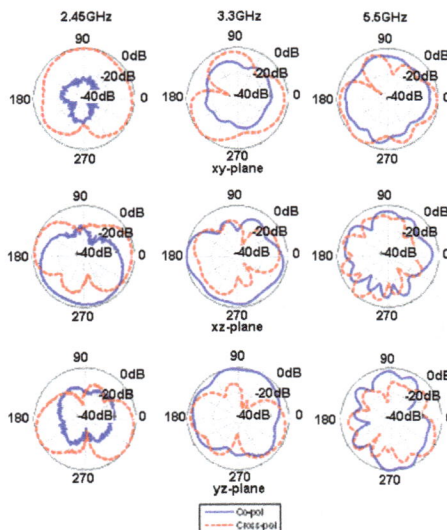

Figure 11. Measured radiation patterns for the first configuration of the CIFA prototype shown in Figures 8 (a) and (b) with a parallel ground slot, at 2.45 GHz, 3.3 GHz and 5.5 GHz.

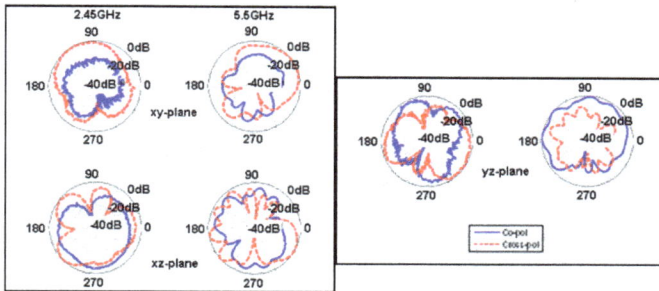

Figure 12. Measured radiation patterns of the second configuration of the CIFA prototype shown in Figures 8 (c) and (d) with a perpendicular ground slot, at 2.45 GHz and 5.5 GHz.

4.2. Coplanar IFA with reconfigurable ground slot

A reconfigurable coplanar inverted-F antenna to operate at the lower bands of wireless services; the GSM, PCS and UMTS service bands is presented. By introducing a dimension-tunable ground slot, the original resonance from the antenna main radiator is not affected. At the same time, frequency reconfigurability is achieved at the ground slot excitations. With multiband capability, reconfigurable antennas can utilize more efficiently radio frequency spectrum, facilitating a better access to wireless services in modern radio transceivers. Several methods have been reported in the literature to achieve reconfigurable antennas. Generally, they are divided into two main categories: frequency tunable and pattern diversity antennas. For frequency tunable antennas, much attention has been given to reconfigurable slot antenna designs [5, 12-14] due to the flexibility of slots in integrating electronic switches. The frequency tuning characteristics of a slot antenna can be achieved by changing the slot effective length [5, 12, 13] or by switching the connection between the feed and the ground [14]. Apart from using ground slots, frequency reconfigurability can also be achieved by changing the induced current distribution [15] or varying the ground plane electrical length [16] supporting a patch antenna. For pattern diversity antennas, reconfigurability can be obtained by adjusting the physical configuration of the antenna radiator to produce tunable radiation patterns [13, 17-22]. Another important element in reconfigurable configuration is the selection of electronic switches. Depending on the type of antennas, the switches such as RF MEMS [23-25], varactors and PIN diodes can be used. The choice is governed by electrical specifications, fabrication complexity, bias requirement, switching time, and price. For instance, RF MEMS switches are very low loss and their other advantages are that they do not require bias lines [23]. However, they are costly. PIN and varactor diodes are low cost and have a simple fabrication process. They require a proper bias network isolating the dc bias current from the RF signal, which usually leads to a complicated biasing network. The complicated dc bias network can sometimes be avoided, and one such solution has been reported in [13]. Furthermore, the limited operating frequency of some commercial low cost PIN diodes can be overcome using solutions proposed in [26]. Most of frequency reconfigurable slot antennas generate only single operating bands at a particular reconfigurable mode. Although many conventional multiband techniques exist such as multi-mode resonator or multi-resonator [27], they are difficult to implement in reconfigurable slot antennas. However, as mentioned in Section 4.1, a coplanar inverted-F antenna (CIFA) with ground slots is capable of generating a new higher-band resonance from a ground slot without affecting the original resonance of the patch radiator. This work on reconfigurable ground slots provides a logical extension to the previous work using a fixed ground slot CIFA, and describes a new solution to achieve a reconfigurable antenna capable of generating multiband operation at each reconfigurable mode. By means of a length-tunable ground slot, a new reconfigurable coplanar inverted-F antenna achieves dual-band operation at four different modes. The generated frequency bands cover several popular wireless services including GSM900, PCS1900 and UMTS2100.

In the antenna configuration, a $\lambda/4$ 900 MHz microstrip radiator of width L_2 = 2 mm is designed in CST Microwave Studio v2010. The radiator is fed by a 50 Ω microstrip feedline of width W_f = 3 mm and placed W_d = 6 mm from the antenna edge. To occupy a small area of L_r = 10 mm × W_g = 40 mm, one end of the radiator is folded twice with L_1 = 3 mm and W_1 = 26 mm to form the configuration shown in Figure 13. To realize an inverted-F antenna, the other end of the radiator is grounded through a via. This is to compensate the large capacitance introduced from the coupling of the folded arm to the ground. The other parameters are the gap between the feedline to the shorting strip G = 4 mm and the L_g = 90 mm × W_g = 40 mm ground plane supporting the antenna. The chosen ground plane size is typical for many wireless transceivers such as a mobile phone.

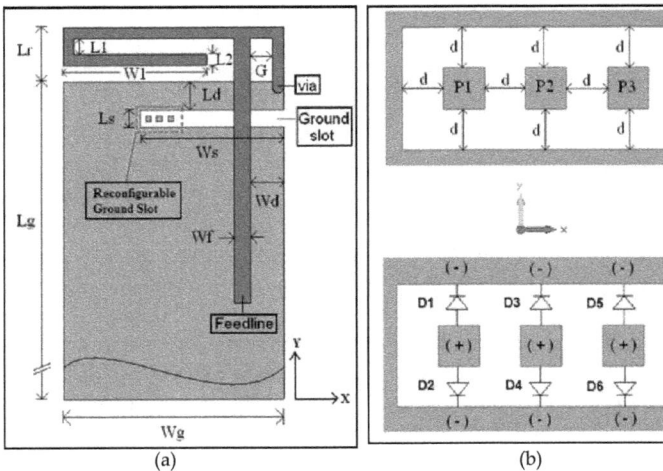

Figure 13. Proposed reconfigurable antenna. (a) Passive antenna configuration. (b) Close-up view of the short-end section of the ground slot and the configuration of the PIN diode switch bias network (marked by the red box in (a)).

Following the technique in [28], a slot is introduced in the ground plane to excite the higher frequency band centred about 1850 MHz. The ground slot is placed $L_d = 5$ mm from the ground top edge and has a dimension of $L_s = 3$ mm × $W_s = 26$ mm as shown in Figure 13(a). In order to achieve an electronically variable (reconfigurable) slot length, three identical 1 mm × 1 mm conducting pads (P1, P2 and P3) and three pairs of PIN diode switches are introduced in the ground slot as presented in Figure 13(b). The gap $d = 1$ mm in Figure 13(b) is chosen to allow uniform decrease of the slot effective length, thus allowing uniform increase in the excited resonant frequency. All final dimensions are achieved through optimization using a parametric study in CST Microwave Studio. The PIN diode used in the antenna is MAP4274-1279T (MACOM). It is a single-diode series with dimension of 1 mm × 0.7 mm and height of 0.6 mm. It is forward biased with a voltage of 0.7 V_{DC} and a current of 10 mA. During forward bias, it exhibits an intrinsic capacitance of 0.1 pF with forward bias resistance of 3 Ω.

Diode Combination	Diode States			
	Mode 1	Mode 2	Mode 3	Mode 4
D1 & D2	OFF	ON	OFF	OFF
D3 & D4	OFF	OFF	ON	OFF
D5 & D6	OFF	OFF	OFF	ON

Table 1. Reconfigurable Antenna Modes of Operation

The switches are designed to operate in three diode pairs. The switch states at each mode are given in Table 1. From the table, the ON-state indicates the diode is forward biased while the OFF-state indicates it is reverse biased. For simplicity in the simulation model, the ON-state diode is represented by a 1 mm × 1 mm PEC. In this case, the effect of the PIN diode forward biased resistance is neglected. The PEC is removed to represent the OFF-state diode. The photograph of the manufactured reconfigurable antenna is shown in Figure 14. The antenna is fabricated on a 1.6 mm FR4 substrate with relative permittivity of $\varepsilon_r = 4.4$. During measurements, the PIN diode switches are biased with a simple DC bias network as proposed in [13]. Each conducting pad (P1, P2 and P3) is initially soldered to a 1.2 kΩ resistor to protect the diodes and the VNA, as shown in Figure 14(d). A GW GPC-3030D dc power supply unit is used to bias the diodes. The S-parameter measurements are carried out using a R&S ZVA24 VNA.

Figure 15 shows both the simulated and measured S11 of the proposed antenna. A close agreement between the measured and simulated results is observed in each mode. The discrepancies are due to the idealized switches used in CST simulations. The measured results show that the proposed reconfigurable antenna covers two frequency bands: between 850 MHz to 960 MHz at the lower band and between 1800 MHz to 2140 MHz at the upper band (at 6 dB reflection coefficient or VSWR 3:1). As a result, the proposed antenna can be considered as a good candidate for wireless services application such as GSM900, PCS1900 and UMTS2100. In fact, the proposed antenna can cover additional bands surpassing mode 4, with extra pairs of switches in the tunable ground slot. In addition, the parameter d can be easily adjusted to obtain the desired band separation between adjacent modes. The measured radiation patterns at the desired frequencies are close to omni-directional. The antenna gain ranges from 1.4 dBi to 3.45 dBi, while the efficiency is between 63% and 80%. The gains are expectedly low due to the small electrical size of the antenna. The obtained efficiencies are due to conduction losses partially incurred in the PIN diodes switches, and are acceptable for portable wireless applications [27].

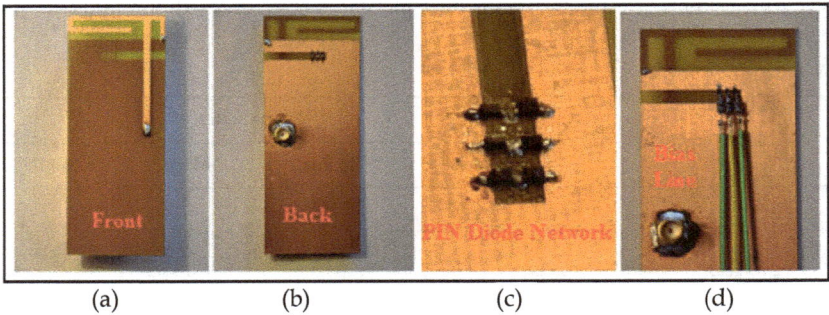

Figure 14. Fabricated prototype of the reconfigurable antenna. (a) Front view. (b) Back view. (c) Integrated PIN diode configuration. (d) Bias line connection to PIN diode switches.

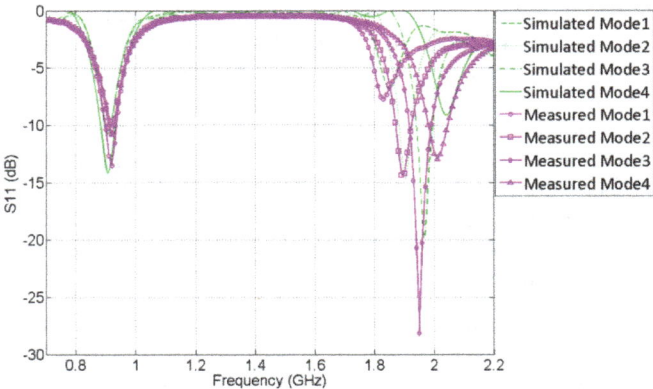

Figure 15. Measured and simulated results for the antenna input refection coefficient, $|S_{11}|$, for the four modes when the individual diodes are turned on or off.

4.3. Slim IFA with a modified ground plane

Apart from the coplanar configurations, inverted-F antennas can also be designed without a dielectric substrate for different transceiver applications. However, without the size-reduction factor by the substrate, the main problem with this antenna configuration is the large size of the main radiator and the ground. In the following work, a novel configuration of an antenna that combines wrapping, folding, ground slots/strips to achieve a super-slim (IFA) with enhanced operational bandwidth is proposed. To fulfill this challenging configuration, multiband radiators have to be miniaturized to fit into the small volume available for modern tranceiver modules. At present, there are many approaches used by antenna designers to reduce the overall projection area of a typical antenna to fit a portable wireless transceiver. Most of them require the use of a single feed structure, as preferred by the majority of wireless device manufacturers. The commonly used antenna for these purposes is a planar inverted-F antenna (PIFA). To meet the multiband operation requirement, its main radiator consists of either an increased single arm/branch or multiple arms/branches to excite multiple resonant modes [7]. The length of each radiating arm has to be at least one quarter wavelength ($\lambda/4$) at the lowest resonance frequency to meet the operation requirements in the lowest frequency band [7]. As a result, this is the longest part of the PIFA which requires miniaturization to fit the available space in a compact transceiver. Nevertheless, folding, meandering and wrapping of the arm and the remaining parts of the PIFA can drastically reduce its projection area. However, these miniaturization techniques have to be carefully applied as they shift resonant frequencies and adversely affect the radiation efficiency [29]. The shift in resonant frequencies is due to the constructive/destructive effects of parasitic reactances which are created during the PIFA structure modification. Usually the length of the electric current path becomes reduced when a thick radiating arm is bent or folded. In turn, a reduction in radiation efficiency is due to the coupling between adjacent parts of radiating arms introduced by folding, meandering or wrapping [29]. The opposite directions of current paths created in this process are responsible for a gradual cancellation of the radiated electromagnetic fields. As a result, the above mentioned miniaturization techniques include a tradeoff between the antenna size and its efficiency. To ease the design challenge, tuning and optimization is usually accomplished with the use of commercial full EM wave simulators such as CST Microwave Studio or Ansoft HFSS. For a typical portable device, such as a cellular phone, the height (thickness) of the primary radiator from 11 mm is reduced down to 7 mm [30]. However with the recent demand for slimmer transceivers, an even smaller height is required and can be as little as just 3 mm. One of the most attractive approaches to meet this miniaturization challenge is to wrap the main radiator in three dimensions. The main advantage of this technique is that it provides extra size reduction on top of folding and meandering of the radiating element. The benefits of this technique in relation to achieving compact antenna designs have been demonstrated in many recent works. For example, the wrapped monopole antenna proposed in [31] features a small volume of 12 x 15 x 7 mm while providing GSM/DCS operation. A similar slim design of a monopole antenna, which is based on wrapping, has been described in [32]. The presented antenna has a height of only 5 mm and delivers GSM, DCS and PCS operation. Another example of a wrapped monopole antenna has been shown in [33] and covers almost all popular the frequency bands between 850 MHz to 6 GHz at 6 dB reflection coefficient (VSWR 3:1). This antenna only occupies a volume of 60 x 13 x 5.6 mm. In [34], a wrapped PIFA capable of operating in frequency bands between 850 MHz and 3 GHz features a volume of 44 x 15 x 8 mm. A further example presented in [35] has shown that the height of the wrapped PIFA can be further reduced to only 4.8 mm while the coverage of GSM/PCS/DCS/UMTS/WLAN bands can still be well maintained. However all these designs do not met the super-slim height of 3 mm. In the following proposed solution, a super slim wrapped inverted-F antenna for GSM/DCS/PCS operation is initially designed by miniaturizing the main radiator. To achieve a reduced-size ground plane, the proposed solution includes the use of a narrow side strip and a small size ground slot. The proposed antenna occupies a low profile volume of just 40 x 10 x 3 mm^3 with a ground plane of 40 x 90 mm^2. The proposed antenna covers several LTE Bands, GSM850 (824-894 MHz) and PCS1900 (1850-1990 MHz) at 6 dB reflection coefficient or 3:1 VSWR.

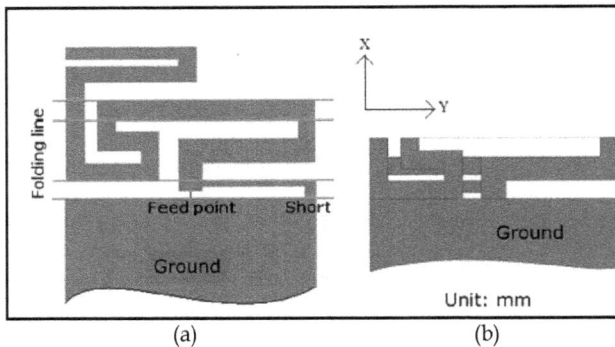

Figure 16. (a) Unwrapped, and (b) wrapped antenna configuration.

Initially, the main radiator configuration of the proposed antenna is shown in Figure 16. The antenna is 40 x 10 mm in area (in the xy-plane) and its height is 3 mm (along the z-axis). The supporting ground plane is 40 x 120 mm. The antenna uses no plastic substrate and is formed by a copper foil. Its very low profile is obtained using a fourth order folding/wrapping technique. As shown in Figure 16(a), it is fed at the middle part of the lowest folded arm with a 50 Ω U.FL mini coaxial cable that is commonly used in many portable devices such as computer notebooks. The lower end of the antenna main radiator arm is shorted to the ground as shown in the figure. The figure also shows the folding transition of the proposed antenna to transform its main radiator into a smaller wrapped dimension. It starts with an inverted-F radiator that has a $\lambda/4$ projection length between its open and short ends at 850 MHz. The arm is folded and meandered as shown in Figure 16(a). Finally the antenna is wrapped using the fourth order folding technique, giving the final height of 3 mm. Figure 17 presents the comparison between the measured and simulated reflection coefficient of the developed antenna. Good agreement between the measured and simulated results is apparent. The measured lower and upper resonant frequency bands of the proposed antenna for the 6-dB reflection coefficient (or VSWR 3:1) reference are 770 MHz to 970 MHz and 1710 MHz to 1990 MHz respectively. They cover the GSM 850/900/1800/1900 MHz, DCS 1800 MHz and PCS 1900 MHz wireless services.

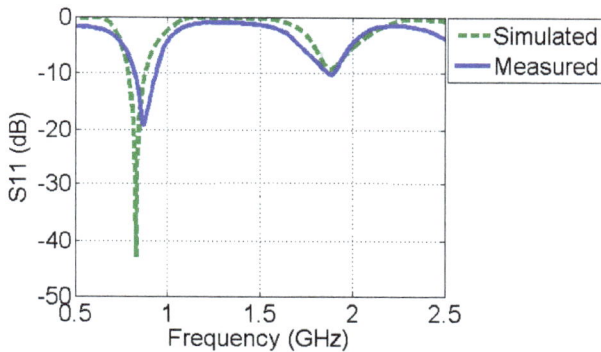

Figure 17. Measured and simulated (CST) reflection coefficient, $|S_{11}|$, for the wrapped antenna of Figure 16 (b).

To achieve a smaller size ground plane, the original 120 mm length is reduced to 90 mm. This reduction is about 25% from the original length. To maintain its original electrical length, the 30 mm long section could be folded. However, this idea does not work because of a large capacitance between the two sections.

A viable alternative is presented in Figure 18 where a 1×10 mm² slot and 2×83 mm² strip are introduced in the ground plane following the earlier work presented in [33]. As observed in Figure 18(b), the introduced modifications do not adversely affect any RF or signal processing modules that are usually placed on the ground plane. The small slot is in the close vicinity of the PIFA while the narrow strip is on the side of the ground plane. The two modifications use less than 5% of the reduced size ground plane area. Photographs of the proposed antenna with the ground modifications are shown in Figure 19. From the photographs, it can be seen that in these structures the plastic foam with air-like dielectric constant is used to support the wrapped radiator and the strip.

In the simulation results, the bandwidth of the lower frequency band reduces with the reduction of the ground plane length. However, the quality of the impedance match improves at the upper band. The reduction of the impedance bandwidth in the lower band is due to the analogous reduction of the optimal electrical length of the ground plane [7, 27].

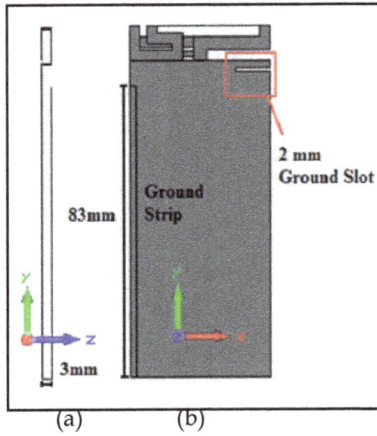

Figure 18. Detailed view of the proposed antenna with a ground plane modification. (a) Side view of the strip configuration. (b) Front view.

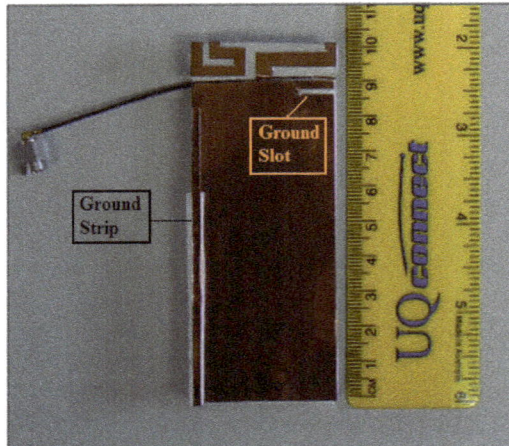

Figure 19. Fabricated prototype of the slim IFA with a modified ground plane.

Figure 20 shows the comparison between the measured and simulated values of the reflection coefficient for the developed antenna that includes the ground plane modifications. It is observed that at 6 dB reflection coefficient (3:1 VSWR), the measured bandwidths cover two bands; from 775 to 925 MHz and from 1800 to 2080 MHz. These bands include several LTE Bands, GSM850 and PCS1900 services.

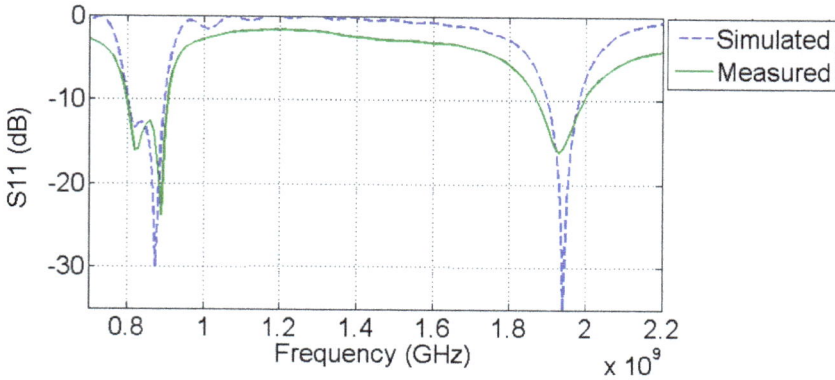

Figure 20. Measured and simulated reflection coefficient, $|S_{11}|$, for the slim IFA with a modified ground plane shown in Figure 19.

The measured radiation patterns in the lower band reveal that the developed antenna features high levels of cross-polarization which are welcome in mobile applications. In this case, the signal reception is unaffected by the orientation of the transceiver. Similar properties of the radiation pattern of the developed antenna are observed in the results for the upper band. The E-theta component dominates in the x-z plane while the E-phi component is predominant in the y-z plane. The cross-polarization is smaller than observed in the lower band. The measured peak gain of the antenna at both bands ranges between -1 to 2.5 dBi in the GSM bands and about 3 dBi in the PCS bands. These obtained gains are adequate for modern portable transceivers [27].

5. Conclusions

This chapter has presented the designs of compact planar multiband antennas for mobile and portable wireless devices. The antennas presented use miniaturization techniques of the main radiator, including meandering, bending, folding and wrapping to achieve compact size features, while the multiband operation of the antennas is generated from ground plane modifications using fixed slots, reconfigurable slots, and a ground strip. All the designs have utilized their ground planes to achieve multiband operation. Following the design guidelines, several novel solutions have been presented. All the presented design models have led to promising configurations for applications in wireless and mobile services.

Author details

Ahmad Rashidy Razali[1], Amin M Abbosh[2] and Marco A Antoniades[2]

1 Faculty of Electrical Engineering, University of Technology MARA, Shah Alam, Malaysia

2 School of Information Technology and Electrical Engineering, The University of Queensland, Brisbane, Australia

References

[1] Wang, Y., Lee, M., & Chung, S. (2007). Two PIFA-related miniaturized dual-band antennas. *IEEE Transactions on Antennas and Propagation*, 55(3), 805-811.

[2] Lee, C., & Wong, K. (2009). Uniplanar printed coupled-fed PIFA with a band-notching slit for WLAN/WiMAX operation in the laptop computer. *IEEE Transactions on Antennas and Propagation*, 57(4), 1252-1258.

[3] Dadgarpour, A., Abbosh, A., & Jolani, F. (2011). Planar multiband antenna for compact mobile transceivers. IEEE Antennas and Wireless Propagation Letters 10, , 651-654.

[4] Abbosh, A., & Dadgarpour, A. (2011). Planar multiband antenna for multistandard mobile handset applications. *Microwave and Optical Technology Letters*, 53(11), 2700-2703.

[5] Peroulis, D., Sarabandi, K., & Katehi, L. (2005). Design of reconfigurable slot antennas. *IEEE Transactions on Antennas and Propagation*, 53(2), 645-654.

[6] Zhang, X., & Zhao, A. (2009). Bandwidth enhancement of multiband handset antennas by opening a slot on mobile chassis. *Microwave and Optical Technology Letters*, 51(7), 1702-1706.

[7] Byndas, A., Hossa, R., Bialkowski, M., & Kabacik, P. (2007). Investigations into operation of single- and multi-layer configurations of planar inverted-F antenna. *IEEE Antennas and Propagation Magazine*, 49(4), 22-33.

[8] Hossa, R., Byndas, A., & Bialkowski, M. E. (2004). Improvement of compact terminal antenna performance by incorporating open-end slots in ground plane. *IEEE Microwave and Wireless Components Letters*, 14(6), 283-285.

[9] Picher, C., Anguera, J., Cabedo, A., Puente, C., & Kahng, S. (2009). Multiband handset antenna using slots on the ground plane: Considerations to facilitate the integration of the feeding transmission line. *Progress In Electromagnetics Research C*, 7-95.

[10] Razali, A. R., & Bialkowski, M. (2009). Design of a dual-band microstrip-fed mean-dered-tail PIFA for WLAN applications. In, *IEEE International Symposium on Antennas and Propagation and USNC/URSI National Radio Science Meeting*, Charleston, SC, 1-4.

[11] CST. (2009). CST Microwave Studio Design Guide. CST Computer Simulation Technology.

[12] Bialkowski, M., Razali, A. R., Boldaji, A., Cheng, K., & Liu, P. (2010). Miniaturization techniques of multiband antennas for portable transceivers. In, *12th International Conference on Electromagnetics in Advanced Applications*, Sydney, NSW, 283-286.

[13] Anagnostou, D., & Gheethan, A. (2009). A coplanar reconfigurable folded slot antenna without bias network for WLAN applications. IEEE Antennas and Wireless Propagation Letters 8, , 1057-1060.

[14] Mak, A., Rowell, C., Murch, R., & Mak, C. (2007). Reconfigurable multiband antenna designs for wireless communication devices. *IEEE Transactions on Antennas and Propagation*, 55(7), 1919-1928.

[15] Lai, M., Wu, T., Hsieh, J., Wang, C., & Jeng, S. (2009). Design of reconfigurable antennas based on an L-shaped slot and PIN diodes for compact wireless devices. *IET Microwaves Antennas and Propagation*, 3(1), 47-54.

[16] Byun, S., Lee, J., Lim, J., & Yun, T. (2007). Reconfigurable ground-slotted patch antenna using PIN diode switching. *ETRI Journal*, 29(6), 832-834.

[17] Wu, S., & Ma, T. (2008). A wideband slotted bow-tie antenna with reconfigurable CPW-to-slotline transition for pattern diversity. *IEEE Transactions on Antennas and Propagation*, 56(2), 327-334.

[18] Kang, W., Park, J., & Yoon, Y. (2008). Simple reconfigurable antenna with radiation pattern. *Electronics Letters*, 44(3), 182-183.

[19] Sarrazin, J., Mahé, Y., Avrillon, S., & Toutain, S. (2009). Pattern reconfigurable cubic antenna. *IEEE Transactions on Antennas and Propagation*, 57(2), 310-317.

[20] Soliman, E., De Raedt, W., & Vandenbosch, G. (2009). Reconfigurable slot antenna for polarization diversity. *Journal of Electromagnetic Waves and Applications*, 23(7), 905-916.

[21] Lai, C., Han, T., & Chen, T. (2009). Circularly-polarized reconfigurable microstrip antenna. Journal of Electromagnetic Waves and Applications 23(2-3), , 195-201.

[22] Li, H., Xiong, J., Yu, Y., & He, S. (2010). A simple compact reconfigurable slot antenna with a very wide tuning range. *IEEE Transactions on Antennas and Propagation*, 58(11), 3725-3728.

[23] Kingsley, N., Anagnostou, D., Tentzeris, M., & Papapolymerou, J. (2007). RF MEMS sequentially reconfigurable Sierpinski antenna on a flexible organic substrate with novel DC-biasing technique. *Journal of Microelectromechanical Systems*, 16(5), 1185-1192.

[24] Ramadan, A. H., Kabalan, K. Y., El -Hajj, A., Khoury, S., & Al-Husseini, M. (2009). A reconfigurable U-Koch microstrip antenna for wireless applications. *Progress In Electromagnetics Research*, 93-355.

[25] Monti, G., de Paolis, R., & Tarricone, L. (2009). Design of a 3 -state reconfigurable CRLH transmission line based on MEMS switches. *Progress In Electromagnetics Research*, 95, 283-297.

[26] Karmakar, N., & Bialkowski, M. (2002). High-performance L-band series and parallel switches using low-cost p-i-n diodes. *Microwave and Optical Technology Letters*, 32(5), 367-370.

[27] Wong, K. (2003). Planar antennas for wireless communications. Wiley New York.

[28] Razali, A. R., & Bialkowski, M. (2009). Coplanar inverted-F antenna with open-end ground slots for multiband operation. IEEE Antennas and Wireless Propagation Letters 8,, 1029-1032.

[29] Marrocco, G. (2008). The art of UHF RFID antenna design: Impedance-matching and size-reduction techniques. *IEEE Antennas and Propagation Magazine*, 50(1), 66-79.

[30] Kabacik, P., Byndas, A., Hossa, R., & Bialkowski, M. (2006). Zero-thickness wideband antennas for small radio transceivers. In, *First European Conference on Antennas and Propagation*, Nice, 1-4.

[31] Sun, B., Liu, Q., & Xie, H. (2003). Compact monopole antenna for GSM/DCS operation of mobile handsets. *Electronics Letters*, 39(22), 1562-1563.

[32] Fang, S., & Shieh, M. H. (2005). Compact monopole antenna for GSM/DCS/PCS mobile phone. In, *Asia Pacific Microwave Conference*, Suzhou, 1-4.

[33] Bialkowski, M. E., Razali, A. R., & Boldaji, A. (2010). Design of an ultrawideband monopole antenna for portable radio transceiver. IEEE Antennas and Wireless Propagation Letters 9, , 554-557.

[34] Park, H., Chung, K., & Choi, J. (2006). Design of a planar inverted-F antenna with very wide impedance bandwidth. *IEEE Microwave and Wireless Components Letters*, 16(3), 113-115.

[35] Bhatti, R., & Park, S. (2007). Hepta-band internal antenna for personal communication handsets. *IEEE Transactions on Antennas and Propagation*, 55(12), 3398-3403.

UWB Printed Antennas

Chapter 6

Recent Trends in Printed Ultra-Wideband (UWB) Antennas

Mohammad Tariqul Islam and Rezaul Azim

Additional information is available at the end of the chapter

1. Introduction

After the Federal Communication Commission (FCC)'s authorization of frequency band of 3.1 to 10.6 GHz for unlicensed radio applications, ultra-wideband (UWB) technology become the most promising candidate for a wide range of applications that will provide significant benefits for public safety, business and consumers, and attracted a lot attention both in industry and academia. The antennas are the key components of UWB system. In wireless communication system, an antenna can take various forms to fulfill the particular requirement. As a result, an antenna may be a piece of conducting wire, an aperture, a patch, a reflector, a lens, an assembly of elements (arrays). A good design of the antenna can fulfill the system requirements and improve overall system performance.

Over the past few years, significant research efforts have been put into the design of UWB antennas and systems for communications. The UWB antenna is essential for providing wideband wireless communications based on the use of very narrow pulses on the order of nanoseconds, covering an ultra-wide bandwidth in the frequency domain, and over short distance at very low spectral power densities. In addition, the antennas required to have a non-dispersive characteristic in time and frequency, providing a narrow, pulse duration to enhance a high data throughput [1]. Different kinds of antennas suitable for use in UWB applications have proposed in past few decade, each with its advantages and disadvantages.

In this paper, a review on printed UWB antennas has been done historically. Then, a technique to miniaturize the antenna's physical size by shrinking the ground plane is proposed. To develop the design technique by which the antennas can be able to achieve both UWB operating bandwidth and the stable radiation pattern across the entire frequency band by reducing the ground plane effect is also described. Finally, the enhancement of operating bandwidth as well as the pattern bandwidth by further modified the ground plane is achieved in order to fulfill the requirements defined by the FCC.

2. History of UWB antennas

The starting of UWB technology was the "spark-gap" transmitter, which broke new grounds in radio technology. The design was first not realized as UWB technology, but then later dug up by investigators. Also, even some of the ideas, which start out as designs for narrowband frequency radio, reveal some of the first concepts of UWB antennas. The concept of "syntony", that is, the received signal can be maximized when both transmitter and receiver are tuned to the same frequency, was presented by Oliver Lodge in 1898. With his new concept, Lodge developed many different types of "capacity areas," or so called antennas. Those antenna designs include spherical dipoles, square plate dipoles, bi-conical dipoles, and triangular or "bow-tie" dipoles. The concept of using the earth as a ground for monopole antennas was also introduced by Lodge. In fact, Lodge's design drawing of triangular or bow-tie elements reproduced in Figure 1(a) clearly shows Lodge's preference for embodied designs. Bi-conical antennas designed by Lodge and shown in Figure 1(b) are obviously used as transmit and receive links [1].

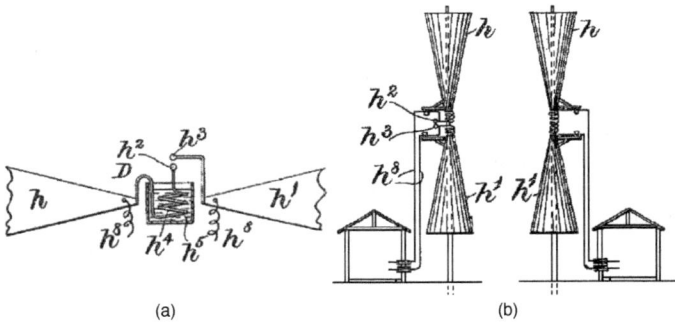

(a)   (b)

Figure 1. Lodge's (a) preferred antennas consisting of triangular "capacity areas," a clear precursor to the "bow tie" antenna (b) biconical antennas [1].

Due to demands of increased frequency band and shorter waves, a "thin-wire" quarter wave antenna dominated the market with its economic advantages over the better performance of Lodge's original designs. Especially, for television antennas, much interest was focused on the ability of handling wider bandwidths due to increased video signals. In 1939, the bi-conical antenna and the conical monopole (Figure 2) were reinvented by Carter to create wideband antennas. By adding a tapered feeding structure, Carter improved Lodge's original designs. Also, Carter was among the first to take the key step of incorporating a broadband transition between a feed-line and radiating elements. This was one of the key steps towards the design of broadband antennas [1]. In 1940, a spherical dipole antenna combined with conical waveguides and feeding structures was proposed by Schelkunoff. Unfortunately, Schelkunoff's dipole antenna does not appear to have seen much use.

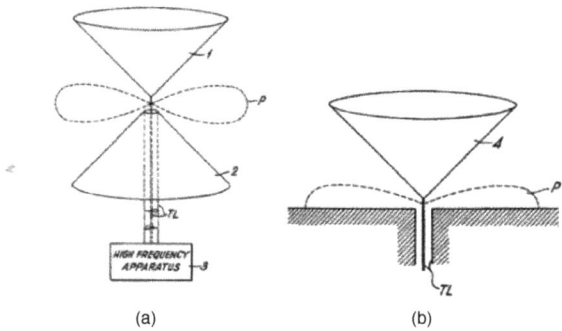

Figure 2. Carter's (a) biconical (b) conical monopole antenna [1].

At that time, the most well-known UWB antenna was the coaxial horn element proposed by Lindenblad [1]. In order to make the antenna more broadband, Lindenblad took the design of a sleeve dipole element and introduced a continued impedance change. In the year of 1941, Lindenblad's elements were used by Radio Corporation of America (RCA) for experiments in television transmission. With the vision of broadcasting multiple channels from a single central station, the need of a wideband antenna was necessary for RCA. On the top of the Empire State Building in New York City, a turnstile array of Lindenblad's coaxial horn elements as an experimental television transmitter were placed by RCA for several years. In 1947, a wideband antenna concept was proposed by the staff of the U.S. Radio Research Laboratory at Harvard University [2]. The concept of a wideband antenna evolves from a transmission line that gradually diverges while keeping the inner and outer conductors ratio constant. Several variations of the concept were developed, such as the teardrop antenna, sleeve antenna and inverted trapezoidal antenna. By the year of 1948, two types of coaxial horn antennas were presented by Brillouin. One of them is omni-directional and the other one is directional [1]. Brilliant results were offered by conventional designs, but other aspects started to grow in significance. In 1968, more complex electric antennas in different variety were developed. Two of those antennas were ellipsoidal monopoles and dipoles which were proposed by Stohr as shown in Figure 3(a).

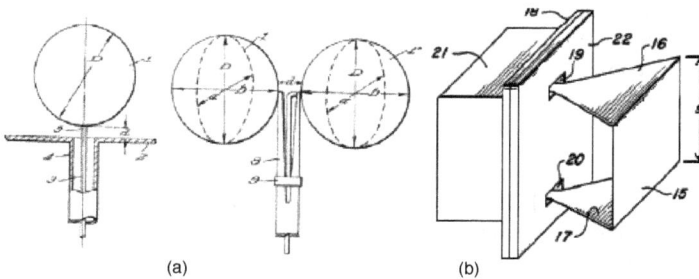

Figure 3. a) Stohr's ellipsoidal monopole (left) and dipole (right) antenna (b) Harmuth's large current radiator [1].

Beside electrical antennas, major progress on magnetic UWB antennas has also been preced-ed. In 1984, an improved magnetic antenna was proposed by Harmuth as illustrated in Fig-ure 3(b). By presenting the idea of the large current radiator surface in the antenna design, the antenna performance had been increased. The concept of this design is to make the mag-netic antenna perform like a large current sheet. However, since both sides of the sheet radi-ate, a lossy ground plane was intentionally constructed to avoid any unwanted resonances and reflections. In this way, the lossy ground plane tends to cause limitations on the anten-na's efficiency and performance.

One of the simplest practical resonant antennas is the dipole antenna. The antenna can only radiate sinusoidal waves on the resonant frequency. Thus, the dipole antennas are not suita-ble for UWB system. On the other hand, a non-resonant antenna can cover a wide frequency range, but special care must be taken in antenna design to achieve sufficient antenna effi-ciency. Moreover, the physical size of available non-resonant antenna is inappropriate for portable UWB devices. Even with appropriate size and sufficient efficiency, until now non-resonant antennas have not been suitable for UWB systems [3].

In 1950's, the spiral antennas were introduced in the class of frequency independent anten-nas. These Antennas whose mechanical dimensions are small compared to the operating wavelength is usually characterized by low radiation resistance and large reactance [4]. Due the effective source of the radiated fields varies with frequency, these antennas tend to be dispersive. The equiangular spiral and archimedean spiral antennas are the most well known spiral antennas. Spiral antennas have about a 10:1 bandwidth, providing the circular polarization in low profile geometry [5]. Transverse electric magnetic (TEM) horns and fre-quency-independent antennas feature very broad well-matched bandwidths and have been widely studied and applied [6-9]. However, for the log-periodic antennas structures, such as planar log-periodic slot antennas, bidirectional log-periodic antennas, and log- periodic di-pole arrays, frequency-dependant changes in their phase centers severely distort the wave-forms of radiated pulses [10]. Biconical antennas are the earliest antennas used in wireless systems relatively stable phase centers with broad well-matched bandwidths due to the ex-citation of TEM modes. The cylindrical antennas with resistive loading also feature broad-band impedance characteristics [11]. However, the antennas mentioned above are rarely used in portable wireless devices due to their bulky size or directional radiation, although they are widely used in electromagnetic measurements. In 1982, R.H. Duhamel patented the sinuous antenna, which exhibits wide bandwidth characteristics with dual linear polariza-tion in a compact, low profile geometry [12]. The sinuous antenna is more complicated than the spiral antenna. However, it provides dual orthogonal linear polarizations so that it can be used for polarization diversity or for transmition and reception.

From 1992, several microstrip, slot and planar monopole antennas with simple structure have been proposed [13-15]. They produce very wide bandwidth with a simple structure such as circular, elliptical or trapezoidal shapes. The radiating elements are mounted or-thogonal to a ground plane and are fed by a coaxial cable. The large ground plane mounted orthogonal to the patch made these antennas bulkier and are difficult to fit into small devi-ces. A novel UWB antenna with combination of two antenna concepts: a slot-line circuit board antenna and a bowtie horn were introduced in 1992 [16]. These two antenna concepts are put together to form a novel antenna type that is wide band with easily controllable E- and H- plane beam-widths. The bowtie horn is known for its broadband radiation pattern; whereas, the slot-line antenna provides a broadband and balanced feed structure. The microstrip and slotline are on opposite sides of the substrate. However, a broadband balun is needed for the transition between the microstrip and slotline transmission line. Proper design of broadband balun is crucial to improve the antenna bandwidth.

In 1998, a new balanced antipodal Vivaldi antenna was proposed for UWB application [17]. The author extended the tapers of the balanced antipodal Vivaldi to make the Vivaldi antenna works as a dipole in lower frequency where the slot can not radiate to extend lower edge frequency. However, the bandwidth of the antenna is limited by the transition from the feed line to the slot line of the antenna. In 1999, Virginia Tech Antenna Group (VTAG) invented and patented the Foursquare antenna. Even though this antenna does not offer as much bandwidth as other elements, it has its unique characteristics such as unidirectional pattern, dual polarization, low profile and compact antenna geometry [18-19]. The compact geometry of the Foursquare antenna is one of a desirable feature for wide scan, phased array antenna.

Several stacked patch antenna have also been proposed for UWB applications. To increase the gain and impedance bandwidth, various arrangements of stacked patch structures have been investigated. A dual layer stacked patch antenna with 56.8% bandwidth was proposed in [20] for UWB applications which does not increase the surface area and has a dimension of $26.5 \times 18 \times 11.5$ mm^3. Elsadek et al. proposed another wide bandwidth by electromagnetically couple the V-shaped patch with the triangular PIFA [21]. UWB operation with 53% bandwidth has been achieved by folding the shorting wall of the triangular PIFA in their research. These techniques can also be applicable to dual-band and wideband applications, albeit more complicated geometrical configurations. An ultra-wideband suspended plate antenna consisting four identical radiating top plates, connected to a common bottom plate is proposed in [22]. This antenna has achieved an impedance bandwidth of 72.7% with a dimension of $45 \times 47 \times 7$ mm^3. More recently, a novel compact stacked patch antenna with a folded patch feed is proposed in [23]. By using a stacked patch fed with a folded patch feed, the antenna achieved an impedance bandwidth of 90 %. Moreover, use of sorting wall significantly reduced the overall antenna size.

Most of the antennas discussed earlier have wider impedance bandwidth, high gain, non-dispersive properties, stable radiation patterns which satisfy the requirement for UWB applications. However, these antenna requires a perpendicular ground plane, which results in increased antenna size, and hence, it is difficult for integration with microwave-integrated circuits. Moreover, their bulky size and directional properties are not suitable for portable devices. When compared with these three-dimensional type of antennas, flat-type UWB antenna printed on a piece of printed circuit board (PCB) is a good option for many applications because it can be easily embedded into wireless devices or integrated with other RF circuitry.

3. Planar UWB antennas

As for portable applications, the planar antennas printed on PCBs are the most suited com-pared to other types of UWB antennas. Mainly printed antennas consist of the planar radia-tor and ground plane etched oppositely onto the dielectric substrate of the PCBs. The radiators can be fed by a microstrip line and coaxial cable.

Several techniques have been suggested to improve the antenna operating bandwidth. First, the radiator may be designed in different shapes. As for example, the radiators may have a bevel or smooth bottom or a pair of bevels to obtain good impedance matching [24]. Second-ly, a different types of slot maybe inserted in the radiators to improve the impedance match-ing, especially at higher frequencies. Besides, use of an asymmetrical strip at the top of the radiator may decrease the height of the antenna and improve the impedance matching [25, 26]. Thirdly, a partial ground plane and feed gap between the ground plane and the radiator may used to enhance and control the impedance bandwidth [27]. In addition, a notch cut from the radiator may be used to control impedance matching and to reduce the size of the radiator. Fourthly, cutting two notches at the bottom portion of the rectangular or square ra-diators can be used to further improvement of impedance bandwidth since they influence the coupling between the radiator and the ground plane [28]. Finally several modified feed-ing structures may used to enhance the bandwidth. By optimizing the position of the feed point, the antenna impedance bandwidth can be widening further since the input impe-dance is varied with position of the feed point [24].

In the past, one major limitation of the microstrip antenna was its narrow bandwidth. It was 15-50% of the centre frequency. This limitation was successfully overcome and now micro-strip planar antennas can attain wider impedance bandwidth by varying parameters like size, height, volume or feeding and matching techniques [29]. To achieve wide band characteris-tics, many bandwidth enhancement techniques have also been suggested, as mention earlier.

Many microstrip line-fed and coplanar waveguide-fed (CPW) antennas have been reported for UWB applications. These antennas use the monopole configuration, such as square, ellip-tical, circular ring, annular ring, triangle, pentagon, and hexagonal antennas [30-35], and the dipole configuration [36-39] such as double-sided printed rectangular and bow-tie antennas. Many of these antennas either have relatively large sizes or do not have a real wide band-width. For example, in [30] an investigation on a small UWB elliptical ring antenna fed by a coplanar waveguide had been carried out. This antenna achieved wideband performance on enlarging the length of the elliptical ring's major axis, and demonstrated a bandwidth from 4.6 to 10.3 GHz. Despite having fairly compact dimensions, the antenna did not cover the entire UWB. In [31], Rajgopal and Sharma proposed an UWB pentagon-shaped planar mi-crostrip slot antenna for wireless communication. Combining the pentagon-shaped slot, feed line, and pentagon stub, the antenna obtained an impedance bandwidth of 124%. However, its use in small wireless devices was limited due to large ground plane. For UWB communi-cation, a new ring antenna adopting a proximity-coupled configuration was proposed in [32]. The antenna had an overall dimension of 44×40 mm^2 with an average gain of 2.93 dBi. In [33], a novel design of printed circular disc monopole antenna with a relatively large size of 42×50 mm^2 was proposed. However, the antenna failed to fulfill the requirement of UWB with an operating bandwidth range of 2.78–9.78 GHz. A miniaturized crescent microstrip antenna for UWB application was proposed in [35]. The antenna had a relatively large size (45×50 mm²) and did not cover the upper edge frequency of the UWB. An improved design of planar elliptical dipole antenna for UWB applications was recently developed in [36]. By using elliptical slots on the dipole arms, the antenna could achieve wideband characteristics having an operating bandwidth of 94.4%. However, the antenna does not possess a physically compact profile, having a dimension of 106 × 85 mm². A double-sided printed bow-tie antenna for UWB application was proposed in [37]. By cutting parts of the rectangular patch, the antenna achieved an impedance bandwidth of 3.1-10.6 GHz to cover the entire UWB frequency band. In [38], Zhang and Wang propose a double printed UWB dipole antenna with two U-shape arms. The antenna exhibit flat amplitude and linear phase responses in 3–8 GHz. However, the operating bandwidth of the antenna is insufficient to cover the entire UWB band.

Compared to the electrical antennas mentioned earlier, slot antennas have relatively large magnetic fields that tend not to couple strongly with near-by objects which make them suitable for applications wherein near-filed coupling is required to be minimized [40]. A conventional narrow slot antenna has limited bandwidth, whereas wide-slot antennas exhibit wider bandwidth. Recently, different printed wide-slot antennas fed by a microstrip line or coplanar waveguide have been reported [41, 42]. Apart from these antennas, monopole like slot antennas have also been reported to have wide bandwidth characteristics [43-45]. By using different tuning techniques or employing different slot shapes such as rectangle, circle, arc-shape, annular-ring [46-49], different slot antennas achieved wideband or ultra-wideband performance.

Many of these proposed antennas either have large physical dimensions or do not have sufficient impedance bandwidth to cover the entire UWB frequency range. Moreover, variation of the electrical length of antennas with frequency causes significant distortion in the radiation patters which posses a challenge to design new antennas for UWB application that achieve physically compact profile and sufficient bandwidth with stable radiation patterns.

After 2003, the trend in UWB antenna was to design antennas with band notch characteristics. This antenna made insensitive to particular frequency band. This technique is useful for creating UWB antennas with narrow frequency notch bands, or for creating multi-band antenna. Since then, many researchers extended their research to investigate the possible interference between UWB system and existing narrow band wireless communication systems such as WiMAX and WLAN. The commonly used techniques to achieve a notched band are cutting a slot on the patch or embedding a quarter wavelength tuning stub within a large slot on the patch. Recently different types of slots such as L-shaped slot [50], U-shaped slot [51], square-shapes slot [52], T-shaped slot [53], pi-shaped slot [54], H-shaped slot [55] and fractal slot [56] have been used to design band notched antenna. Another simple way is to put parasitic elements near printed monopole, playing a role as filters to reject the limited band. By adding either a split-ring resonator [57] or a multi-resonator load [58] in the antenna structure, the undesired frequencies can also be stopped with better system performance. An isolated slit inside a patch, two open-end slits at the top edge of a T-stub, two parasitic strips [59] and a square ring resonator embedded in a tuning stub [60] have also been used to design band-notched antennas.

4. Reduction of ground plane effect on antenna performance

Antennas play a vital role in wireless communication systems and to fulfill UWB technology requirement, various monopole-like UWB antennas have been proposed due to their attractive features of simple configuration and ease of fabrication and numerous techniques have been exploited to broaden their bandwidth as well as improving their performance. Several broadband antennas such as vertical monopole, Vivaldi, log-periodic, cavity-backed, waveguide, bow-tie, TEM horn and dielectric loaded rod antennas have been proposed to support UWB communications. Recently, several broadband configurations, such as stack patch, plate, elliptical, pentagonal and planar unidirectional with broadband feeding structure have been proposed for UWB applications. These antennas characterizes with wide bandwidth, stable radiation patterns, simple structures and ease of fabrication. However, these antennas are relatively large and their structures make them difficult for low profile system integration. Moreover, in these antennas, the radiators are perpendicular to the ground plane resulted in increased antenna size and are difficult to be integrated with microwave circuitry.

Compared with the three dimensional type of antennas, planar structure in which the antenna can be printed onto a piece of printed circuit board is one of the possible options to satisfy the requirements for small UWB antennas. For to this advantage, industry and academia have put enormous efforts on researches to study, design and develop planar antennas for UWB communication system. In the design of planar UWB antenna, the patch and the size and shape of the ground plane as well as the feeding structure can be optimized to achieve a wide operating bandwidth. However, the planar antennas consist of a radiator and ground plane essentially an unbalanced design. The electric currents in these antennas are distributed both on the radiating element and on the ground plane, and the radiation from the ground plane is unavoidable. Therefore, the performance of the printed UWB antenna is considerably affected by the size and shape of the ground plane in terms of operating bandwidth, gain and radiation patterns [61, 62]. Moreover, due to large lateral size or asymmetric geometry of the radiator, the planar monopole antennas suffer high cross-polarization level in the radiation patterns.

4.1. Planar antenna geometry

The geometries of square patch planar monopole antennas that are considered in this chapter are shown in Figure 4. The planar monopole antennas is chosen due to their remarkably compact size, low spectral power density, simplicity, stable radiation characteristics and easy to fabricate and very easy to be integrated with microwave circuitry for low manufacturing cost. A shortcoming of this structure is limited bandwidth and high cross polarization levels. The objectives of this study are to modify the structure of the ground plane and incorporate the techniques to increase the bandwidth. The initial antenna in this study consists of an almost square radiating patch which is feed by microstrip line and ground plane. W_P and L_P

denote the width and length of the patch respectively while the ground plane has a dimension of $W \times L$. The width of the microstrip line is chosen as w_f to achieve 50 Ω characteristics impedance and is d_f mm away from the left edge of the substrate. The radiating patch and the microstrip feed line is printed on one side of a low cost FR4 PCB substrate of thickness 1.6 mm, with relative permittivity 4.6 and loss tangent 0.02 while the ground plane is printed on the other side. The radiating patch is 3.75 mm away from the left edge of the substrate.

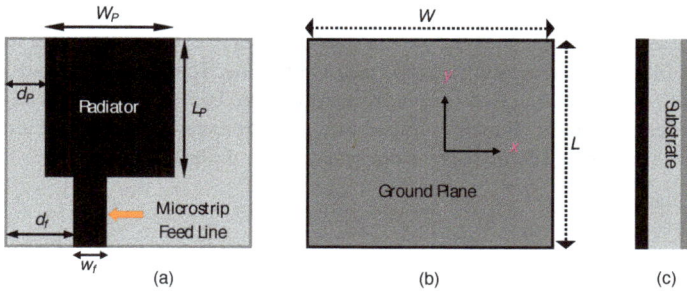

Figure 4. Geometry and dimensions of the proposed antenna (a) top view (b) bottom view and (c) side view.

4.2. Performance and characterization

The performance of the proposed initial design has been analyzed by commercially available full-wave electromagnetic simulator IE3D from Zeland which utilize the methods of moment (MoM) for electromagnetic computation. Figure 5 illustrates the simulated return loss curve for the initial design with a ground plane of dimension 30 mm × 22 mm and other parameters are W_p = 14.5 mm, L_p = 14.75 mm, d_f = 6.75 mm, w_f = 3 mm and d_p = 3.75 mm.

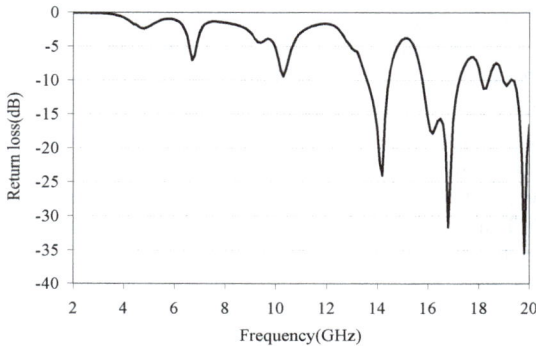

Figure 5. Simulated return loss curve of the initial antenna with W = 30 mm and L = 22 mm.

Its simulated input impedance curve is depicted in Figure 6. It is noticed in Figure 5 that first -10 dB bandwidth rages from 13.6 - 14.5 GHz, much away from UWB band. Although at higher frequencies the -10 dB bandwidth is much wider, still it is out of our desire band. This may be due to the impedance mismatching over an extremely wide frequency range resulting from large ground plane below the radiating element. From input impedance curve in Figure 6, it is seen that both the resistance and reactance fluctuate substantially in the frequency range of 0 - 13 GHz. From 0 - 6 GHz, the fluctuation in both resistance and reactance is quite high and has a peak value of around 1000 Ω. Furthermore, at the frequencies where resistance is close to 50 Ω, reactance is far from 0 Ω; when reactance reaches 0 Ω, resistance is either in its maximum or near its lowest value. Therefore, the input impedance mismatched to 50 Ω resulting in a very narrow impedance bandwidth. So it can be concluded from Figures 5 and 6 that the input impedance characteristics of the printed antenna with a ground size of 30 mm × 22 mm suffer from strong ground plane effects.

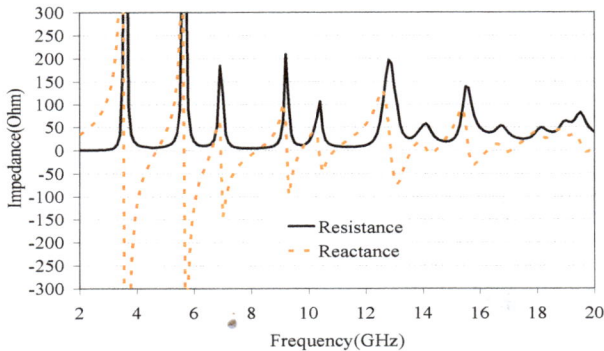

Figure 6. Simulated input impedance of the initial antenna.

This phenomenon can be explained when the ground plane is treated as a part of the antenna. When the ground plane size is large, the current flow on the top edge of the ground plane is increases. This corresponds to an increase of the inductance of the antenna if it is treated as a resonating circuit, which causes the first resonance mode either up-shifted or down-shifted in the spectrum. Also, this change of inductance causes the frequencies of the higher harmonics to be unevenly shifted. Therefore, the size of the ground plane makes some resonances become not so closely spaced across the spectrum and reduces the overlapping between them. Thus, the impedance matching becomes worse (return loss ≥ -10 dB) in ultra-wide frequency band.

The current distributions on the top (Patch) and bottom (Ground plane) surface of the initial antenna at 3.5, 6 and 14 GHz are shown Figure 7 and radiation patterns at these frequencies are depicted in Figure 8. As shown in Figure 7, at all frequencies the current is mainly distributed along the edge the radiating patch. This is due to the reason that the first resonance

frequencies are associated with the size of the patch. On the ground plane, the surface current is strongly directed towards x-axis which assures that the antenna characteristic is critically dependent on ground plane size. At the low frequency of 3.5 GHz, Figure 7 (a)& (b)
shows the current is evenly distributed on the radiator as well as in the ground plane, thus
the radiation pattern in the *H*-plane as shown in Figure 8(a) is omnidirectional. At a higher
frequency of 6.5 GHz, the current shown in Figure 7 (c) is still roughly evenly distributed on
the radiator, and so the radiation pattern is still approximately omnidirectional. At these frequencies, both *E*- and *H*- plane radiation patterns are roughly the same as that of a monopole antenna. At the higher frequencies of 14 and 16.5 GHz, higher order current modes are
excited, and the surface current density is no longer evenly distributed on the radiator as
well as on ground plane. The radiation patterns, as can be seen in Figure 8 (c), become directional with some nulls.

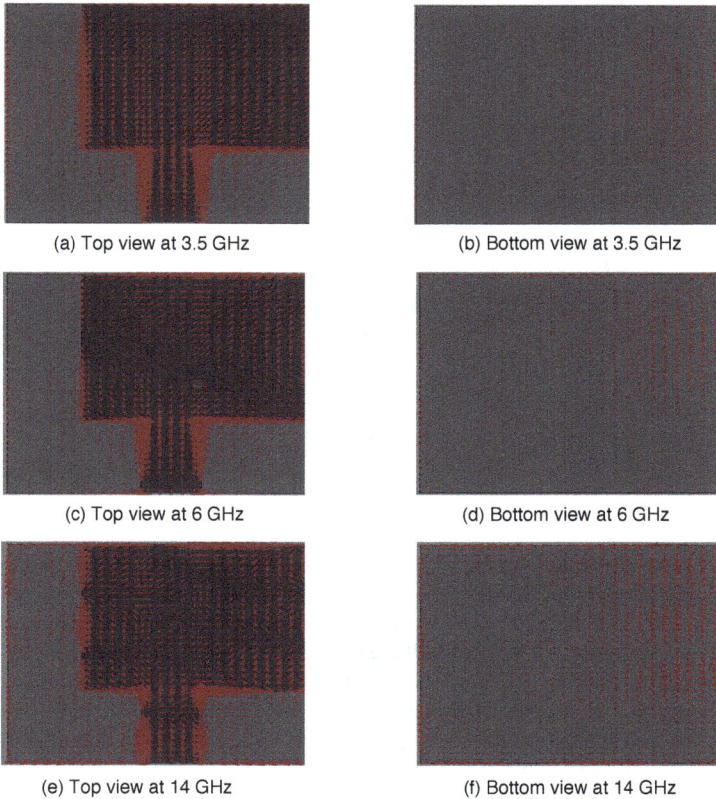

(a) Top view at 3.5 GHz

(b) Bottom view at 3.5 GHz

(c) Top view at 6 GHz

(d) Bottom view at 6 GHz

(e) Top view at 14 GHz

(f) Bottom view at 14 GHz

Figure 7. Surface current distributions at different frequencies.

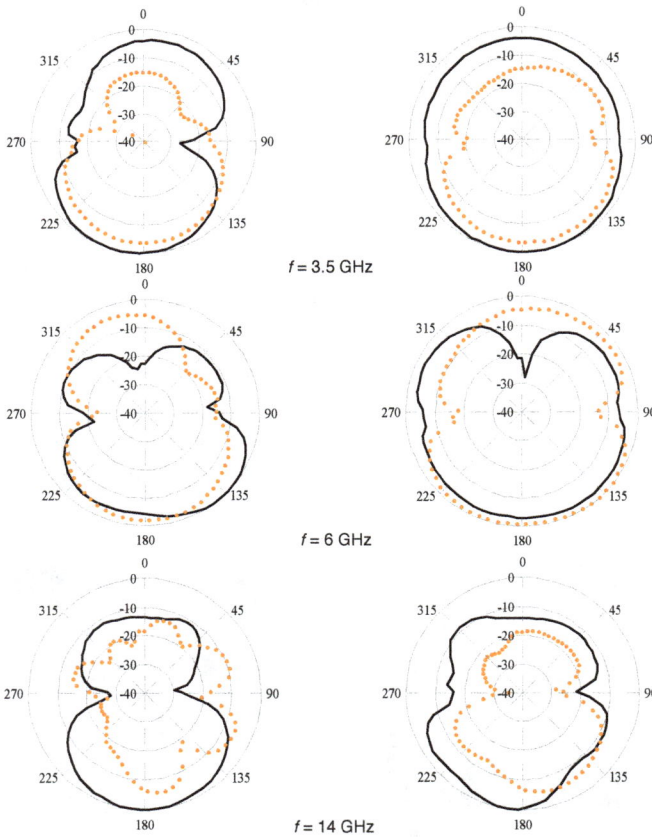

Figure 8. Simulated E (Left column) - and H (Right column)-field patterns at (a) 3.5, (b) 6 and (c) 14 GHz [solid line: co-polarization; dotted line: cross-polarization].

From the current distribution in Figure 7 it is observed that the electric currents are uniform-ly distributed in the ground plane of the initial design as like as radiating element and it be-come stronger at higher frequencies. However, such a radiation from the ground plane of planar antennas is undesirable because it creates an unbalanced structure between radiator and ground plane resulting in degradation of the antenna performance in terms of operating bandwidth and radiation patterns. Moreover, from radiation patterns display it is observed that the initial antenna suffer high cross-polarization levels in both E- and H-plane. This is due to large size of the ground plane. These sorts of ground plane effect may cause severe practical engineering problem such as deployment difficulties and design complexity [61]. That is why, it is necessary to introduce a technique to reduce the effect of ground plane on compact planar UWB antennas.

5. New bandwidth enhancement technique in UWB antenna

As the operating bandwidth of the initial antenna does not fulfill the requirements by FCC, i.e. 3.1 - 10.6 GHz, its ground plane (30 mm × 22 mm) is to be modified to improve the input impedance characteristics at the lower frequency band. Moreover, the lateral size of the ground plane has to be minimized to reduce the high cross-polarization level as well to compact the antenna which is desirable for many portable devices. For reducing the effect of ground plane on antenna performance, the ground plane is modified by using the following techniques.

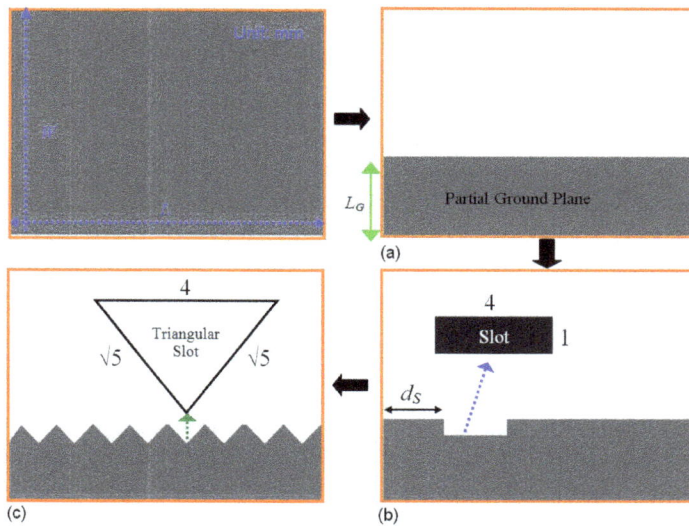

Figure 9. Modification of initial ground plane.

First, the length of ground plane decreased to L_G as shown in Figure 9, which is equal to the length of microstrip feeding line, l_f that is, the ground plane is printed only beneath of the feeding line and the size and position of the radiating patch remain unchanged. As the ground plane serves as an impedance matching circuit, it is seen the from the from the Figure 10 that the antenna with a partially shrunken ground plane of $W \times L_G$ can now achieved an impedance bandwidth (return loss ≥ -10 dB) of 8.8 GHz (3.1 - 11.9 GHz) which covers the entire ultra-wideband assigned by FCC. The overlapping of multiple resonance modes which are closing distributed across the spectrum resulting in such an ultra-wide operating bandwidth.

To further reduce the effect of ground plane on antenna performance and to improve operating bandwidth, a rectangular shape slot is introduced at the top side of the ground plane.

The slot is placed at a distance of d_s mm from left edge of the substrate. The resultant ground plane is shown in Figure 9(b). The return losses in Figure 10 shows that the rectangular slot of dimension 4 mm × 1 mm has small effect on the lower edge frequency while it increase the upper edge frequency of the operating band and the antenna can provide an impedance bandwidth of 9.3 GHz operating from 3.1 to 12.42 GHz. Compared to the partial ground plane without any slot, the antenna with single slot on the top edge of the ground plane can enhance the bandwidth by 520 MHz.

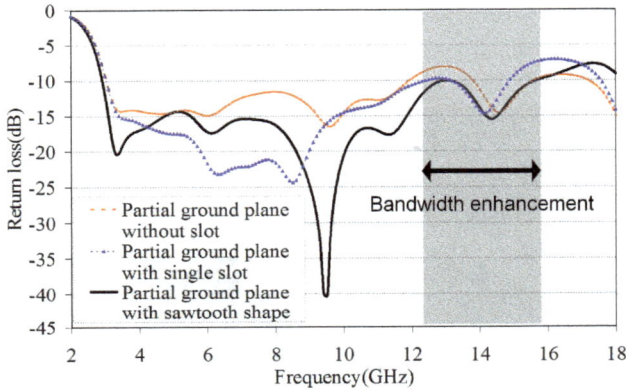

Figure 10. Comparison of return loss curves of the antenna with partial ground plane, partial ground plane with single slot and partial ground plane with sawtooth top edge.

To enhance the bandwidth and reduce the ground plane effect further, the top edge of the partial ground plane is reshaped to form a symmetrical sawtooth shape top edge by cutting of triangular shape slot as shown in Figure 9(c). This technique alter the distance between the ground plane and lower part of planar monopole antenna and tune the capacitive coupling between them resulting in wider operating bandwidth. The optimized dimension of the triangular shape slot is 4 mm × √5 mm × √5 mm. From the return loss curve shown in Figure 10 it is seen that the modified ground plane with sawtooth shape top edge has a little effect on lower edge frequency while it significantly influence the upper edge frequency of the operating band as expected. It is also seen from the Figure that the antenna with modified ground plane can be operated from 2.92 GHz to 15.70 GHz providing a -10 dB impedance bandwidth of 12.78 GHz. It is also observed that, introduction of triangular shaped slots not only widens the bandwidth but also reduces the return loss. The insertion of slots in the top edge of the ground plane increases the gap between the radiating patch and the ground plane and as a result the impedance bandwidth increases further due to extra electromagnetic coupling in between radiating element and the ground plane. Compared to the result associated with the initial design, the antenna with modified sawtooth shape ground plane can increase the bandwidth by 45.25% (3.98 GHz) as depicted in Figure 10.

The simulated input impedance curve with different types of partial ground plane is shown in Figure 11. It is seen that compared to partial ground plane without any slot and with a rectangular slot, the antenna with sawtooth shape ground plane exhibit less capacitive load to the antenna especially at higher frequencies of the operating band, which means the impedance match is getting better, thus leading to a wider operating bandwidth as illustrated in Figure 10.

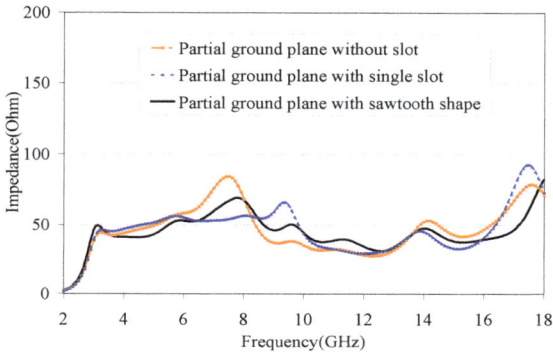

Figure 11. Simulated input impedance for different types of ground plane.

5.1. Operating principle of UWB characterization

It has been observed from the impedance characteristics that first resonance is occurs at 3.3 GHz when the ground plane is shrunk to 30 mm × 7.5 mm and modify its top edge. In input impedance characteristics, the resonance frequencies are defined where the dips are located. When the radiating patch is backed by a large ground plane of whole substrate size, the first resonance frequency is slightly shifted towards higher frequencies. If the ground plane is shrunk in length, the second and third resonance is also shifted slightly, but still not faraway from initial one as shown in Figure 5 and Figure 10. This demonstrates that these resonant frequencies is mostly determined by the radiating patch and slightly detuned by the ground plane dimension. Furthermore, it is observed that the first resonance frequency is dependent on the size of the patch while the second and third resonance frequencies as well as the bandwidth obey the size of the triangular slots that cut the top edge of the partial ground plane. However, the fourth and fifth resonance frequencies are strongly dominated by the size of the ground plane as seen from Figure 5 and Figure 10.

At lower frequency of the operating band (first resonance) where the corresponding wavelength is larger than the antenna dimension, the electromagnetic signal can couple easily into the antenna configuration therefore it act as an oscillator, i.e. a stationary wave as presented earlier in [33]. As the frequency increases, the antenna starts to operate in a dual mode of stationary and travelling waves. At the upper edge frequencies, the travelling wave becomes more influential to antenna operation since the electromagnetic signal required to go down to the antenna structure which is large in terms of wavelength.

The rectangular radiating patch of the printed planar antenna and the slots formed at top edge of the partial ground plane with an appropriate dimension can support travelling wave very well. Therefore the planar monopole antenna with modified sawtooth shape ground plane can exhibit an ultra-wide operating bandwidth (return loss ≥ -10 dB) with optimal design parameters. Furthermore, it is depicted from Figure 10 that the proposed antenna is capable of supporting multiple resonance modes and the higher order modes are the harmonics of the fundamental mode of the patch. It is also observed that these higher order modes are very much spaced. Therefore, the overlapping of these resonance modes leads to the characterization of ultra-wideband, as depicted in Figure 12.

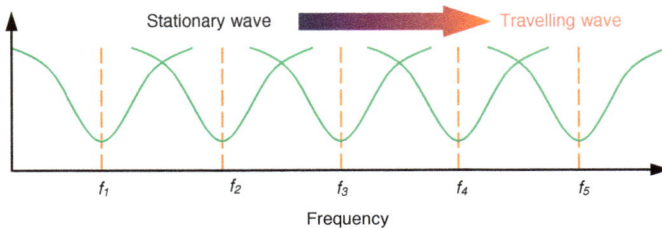

Figure 12. Overlapping of multiple resonances leading to UWB characterization.

5.2. Current distributions analysis

The current distributions usually present an insight into the physical behavior of the antenna. The simulated current distributions of the final design with modified ground plane at different frequencies are depicted in Figure 13. The current pattern at first resonance frequency of 3.1 GHz is illustrates at Figure 13(a). Figure 13(b) show the current distribution pattern at 6.1 GHz, representing a second order harmonics. A more complicated current distribution at third resonance frequency of 9.5 GHz, corresponding to third order harmonics is depicted in Figure 13(c). Figure 13(d) presents the fourth order harmonics at 11.3 GHz. These current distributions support the principle that the overlapping of closely spaced resonances resulting in the UWB characterization. At these frequencies the resonances are clearly observed on the edge of the radiating patch as well as in the ground plane.

It can be observed from the current distribution pattern that majority of the electric currents is concentrated around the edge of the radiating patch and the ground plane while the currents at the centre of the patch and ground plane are very weak. It is also seen that the current is coupled from the top and bottom edge of the ground plane to the patch through microstrip feed line and radiates to the free space. At lower frequencies, the current path length in antenna with modified sawtooth shape ground plane is much smaller than the antenna with large ground plan. In the antenna with modified ground plane, the currents at the junction between ground plane and patch is weaker than the currents in antenna large ground plane. Therefore, very small amount of current is flow into the ground plane and thus its effects on antenna performance reduced significantly. Moreover, triangular slots on the top edge of the partial ground plane effectively alleviate the changes in antenna impedance by altering the current path and creating a symmetrical current distribution to a small ground plane which reduce the effect of ground plane on antenna performances. However, at higher frequencies, the currents are mainly distributed on the microstrip line and the junction between patch and ground plane. As a result, the currents on the ground plane become stronger than lower frequencies and impedance matching becomes worse for travelling wave dependent modes.

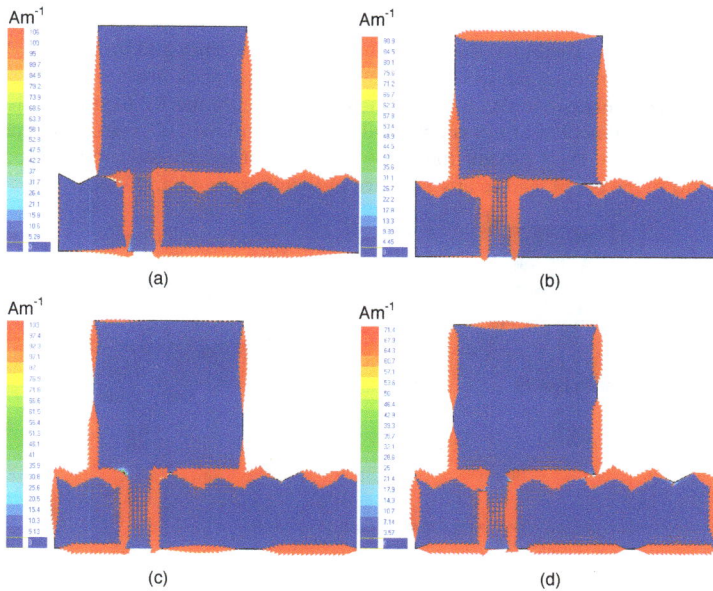

Figure 13. Surface current distributions at (a) 3.3 GHz, (b) 6.1 GHz, (c) 9.5 GHz and (d) 11.3 GHz.

5.3. Experimental verification

After a comprehensive investigation of the effect of different parameters and ground plane on antenna performance it is found that the final design had the optimized structural parameters of $W = 30$ mm, $L = 22$ mm, $L_G = 7.5$ mm, $W_P = 14.5$ mm, $L_P = 14.75$ mm, $w_f = 3$ mm, $d_f = 6.75$ mm, $d_S = 3.5$ mm and $d_p = 3.75$ mm. The very small sized ground plane would be able to cope with the increasing demand for compact antenna for portable devices. Moreover, since the antenna is printed on substrate, no additional space is required for height of the antenna. A set of prototype of the final design of the proposed antenna with optimal parameters was fabricated for experimental verification as shown in Figure 14. The prototype consist of two 35μm-thick copper layers with the antenna printed on the top side for better radiation performance while the modified partial ground is etched out at the bottom side. The antenna is printed on a 1.6 mm-thick FR4 dielectric substrate with relative permittivity of 4.6 and loss tangent of 0.02. An SMA connector is connected to the port of the microstrip feed line.

The input impedance characteristic of the realized antenna has been measured in an anechoic chamber using Agilent E8362C vector network analyzer. Figure 15 plotted the measured and simulated return loss curves. The simulated -10 dB return loss bandwidth ranges from 2.92 GHz to 15.70 GHz which is equivalent to a fractional bandwidth of 137.3%. This UWB characteristic of the compact planar monopole antenna is confirmed in measurement, with only a small shift of the lower and upper edge frequency to 2.95 GHz and 15.45 GHz respectively. Despite very compact size, the performance of the proposed antenna exceeds the UWB as defined by FCC. Although there is a disparity between the measured and simulated resonances which possibly attributed due to manufacturing tolerance and imperfect soldering effect of the SMA connector, the measured resonance frequencies are nearly identical to the simulate one. This mismatch also may be due to the effect of the RF feeding cable, which is used in the measurements but not considered during simulation. Despite some mismatched as observed in Figure 15, it is confirmed that reduction of length and modification of shape of the ground plane is not lead to any sacrifice of the operating bandwidth.

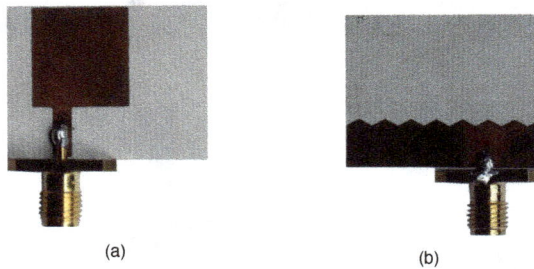

(a)   (b)

Figure 14. Photograph of realized antenna (a) Top view, (b) Bottom view.

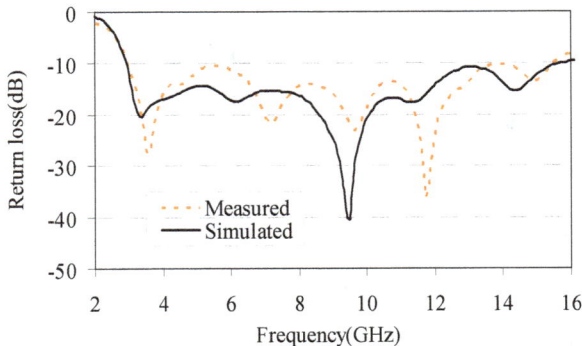

Figure 15. Measured and simulated return loss curves of the final design.

The phase of the input impedance is another important parameter of the planar UWB antenna. Since UWB antenna operates in wide range of frequencies, the phase across the operating band should be linear for preventing the pulse distortions. Figure 16 illustrates the measured phase variation of the input impedance of the final design measured in anechoic chamber using Agilent E8362C PNA series network analyzer. The phase variation across the entire operating band is reasonably linear except at around 10 and 14 GHz. This linear variation in the phase with frequency ensures that all the frequency components of signal have same delay leading to less pulse distortion.

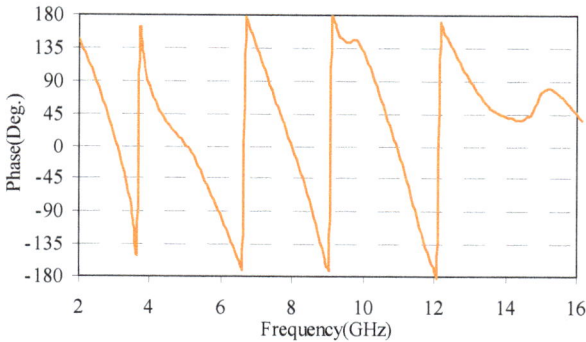

Figure 16. Measured phase variation of the input impedance.

The final design with symmetrical sawtooth shape partial ground plane was characterized in the anechoic chamber using SATIMO's StarLab antenna measuring equipment in order to measure the radiation patterns, peak gain and radiation efficiency. To achieve untruncated extent near field sampling using a probe array, the spherical scanning system was utilized for this near-field antenna measurement system.

Based on the antenna orientation with respect to the axes in Figure 4, yz-plane is the E-plane while xz-plane represents the H-plane and $\theta = 0^0$ corresponds to z-axis, while $\theta = 90^0$ corresponds to y-axis and x-axis for E-and H-planes respectively. The radiation patterns of the proposed antenna in E- and H- planes are measured at frequencies that are very close to resonance frequencies. The measured 2D radiation patterns at 3.3 GHz, 6.2 GHz and 9.4 GHz in E- and H- planes are presented in Figures 17. The radiation patterns are normalized by the taking the highest value as reference. It is observed that, in both E- and H-plane, the co-polarized field is omni-directional at lower frequencies and retain a good omni-directional patterns even at higher frequencies. In, E-plane the cross-polarized fields are much lower than that of the co-polarized one especially at lower frequencies. Although some harmonic is introduced at higher frequencies, the proposed antenna with modified partial ground plane exhibits a symmetric omni-directional radiation patterns that are same as that of a monopole antenna. Compared to radiation patterns of the antenna with large ground plane, the radiation pattern of the antenna with partial ground plane is more omni-directional and size reduction does not deteriorate the radiation characteristics.

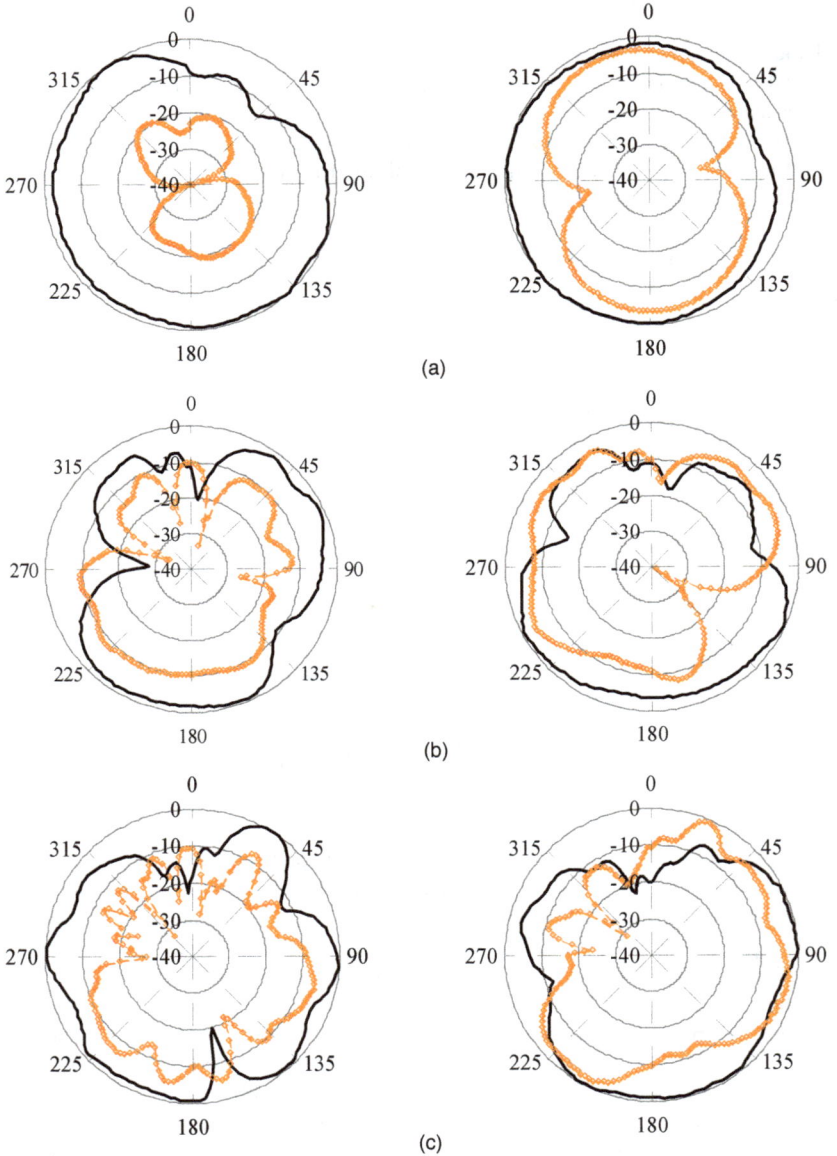

Figure 17. Measured E(left)-and H-(right)-plane radiation patterns at (a) 3.3 GHz, (b) 6.2 GHz and (c) 9.4 GHz[co-polarization :solid line; cross-polarization: cross line].

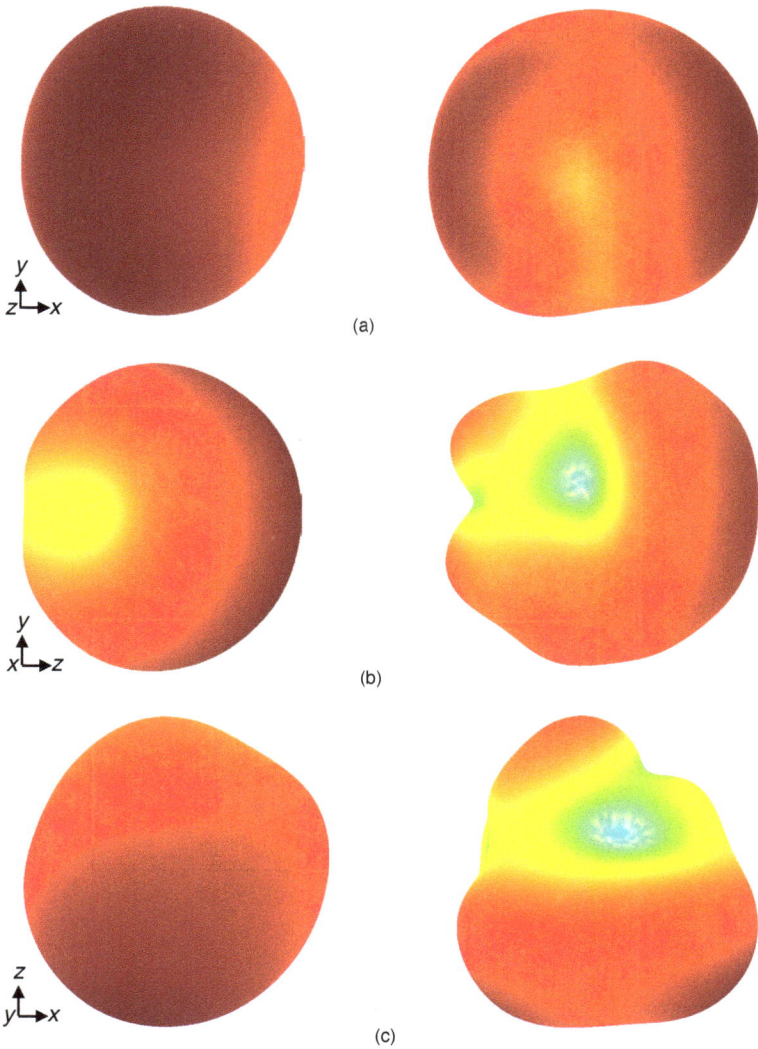

Figure 18. Measured 3D radiation patterns at 3 GHz (left column) and 8(right column GHz (a) *xy*-, (b) *yz*- and (c) *xz*-plane.

The measured 3D radiation patterns for total electric field at 3 and 8 GHz is shown in Figure 18. In the patterns the red color indicates the stronger radiated E-field and the sky blue is the weakest ones. The radiation is slightly weak in *z*-direction. It is observed from the figures that the proposed antenna exhibits almost omni-directional radiation patterns which similar like a typical monopole antenna. This 3D omni-directional radiation pattern is required for many wireless applications such as mobile communication.

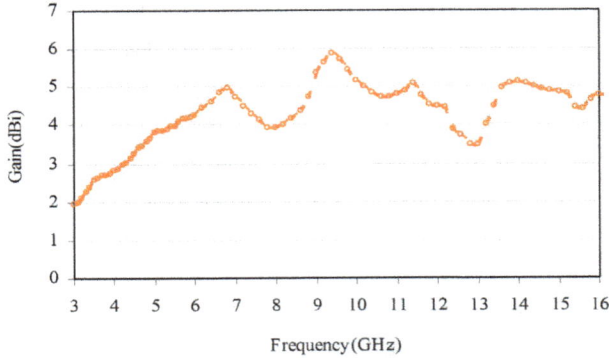

Figure 19. Measured peak antenna gain in UWB frequency range.

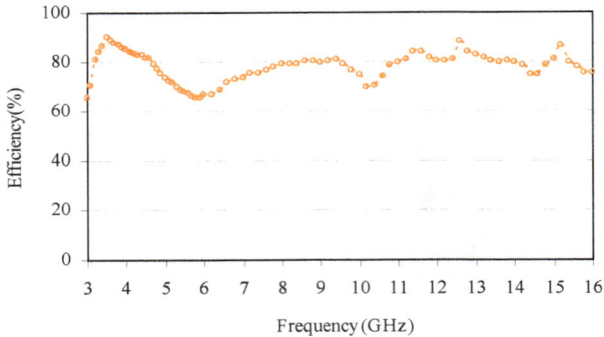

Figure 20. Measured radiation efficiency with sawtooth shape ground plane.

The measured peak gain of the final realized antenna at boresight (+z direction) is shown in Figure 19. It is observed that the antenna has a good gain with a maximum value of 5.9 dBi at 9.4 GHz. The average gain in the operating band is 3.97 dBi and the measured gain variation is less than ± 2 dBi. The measured radiation efficiency of the proposed antenna at boresight is shown in Figure 20. The antenna has a maximum of 90.2% radiation efficiency with an average of 76.79%. As the antenna was fabricated on FR4 dielectrics substrate with modified partial ground plane, the dielectric loss was high, which affected the gain as well as the efficiency.

Since UWB systems directly transmit narrow pulses rather than continuous wave, the time domain performances of the UWB antenna is very crucial. A good time domain performances is a primary requirement of UWB antenna. The antenna features can be optimized to avoid undesired pulse distortions. The group delay is defined as the negative derivative of the phase response with respect to frequency. The group delay gives an indication of the time delay of an impulse signal at different frequencies. Since UWB technology employed in short range communication systems, in the measurements the transmitting and receiving antennas are placed at distance 50 cm apart in face to face orientation.

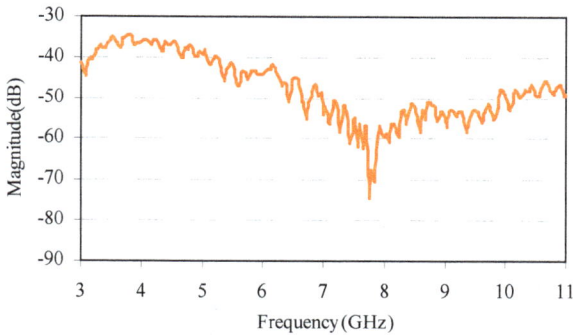

Figure 21. Magnitude of the measured transfer function in UWB range.

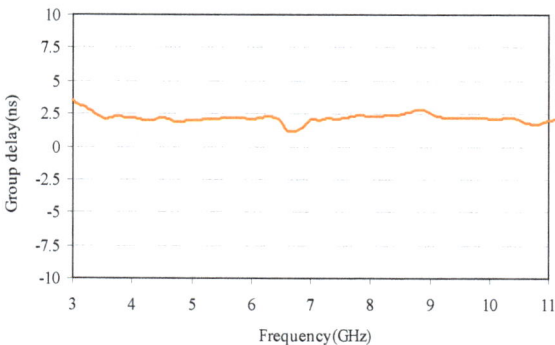

Figure 22. Measured group delay of the proposed antenna in UWB range.

The measured transfer functions of the proposed planar antenna with modified ground planes are shown in Figures 21 and 22. It is observed from Figure 21 that the magnitude curve has little variation with some ripples due noise. The group delay characteristics are almost smooth across the UWB frequency band as shown in Figure 22. The measured group delay is flat at about 2.2 ns and variation is only about 0.17 ns. This small variation in group delay indicates that the proposed antenna has good linear transmission function characteristics and could be useful for UWB impulse radios applications.

6. Conclusion

UWB is a promising technology that brings the convenience and mobility of communications to high-speed interconnections in wireless devices. As a vital part of the communication system, the design of antenna is challenging since there are more particular requirements for UWB antenna compared to its narrow band counterpart. In addition, in order to be easily integrated into portable devices, printed UWB antennas with small size and compact planar profile are highly desirable and essential for a wide variety of applications. Therefore, the design of printed planar antenna for UWB applications and its trends has been analyzed in this article. Rectangular planar antenna is initially chosen as conventional structure due to its low profile and ease of fabrication. However, its performance is significantly affected by the large ground plane. A technique, reducing the size of the ground plane and cutting of different slots has then been applied to reduce the ground plane dependency. It has been revealed that modification of large ground plane reduces the ground plane dependency as well the cross-polarization component and increases the impedance bandwidth while the addition of different types of slots enhances the bandwidth further. It has been observed that shortening of current path due to removal of the upper portion of the ground plane and insertion of the slots contributes to the wider bandwidth at low frequency end. Studies indicate that the rectangular antenna with modified sawtooth shape ground plane is capable of supporting closely spaced multiple resonant modes and overlapping of these resonances leads to the UWB characteristic. It is observed that the cutting triangular shape slots on the ground plane help to increase the bandwidth by 45.25% and in overall the antenna achieve an impedance bandwidth of 137.3% (2.92 GHz to 15.70 GHz). Moreover, it exhibits stable radiation patterns with satisfactory gain, radiation efficiency and good time domain behavior.

Author details

Mohammad Tariqul Islam[1*] and Rezaul Azim[1,2]

*Address all correspondence to: titareq@yahoo.com

1 Institute of Space Science (ANGKASA), Universiti Kebangsaan, Malaysia

2 Department of Physics, University of Chittagong, Bangladesh

References

[1] Schantz, H. G. (2004). A brief history of UWB antennas. *IEEE Aerospace and Electronic Systems Magazine*, 19(4), 22-26.

[2] Schelkunoff, S. A. (1943). *Electromagnetic Waves*, New York: D. Van Nostrand Company, Inc.

[3] Yazdandoost, K. Y., & Kohno, R. (2004). Ultra Wideband Antenna. *IEEE Radio Communications Magazine*, 42(6), S29-S32.

[4] Rumsey, V. (1957). Frequency independent antennas. *IRE International Convention Record*, 5(1), 114-118.

[5] Dyson, J. D. (1959). The Equiangular Spiral Antenna. *IRE Transactions on Antennas and Propagation*, 7, 181-187.

[6] Kanda, M. (1980). Transients in a resistively loaded linear antenna compared with those in a conical antenna and a TEM horn. *IEEE Transactions on Antennas and Propagation*, 28(1), 132-136.

[7] Shlager, K. L., Smith, G. S., & Maloney, J. G. (1996). Accurate analysis of TEM horn antennas for pulse radiation. *IEEE Transactions on Electromagnetic Compatibility*, 38(3), 414-423.

[8] Chang, L. T., & Burnside, W. D. (2000). An ultrawide-bandwidth tapered resistive TEM horn antenna. *IEEE Transactions on Antennas and Propagation*, 48(12), 1848-1857.

[9] Lee, R. T., & Smith, G. S. (2004). A design study for basic TEM horn antenna. *IEEE Antennas and Propagation Magazine*, 46(1), 86-92.

[10] Hertel, T. W., & Smith, G. S. (2003). On the dispersive properties of the conical spiral antenna and its use for pulsed radiation. *IEEE Transactions on Antennas and Propagation*, 51(7), 1426-1433.

[11] Wu, T. T., & King, R. W. P. (1965). The cylindrical antenna with non-reflecting resistive loading. *IEEE Transactions on Antennas and Propagation*, 13(3), 369-373.

[12] Duhamel, R. H. (1987). Dual Polarized Sinuous Antennas. U.S. Patent 4658262.

[13] Hammoud, M., Poey, P., & Colombel, F. (1993). Matching the input impedance of a broadband disc monopole. *Electronics Letters*, 29(4), 406-407.

[14] Evans, J A, & Amunann, M. J. (1999). Planar trapezoidal and pentagonal monopoles with impedance bandwidth in excess of 10:1. *In proceedings of the IEEE Antenna Propagation Society International Symposium*, 11-16 July, FL, USA, 1558-1561.

[15] Taylor, R. M. (1994). A broadband omni-directional antenna. *In proceedings of the Antennas and Propagation Society International Symposium*, 19-24 June, CA, USA,, 1294-1297.

[16]   Lai, A. K. Y., Sinopoli, A. L., & Burnside, W. D. (1992). A novel antenna for ultra-wideband applications. *IEEE Transactions on Antennas and Propagation*, 40(7), 755-760.

[17]   Guillanton, E., Dauvignac, J. Y., Pichot, C., & Cashman, J. (1998). A new design tapered slot antenna for ultra-wideband applications. *Microwave and Optical Technology Letters*, 19(4), 286-289.

[18]   Nealy, J. R. (1999). Foursquare Antenna Radiating Element. US patent 5926137.

[19]   Nealy, J. R., Monkevich, J. M., Stutzman, W. L., & Davis, W. A. (2000). Improvements to the Foursquare Radiating Element-Trimmed Foursquare. US patent 6057802.

[20]   Matin, M. M., Sharif, B. S., & Tsimenidis, C. C. (2007). Dual layer stacked rectangular microstrip patch antenna for ultra wideband applications. *IET Microwave and Antennas Propagation*, 1(6), 1192-1196.

[21]   Elsadek, H., & Nashaat, D. M. (2008). Multiband and UWB V-shaped antenna configuration for wireless communications applications. *IEEE Antennas and Wireless Propagation Letters*, 7, 89-91.

[22]   Low, X. N., Chen, Z. N., & Toh, W. K. (2008). Ultrawideband suspended plate antenna with enhanced impedance and radiation performance. *IEEE Transactions on Antennas and Propagation*, 56(8), 2490-2495.

[23]   Shakib, M. N., Islam, M. T., & Misran, N. (2010). Stacked patch antenna with folded patch feed for ultra-wideband application. *IET Microwave Antennas Propagation*, 4(10), 1456-1461.

[24]   Amman, M. J., & Chen, Z. (2003). A wide-band shorted planar monopole with bevel. *IEEE Transactions on Antennas and Propagation*, 51(40), 901-903.

[25]   Chen, C. C., Rao, K. R., & Lee, R. (2003). A new ultrawide-bandwidth dielectric-rod antenna for ground-penetrating radar applications. *IEEE Transactions on Antennas and Propagation*, 51(3), 371-377.

[26]   Cai, A., See, T., & Chen, Z. (2005). Study of human head effects on UWB antenna. *In proceeding of the IEEE International Workshop on Antenna Technology*, Singapore, 310-313.

[27]   Agrawell, N., Kumar, G., & Ray, K. (1998). Wide-band planar monopole antennas. *IEEE Transactions on Antennas and Propagation*, 51(4), 294-295.

[28]   Lee, S., Park, J., & Lee, J. (2005). A novel CPW-fed ultra-wideband antenna design. *Microwave and Optical Technology Letters*, 44(5), 393-396.

[29]   Pramanick, P., & Bhartia, P. (1985). CAD models for millimeter-wave fin-lines and suspended substrate microstrip lines. *IEEE Transactions on Microwave Theory and Techniques*, 33, 1429-1435.

[30]   Ren, Y. J., & Chang, K. (2006). Ultra-wideband planar elliptical ring antenna. *Electronics Letters*, 42(8), 447-449.

[31] Rajgopal, S. K., & Sharma, S. K. (2009). Investigations on ultrawideband pentagon shape microstrip slot antenna for wireless communications. *IEEE Transactions on Antennas and Propagation*, 57(5), 1353-1359.

[32] Ren, Y. J., & Chang, K. (2006). An annual ring antenna for UWB communications. *IEEE Antennas and Wireless Propagation Letters*, 5, 274-276.

[33] Liang, J., Chiau, C. C., Chen, X., & Parini, C. G. (2005). Study of a printed circular disc monopole antenna for UWB systems. *IEEE Transactions on Antennas Propagation*, 53(11), 3500-3504.

[34] Xuan, H. W., & Kishk, A. A. (2008). Study of an ultrawideband omnidirectional rolled monopole antenna with trapezoidal cuts. *IEEE Transactions on Antennas and Propagation*, 56(1), 259-263.

[35] Azenui, N. C., & Yang, H. Y. D. (2007). A printed crescent patch antenna for ultrawideband applications. *IEEE Antennas and Wireless Propagation Letters*, 6, 113-116.

[36] Nazli, H., Bicak, E., Turetken, B., & Sezgin, M. (2010). An improved design of planar elliptical dipole antenna for UWB applications. *IEEE Antennas and Wireless Propagation Letters*, 9, 264-267.

[37] Kiminami, K., Hirata, A., & Shiozawa, T. (2004). Double-sided printed bow-tie antenna for UWB communications. *IEEE Antennas and Wireless Propagation Letters*, 3, 152-153.

[38] Zhang, J. S., & Wang, F. J. (2008). Study of a double printed UWB dipole antenna. *Microwave and Optical Technology Letters*, 50(12), 3179-3181.

[39] Xiao, J. X., Yang, X. X., Gao, G. P., & Zhang, J. S. (2008). Double-printed U-shape ultra-wideband dipole antenna. *Journal of Electromagnetic and Waves Applications*, 22(8-9), 1148-1154.

[40] Schantz, H. G. (2003). UWB Magnetic Antennas. , *In proceedings of the IEEE Antennas and Propagation Society International Symposium*, Columbus, Ohio, USA, 604-607.

[41] Liu, Y. F., Lan, K. L., Xue, Q., & Chan, C. H. (2004). Experimental studies of printed wide-slot antenna for wide-band applications. *IEEE Antennas and Wireless Propagation Letters*, 3(1), 273-275.

[42] Qu, S. W., Ruan, C., & Wang, B. Z. (2006). Bandwidth enhancement of wide-slot antenna fed by CPW and microstrip line. *IEEE Antennas and Wireless Propagation Letters*, 5(1), 15-17.

[43] Latif, S. I., Shafai, L., & Sharma, S. K. (2005). Bandwidth enhancement and size reduction of microstrip slot antenna. *IEEE Transactions on Antennas and Propagation*, 53(3), 994-1003.

[44] Lui, W. J., Cheng, C. H., & Zhu, H. B. (2007). Experimental investigation on novel tapered microstrip slot antenna for ultra-wideband applications. *IET Microwave and Antennas Propagation*, 1(2), 480-487.

[45] Qing, X., & Chen, Z. N. (2009). Compact coplanar waveguide-fed ultra-wideband monopole-like slot antenna. *IET Microwave and Antennas Propagation*, 3(5), 889-898.

[46] Chang, D. C., Liu, J. C., & Liu, M. Y. (2005). Improved U-shaped stub rectangular slot antenna with tuning pad for UWB applications. *Electronics Letters*, 41(20), 1095-1097.

[47] Chair, R., Kishk, A. A., Lee, K. F., Smith, C. E., & Kajfez, D. (2006). Microstrip line and CPW fed ultra wideband slot antennas with U-shaped tuning stub and reflector. *Progress In Electromagnetics Research*, 56, 163-182.

[48] Yang, S. L. S., Kishk, A. A., & Lee, K. F. (2008). Wideband circularly polarized antenna with L-shaped slot. *IEEE Transactions on Antennas and Propagation*, 56(6), 1780-1783.

[49] Gopikrishna, M., Krishna, D. D., Anandan, C. K., Mohanan, P., & Vasudevan, K. P. (2009). Design of a compact semi-elliptic monopole slot antenna for UWB systems. *IEEE Transactions on Antennas and Propagation*, 57(6), 1834-1837.

[50] Zaker, R., Ghobadi, C., & Nourinia, J. (2009). Bandwidth enhancement of novel compact single and dual band-notched printed monopole antenna with a pair of L-shaped slots. *IEEE Transactions on Antennas Propagation*, 57(12), 3978-3983.

[51] Lee, W. S., Kim, D. Z., Kim, K. J., & Yu, J. W. (2006). Wideband planar monopole antennas with dual band-notched characteristics. *IEEE Transactions on Microwave Theory and Techniques*, 54(6), 2800-2806.

[52] Hu, S., Chen, H., Law, C. K., Shen, Z., Zhu, L., Zhang, W., & Dou, W. (2007). Backscattering cross section of ultrawideband antennas. *IEEE Antennas and Wireless Propagation Letters*, 6, 70-73.

[53] Ojaroudi, M., Ghobadi, C., & Nourinia, J. (2009). Small square monopole antenna with inverted T-shaped notch in the ground plane for UWB application. *IEEE Antennas Wireless Propagation Letters*, 8, 728-731.

[54] Zhao, Y. L., Jiao, Y. C., Zhao, G., Zhang, L., Song, Y., & Wong, Z. B. (2008). Compact planar monopole UWB antenna with band-notched characteristic. *Microwave and Optical Technology Letters*, 50(10), 2656-2658.

[55] Deng, J. Y., Yin, Y. Z., Zhou, S. G., & Liu, Q. Z. (2008). Compact ultra- wideband antenna with tri-band notched characteristic. *Electronics Letters*, 44(21), 1231-1233.

[56] Lui, W. J., Cheng, C. H., & Zhu, H. B. (2006). Compact frequency notched Ultra-wideband fractal printed slot antenna. *IEEE Microwave and Wireless Components Letters*, 16(4), 224-226.

[57] Kim, J., Cho, C. S., & Lee, J. W. (2006). 5.2 GHz notched ultra-wideband antenna using slot-type SRR. *Electronics Letters*, 42(6), 315-316.

[58] Ma, T. G., Hua, R. C., & Chou, C. F. (2008). Design of a multiresonator loaded band-rejected ultrawideband planar monopole antenna with controllable notched bandwidth. *IEEE Transactions on Antennas and Propagation*, 56(9), 2875-2883.

[59] Ryu, K. S., & Kishk, A. A. (2009). UWB antenna with single or dual band-notches for lower WLAN band and upper WLAN band. *IEEE Transactions on Antennas and Propagation*, 57(2), 3942-3950.

[60] Lui, W., Cheng, C., & Zhu, H. (2007). Improved frequency notched ultrawideband slot antenna using square ring resonator. *IEEE Transactions on Antennas and Propagation*, 55(9), 2445-2450.

[61] Chen, Z. N., Terence, S. P., & Xianming, Q. (2007). Small printed ultrawideband antenna with reduced ground plane effect. *IEEE Transactions on Antennas Propagation*, 55(2), 383-388.

[62] Lu, Y., Huang, Y., Chattha, H. T., & Cao, P. (2011). Reducing ground-plane effects on UWB monopole antennas. *IEEE Antennas and Wireless Propagation Letters*, 10, 147-150.

Chapter 7

Printed Wide Slot Ultra-Wideband Antenna

Rezaul Azim and Mohammad Tariqul Islam

Additional information is available at the end of the chapter

1. Introduction

With the beginning of the new information era, necessity of wideband wireless communications technology is increasing rapidly due to the need to support more users and to provide information with higher data transmitting rates. Ultra-wideband (UWB) technology could be the most suitable technologies that promise to revolutionize high data rate transmission and enable the personal area networking industry leading to new innovations and greater quality of services to the end users. A UWB system is found to be extremely useful and consists of various satisfying features such as high data rate, high precision ranging, fading robustness, and low cost transceiver implementation. UWB is regarded as a very promising and fast emerging low-cost technology with uniquely attractive features inviting major advances in wireless communications, sensor networking, radar, imaging, and positioning systems [1, 2].

Antennas are indispensable elements of any wireless communication systems. For UWB communication systems, the antennas must be of low profile, compact size, light weight, low cost and conformable to the architecture of the mounting devices. Amongst various types of antennas such as log periodic, TEM horn, stacked patch, spiral and planar structure, the antenna with planar profile seems to be the most preferred choice [3-5]. It has the advantage of low profile in size, compactness, and easily embeddable into wireless devices or integratable with other RF circuitry

In recent years, printed slot antennas are under consideration for use in UWB applications and are getting more and more popular because of the merits of wide frequency bandwidth, low profile, lightweight, ease of fabrication and integration with other devices or RF circuitries. Compared to the electrical antennas, slot antennas have relatively large magnetic fields that tend not to couple strongly with near-by objects, which make them suitable for applications wherein near-filed coupling is required to be minimized [6]. A conventional narrow slot antenna has limited bandwidth, whereas wide-slot antennas exhibit wider bandwidth. Recently, different printed wide-slot antennas fed by a microstrip line or coplanar wave-guide have been reported [7, 8]. Apart from these antennas, monopole like slot antennas have also been reported to have wide bandwidth characteristics [9-11]. By using different tuning techniques or employing different slot shapes such as rectangle, circle, arc-shape, an-nular-ring, U-shaped [12-16], different slot antennas achieved wideband or ultra-wideband performance. A square slot antenna excited by a CPW-fed widened tuning stub was pro-posed in [17]. By properly choosing the location and size of the tuning stub, the proposed antenna achieved a bandwidth of 60% with an overall dimension of 72 mm × 72 mm. In [18], a novel broadband design of a CPW-fed square slot antenna loaded with conducting strips has been introduced. The -10 dB impedance bandwidth of the proposed slot antenna is more than 60%. In [19], a printed wide-slot antenna fed by a microstrip line is introduced. By em-ploying an arc-shaped slot and a square-patch feed, the antenna achieved an impedance bandwidth ranging from 1.82 GHz to 7.23 GHz. Although the antenna achieved a good im-pedance bandwidth with an overall dimension of 110 mm × 110 mm, it does not operate within the entire UWB. The design of a printed wide-slot antenna with a rotated slot is pre-sented in [20]. The impedance bandwidth of the antenna varies with the rotation angle of the slot and can maintain 50.2% with suitable angle. More recently, the design of a printed wide-slot antenna for wideband applications is proposed in [21]. The antenna consists of an E-shaped patch and E-shaped slot and achieves an impedance bandwidth of 120% (2.8 - 11.4 GHz). However, the antenna does not possess a compact profile having a dimension of 85 mm × 85 mm. A new CPW-fed tapered ring slot antenna was presented in [22]. With an overall size of 66.1 mm × 44 mm, the proposed antenna achieved an impedance bandwidth range of 8.9 GHz (ranging from 3.1 - 12 GHz). The actual bandwidth was, however, limited by the distortion of radiation patterns.

In this chapter, a printed wide slot antenna that achieves a physically compact planar profile having sufficient impedance bandwidth and omnidirectional radiation pattern is proposed for UWB communication systems. By etching a microstrip fed rectangular tuning stub as ra-diating element and a tapered shape slot in the ground plane, the proposed antenna ach-ieved a UWB characteristics. The antenna structure is flat, and its design is simple and easy to fabricate.

2. Antenna configuration

The geometry and configuration of the proposed antenna is illustrated in Figure 1. The an-tenna consists of a tapered shape slot etched out of the ground plane and a microstrip line fed rectangular tuning stub for excitation. The tuning stub fed by microstrip line of 50 Ω characteristics impedance is printed on one side of an inexpensive FR4 substrate of thickness 1.6 mm, with relative permittivity 4.6 and loss tangent 0.02 while the slot is etched out on the other side. The reason for choosing FR4 substrate material is its low cost. Despite of rela-tively high loss tangent, the antenna fabricated on FR4 achieved moderate gain and efficient, which are sufficient for UWB wireless communications. The slot in the ground plane consists of two sections: the rectangular section of dimension $W_3 \times L_1$ and the triangular section, which is tapered with a slant angle $\alpha = 90^0$ for a length L_2 and has strong coupling to the feeding structure. The distance between bottom edge of the tuning stub and lower edge of the tapered shape slot is h. Therefore by properly selecting the slot shape and tuning stub, a good impedance bandwidth and radiation characteristics can be achieved. The overall size of the proposed antenna is only 22 mm × 24 mm, which can be considered as one of the smallest UWB slot antenna found in open literature.

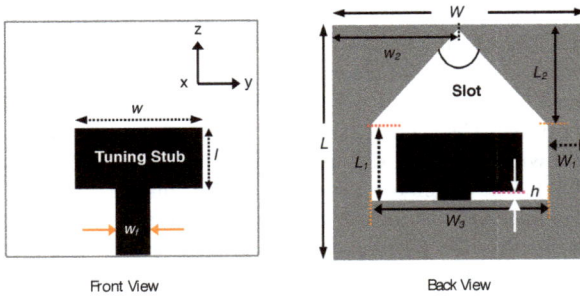

Figure 1. Geometry and detailed view of the proposed slot antenna.

3. Effects of designed parameters

Based on this design, some sensitive parameters are studied numerically in order to investigate the influence of the parameters on antenna performance. In the simulation only one parameter was varied each time, where as the others were kept constant. All simulation was carried out by employing Zeland's IE3D based on method of moment [23].

3.1. Effect of the tuning stub

Usually a large slot is used in a wide-slot antenna to achieve a high level of electromagnetic coupling to the tuning stub. Therefore variation of the tuning stub shape and slot shape will change the coupling; and thus control the impedance matching. In order to optimize the coupling between the microstrip-line and the tapered slot, different stub shapes are studied. The rectangular shape tuning stub is compared with four other stubs as shown in Figure 2. Figure 3 shows the simulated return loss curves for the five different stubs. It is observed that, for elliptical and circular shape tuning stubs, the impedance matching become very poor due to poor electromagnetic coupling between the feed-line and tapered slot. The rectangular shape tuning stub shows a good coupling with tapered shape slot proving a wider impedance matching for UWB application.

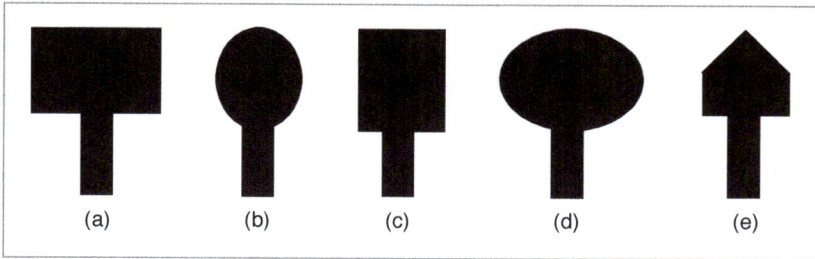

Figure 2. Different tuning stub shape (a) Rectangular (b) Circular, (c) Square, (d) Elliptical and (e) Tapered.

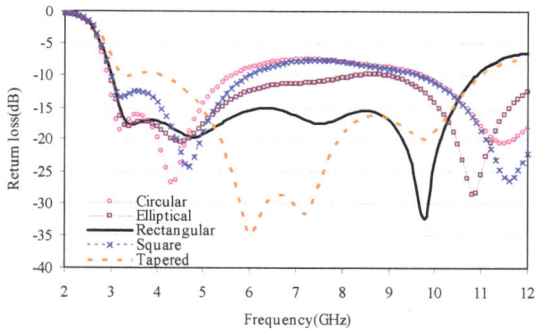

Figure 3. Simulated return loss curves for different tuning stub shape

3.2. Effect of slot shape

The wide-slot antenna is well-known to have wide impedance bandwidth though its operating bandwidth is limited due to the degradation of the radiation patterns at higher frequencies [7]. Through the numerical study on different slot shapes as shown in Figure 4, it is seen that currents flowing on the edge of the slot will increase the cross-polarization component in the yz-plane and cause the main beam to tilt away from the broadside direction in the xz-plane. Unlike the conventional wide-slot antenna proposed in [17], the slot in the ground plane of the proposed antenna with tapered shape is surrounded by ground strips of small width, which makes the antenna very compact. Moreover, introduction of the tapered slot instead of the rectangular slot changes the electric field distribution by reducing the longest current path and reducing the slot size. As a result, the impedance matching is much improved, especially at lower frequencies, resulting in overall enhancement of operating bandwidth as shown in Figure 5. It is also observed that high-frequency performance can also be improved by employing tapered slot structure, and a tapered-shape slot matched with a rectangular tuning stub can produce wider bandwidth than with a circular, elliptical, and square-shaped slot.

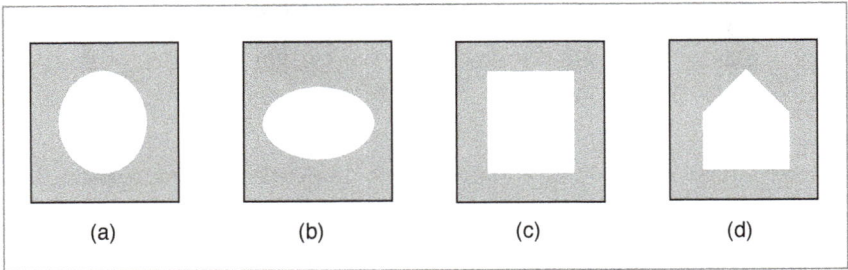

Figure 4. Different slot shape (a) Circular, (b) Elliptical, (c) Square and (d) Tapered.

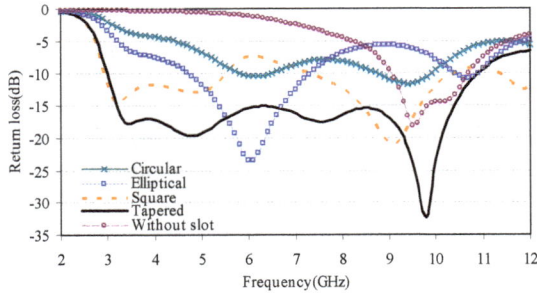

Figure 5. Simulated return loss curves for different tuning slot shape.

3.3. Effect of feed gap

The gap between the slot and the ground plane determines the matching between the feed line and slot antenna. The effect of feed gap on the impedance matching was investigated in [7] and [21]. It was found that by enhancing the coupling between the slot and microstrip feed line, good impedance matching can be obtained. An optimum value of the impedance bandwidth can be obtained for a certain optimum value of coupling. However, if the coupling increases further from this optimum value, the impedance matching becomes worse due to over-coupling. Figure 6 shows the simulated results of the proposed antenna for different feed gaps of -0.25, 0, 0.75 and 1.25 mm. It can be observed from the Figure that lower edge frequency of the operating band is highly dependent on the feed gap, while the feed gap has a little effect on the upper edge frequencies. It is also observed that a feed gap of 0.75 mm can give the widest operating band with good return loss values. Table 1 is therefore represents a good summary of the optimized parameters of the proposed antenna for achieving the ultra-wide impedance bandwidth.

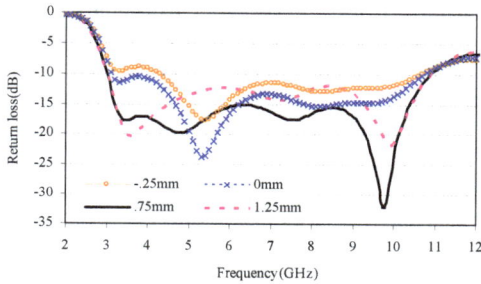

Figure 6. Simulated return loss curves for different feed gap.

Parameter	Value(mm)
W	22
L	24
W_1	2
L_1	10.75
W_2	11
L_2	7
W_3	18
w	13
l	7
W_f	3
a	90°
h	0.75

Table 1. Optimized antenna parameters

4. Antenna performance and characteristics

A prototype of tapered shape slot antenna with optimal parameters tabulated in Table 1 is fabricated for experimental verification as shown in Figure 7. The antenna performance is measured in an anechoic chamber using Satimo's antenna measurement system and Agilent E8362C vector network analyzer.

4.1. Input impedance characteristics and current distribution

The measured and simulated return loss curves of the proposed antenna are depicted in Figure 8. It is seen that the proposed antenna exhibits a wideband performance from 3 to 11.2 GHz (115.5%) for -10 dB return loss value. The measured result agrees reasonably with the simulated one across the whole operating band. The disagreement between simulation and measurement is mainly due to the fabrication tolerance. It may also be due to the effect of the feeding cable as the antenna is small. Despite being physically small than the antenna proposed in [7, 13, 17, 20, 21], the antenna still achieved wide bandwidth to cover the entire ultra-wide frequency band.

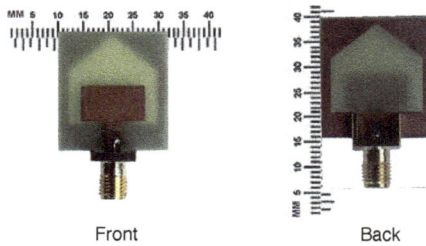

Front   Back

Figure 7. Photograph of the realized antenna.

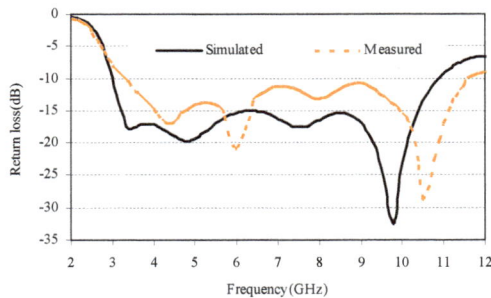

Figure 8. Simulated and measured return loss curves of the proposed antenna.

It is observed from the return loss curve that the proposed antenna is capable of supporting multiple resonances. The first resonance emerges at around 3.4 GHz, the second resonance at 6 GHz, third resonance at 8 GHz and fourth one at 10 GHz. The overlapping of these resonances, which are closely spaced over the spectrum leads to an ultra-wide operating band, which support the principle presented in [24].

The input impedance of proposed antenna is shown in Figure 9. Though there is variation in the frequency range from 5 - 8 GHz, it is seen in the Figure that the resistance is nearly flat and tends to 50 Ω values while the reactance is relatively constant at 0 Ω. Moreover, at these frequencies the measured phase of the input impedance is almost linear, which ensure that all the frequency components of the signal have the same delay leading to less pulse distortions. It is also seen that at the higher frequency end the resistance and reactance are getting away from 50 Ω and 0 Ω lines, respectively, i.e. the impedance matching is getting worse.

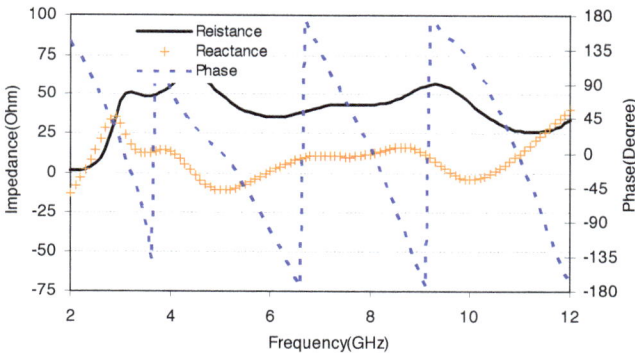

Figure 9. Input impedance and phase of the proposed slot antenna.

The return loss curve or the input impedance can only illustrate the antenna performance as a lumped load at the end of microstrip line [25]. The electromagnetic characteristics of the antenna can only be understood by examining the current distributions behavior at resonance frequencies. Simulated surface current distributions on the antenna close to the resonance frequencies are depicted in Figure 10. Figure 10(a) shows the current distribution at first resonance frequency of 3.4 GHz. The current pattern near the second resonance at around 6 GHz is shown in Figure 10(b), representing approximately a second order harmonic. Figure 10(c) present third order harmonic at 8 GHz. Figure 10(d) plots a more complicated current pattern at 10 GHz, corresponding to the fourth order harmonic. These current distributions is also support the principle that the overlapping of closely spaced resonances resulting in UWB characterization of the proposed antenna. At these four frequencies the resonances are clearly observed on the edges of both the tapered shape slot and rectangular tuning stub.

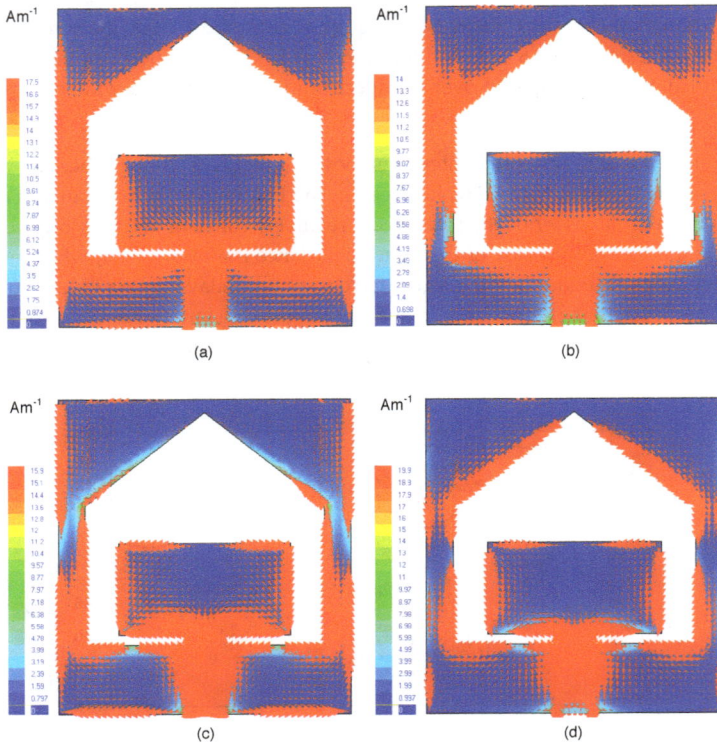

Figure 10. Simulated current distributions at (a) 3.4, (b) 6, (c) 8 and (d) 10 GHz.

4.2. Radiation characteristics

Satimo Starlab 0.6–18 GHz anechoic chamber at University of Hong Kong is used for the measurements of gain, total antenna efficiency, and radiation pattern [27]. This system uses the near-field measurement techniques that allow measurement of electric fields within the near-field of the antenna to calculate the equivalent far-field data of the antenna under test. The near-field of an antenna is the area close to the antenna, where the electric charge and electromagnetic induction effects occur. These effects fade out far more rapidly with increasing distance from the antenna (proportional to the cube of the distance) than the radiated electromagnetic far-field that fades out proportional to the distance. Near-field effects become negligible more than a few wavelengths away from the antenna. Once the near-field data have been measured, a Fourier transformation is used to calculate the equivalent far-field data. The antenna, mounted on the test board, is positioned in the center of a circular "arch" that contains 16 separate measurement probes. These probes are spaced equally apart along the circular surface. The antenna is rotated horizontally through 360°, and the combination of this rotation and the array of probes allows a full 3D scan of the antenna to be carried out, allowing full 3D radiation patterns to be measured, plotted, and analyzed. Information about antenna gain and efficiency can then be calculated from the far-field radiation pattern data. A coaxial cable was incorporated in the measurement system and the system was calibrated.

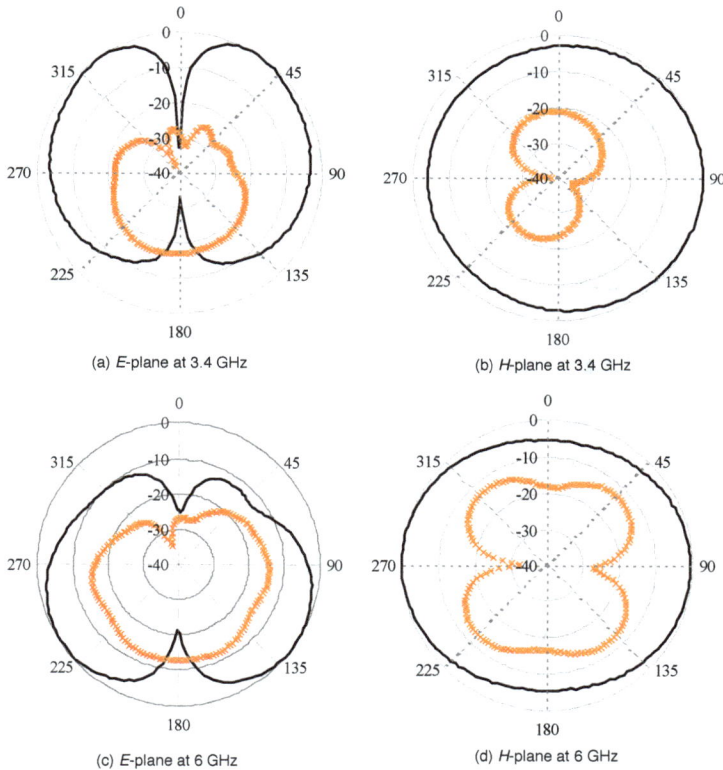

(a) E-plane at 3.4 GHz

(b) H-plane at 3.4 GHz

(c) E-plane at 6 GHz

(d) H-plane at 6 GHz

Figure 11. Measured radiation patterns at 3.4 GHz and 6 GHz [solid line: co-polarization, crossed line: cross-polarization].

Figures 11 and 12 show the measured 2D radiation patterns in two principal planes-namely, the E-(xz) and H-(xy) planes for four resonant frequencies of 3.4, 6, 8 and 10 GHz. It can be observed that at lower frequencies the antenna exhibit an omni-directional radiation patterns for H-plane and donut shape for E-plane with low cross-polarization field and patterns are about the same as that of a typical monopole antenna. As the frequency increases, higher order harmonic introduced to patterns and both H- and E-plane become more directional but still retain omni-directionality. Polarization purity can only be seen at low frequency region where the cross-to-co polarization ratio is around -20 dB, in contrast to higher frequencies, where the cross-polarization is dominant especially in H-plane. The slight asymmetry observed at higher frequencies in both in H -and E -plane may be due to the fact that microstrip feed line itself act as a radiator.

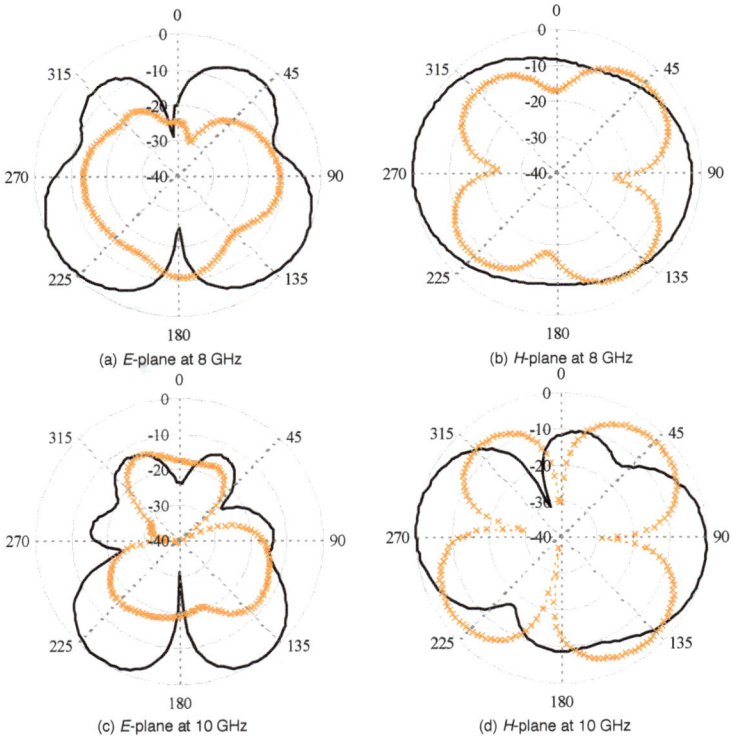

Figure 12. Measured radiation patterns at 8 GHz and 10 GHz [solid line: co-polarization, crossed line: cross-polarization].

Figure 13 depict the measured 3D radiation patterns at 3.4 and 8 GHz. In the patterns the red color indicates the stronger radiated E-field and the blue is the weakest ones. At low frequency of 3.4 GHz, the radiation patterns are almost omni-directional similar to a typical monopole antenna. The radiation is slightly weak in z-direction. At the higher frequency of 8 GHz, the radiation becomes slightly directional with a null in the z-direction due to higher order harmonics. The 3D omni-directional radiation patterns of the proposed antenna make it suitable for being used in different wireless communications especially in mobile communication.

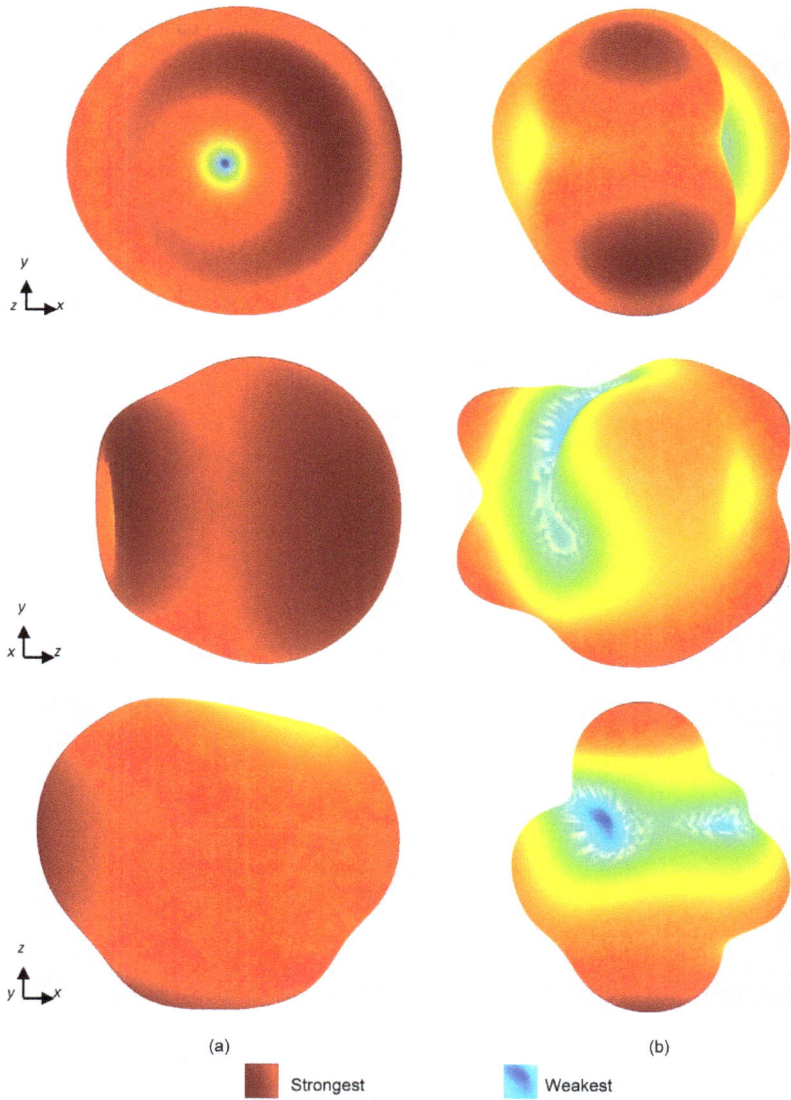

Figure 13. Measured 3D radiation pattern in *xy*, *yz* and *xz*-planes at (a) 3.4 and (b) 8 GHz.

4.3. Gain and radiation efficiency

The measured peak gain of the proposed slot antenna is shown in Figure 14. From the Figure, it can be seen that the proposed antenna achieves an average peak gain of 3.81 dBi. The maximum realized gain is 5.4 dBi at 9.8 GHz, where the radiation patterns become slight directional.

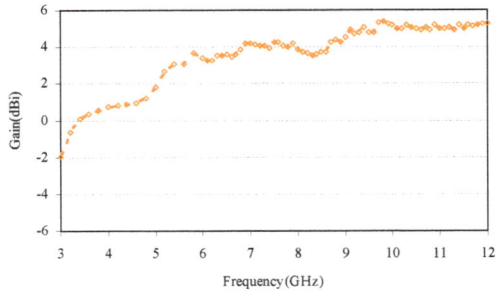

Figure 14. Measured peak antenna gain.

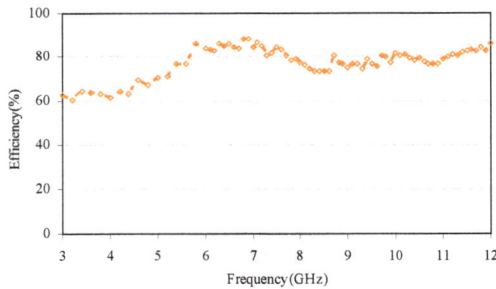

Figure 15. Measured radiation efficiency.

The realized radiation efficiency of the proposed antenna is shown in Figure 15. It is seen that the fabricated antenna achieves an average radiation efficiency of 77.9% and the maximum efficiency is 88.2%. Despite of fluctuations observed in the curves due to wider bandwidth, the proposed slot antenna achieves good gain and radiation efficiency with a compact profile in comparison with the other reported microstrip line fed planar antennas and is similar to those proposed in [17] and [26].

5. Time domain behavior

Since UWB systems directly transmit narrow pulses rather than continuous wave, the time domain performances of the UWB antenna is very crucial. A good time domain performances is a primary requirement of UWB antenna. The antenna features can be optimized to avoid undesired pulse distortions. For a transmitting/receiving antenna system as shown in Fig. 16(a), the transfer function (S_{21} parameter) is required to have flat magnitude and linear phase response over the operating band to minimize the distortions in the received signal waveform and is defined as [28-30]

$$T(\theta,\varphi,\omega) = \frac{\sqrt{\eta_0 Z_0}}{Z_0 + Z_A(\omega)} \vec{h}_{eff}(\theta,\varphi,\omega) \tag{1}$$

where $\vec{h}_{eff}$ is the complex vector effective height and $Z_A(\omega)$ is the antenna input impedance. In terms of this transfer function the port-to-port S_{21} between transmitting and receiving antennas is

$$S_{21}(\omega) = j\omega \vec{T}_{TX}(\omega) \bullet \vec{T}_{RX}(\omega) \frac{e^{-jkR}}{2\pi RC_0} \tag{2}$$

where $\omega = 2\pi f$, f is the operating frequency, C_0 is the velocity of light, (θ, φ) is the orientation and T_{TX} and T_{RX} are the transfer functions of the transmitting and receiving antennas respectively. If S_{21} is measured using two identical antennas, for single polarization, the transfer can be calculated as

$$T(\omega) = \sqrt{\left(\frac{2\pi RC_0}{j\omega} S_{21}(\omega) e^{j\omega R} \middle/ C_0 \right)} \tag{3}$$

The distance, R is to be derived from the S_{21} data itself.

The group delay is defined as the negative derivative of the phase response with respect to frequency and usually used to evaluate the phase response of the transfer function. The group delay gives an indication of the time delay of an impulse signal at different frequencies. Ideally, when the phase response is strictly linear, the group delay variation is zero. The transfer functions and group delay between a pair of proposed antennas had been measured inside an anechoic chamber with dimension of 4 m × 4 m × 8 m using Satimo's StarLab antenna measuring equipment. Since UWB technology employed in short range communication systems, in the measurements the transmitting and receiving antennas are placed face-to face at distance 0.5 m apart as illustrates in Figure 16. The measurements were taken at different azimuth angle in xz-plane.

(a)

(b)

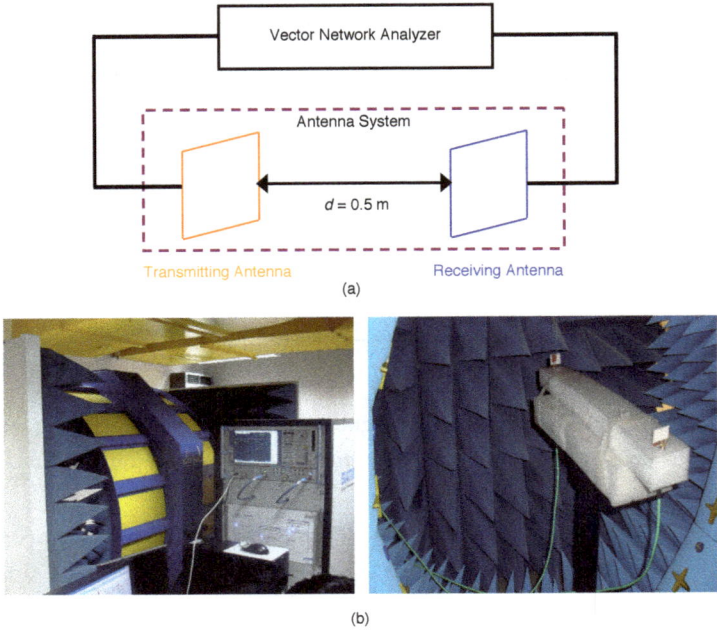

Figure 16. Setup for transfer function and group delay measurement (a) schematic diagram and (b) in anechoic chamber.

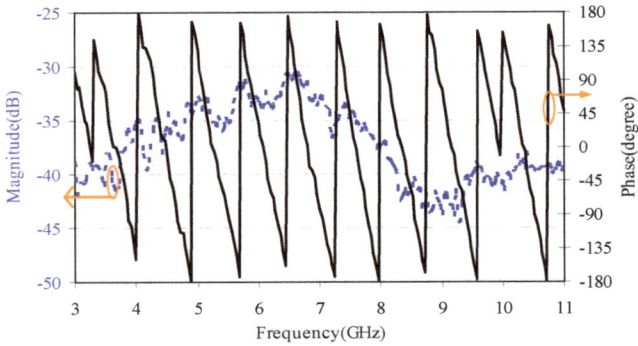

Figure 17. Magnitude and phase of the measured transfer function.

The magnitude and phase of the measured transfer function of the proposed antenna are shown in Figure 17. It is observed that the magnitudes of the transfer function are relatively smooth over the whole UWB frequency range and the variation is less than 10 dB. Linear phase response is also observed within the frequency range from 3 - 10 GHz as depicted in Figure 17. The measured group delay as shown in Figure 18 demonstrates relatively constant responses over the entire UWB frequency band. The average variation in the group delay is less than 1.3 ns, which corresponds very well to the phase of the transfer functions. This small variation in transfer function indicates that the proposed antenna does not distort the phase of the transmitted/received signals, which is a primary requirement of UWB applications.

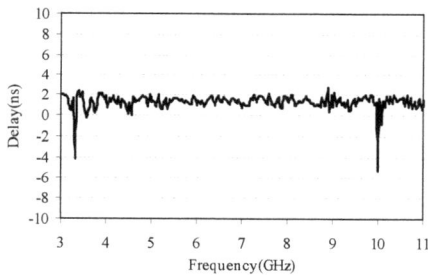

Figure 18. Measured group delay of the proposed antenna.

6. Conclusion

The design of a compact printed wide slot antenna has been proposed and implemented for ultra-wideband applications. The proposed antenna consist of a tapered shape slot and rectangular tuning stub, and fabricated onto a 22 mm × 24 mm× 1.6 mm size FR4 dielectric substrate. The measured results show that the proposed antenna achieves good impedance matching constant gain, stable radiation patterns over an operating bandwidth of 3 to 11.2 GHz (115.5%) to cover the entire UWB. The stable radiation pattern with a maximum gain of 5.4 dBi and good time domain behaviors makes the proposed antenna a suitable candidate for practical UWB applications.

Acknowledgements

This work is funded by Universiti Kebangsaan Malaysia under the grant DIP-2012-06. The authors would like to thank Associate Professor S. W. Cheung for allowing using the SATI-

MO's StarLab antenna measurement equipment of Department of Electrical and Electronic Engineering, University of Hong Kong.

Author details

Rezaul Azim[1] and Mohammad Tariqul Islam[1,2]

1 Institute of Space Science (ANGKASA), Universiti Kebangsaan Malaysia, Malaysia

2 Department of Physics, University of Chittagong, Bangladesh

References

[1] Allen, B., Dohler, M., Okon, E., Malik, W., Brown, A., Edwards, D. *Ultra-wideband Antennas and Propagation for Communications, Radar and Imaging*. New York: John Wiley & Sons, Ltd. (2006).

[2] Chen, C. C., Rao, K. R., Lee, R. A New Ultrawide-bandwidth Dielectric-rod Antenna for Ground-Penetrating Radar Application. *IEEE Transactions on Antennas & Propagation*, 51, 371-377(2003).

[3] Chang, L. T., Burnside, W. D. An Ultrawide-Bandwidth Tapered Resistive TEM Horn Antenna. *IEEE Transactions on Antennas & Propagation*, 48, 1848–1857(2000).

[4] Shakib, M. N., Islam, M. T., Misran, N. Stacked Patch Antenna with Folded Patch Feed for Ultra-wideband Application. *IET Microwave Antennas & Propagation*, 4, 1456–1461(2010).

[5] Azim, R., Islam, M. T., Misran, N., Cheung, S. W., Yamada, Y. Planar UWB Antenna with Multi-Slotted Ground Plane. *Microwave & Optical Technology Letters*, 53, 966-968 (2011).

[6] Schantz, H. G. UWB Magnetic Antennas. *Proceedings of the IEEE Antennas & Propagation Society International Symposium*, 22-27 June (2003), Columbus, Ohio, USA, 604 - 607.

[7] Liu, Y. F., Lan, K. L., Xue, Q., Chan, C. H. Experimental Studies of Printed Wide-slot Antenna for Wide-band Applications. *IEEE Antennas & Wireless Propagation Letters*, 3, 273 – 275 (2004).

[8] Qu, S. W., Ruan, C., Wang, B. Z. Bandwidth Enhancement of Wide-slot Antenna fed by CPW and Microstrip Line. *IEEE Antennas & Wireless Propagation Letters*, 5, 15 – 17 (2006).

[9]   Latif, S. I., Shafai, L., Sharma, S. K. Bandwidth Enhancement and Size Reduction of Microstrip Slot Antenna. *IEEE Transactions on Antennas & Propagation*, 53, 994 – 1003 (2005).

[10] Lui, W. J., Cheng, C. H., Zhu, H. B. Experimental Investigation on Novel Tapered Microstrip Slot Antenna for Ultra-wideband Applications. *IET Microwave & Antennas Propagation*, 1, 480 – 487 (2007).

[11] Qing, X., Chen, Z. N. Compact Coplanar Waveguide-fed Ultra-wideband Monopole-like Slot Antenna. *IET Microwave & Antennas Propagation*, 3, 889 – 898 (2009).

[12] Chang, D. C., Liu, J. C., Liu, M. Y. Improved U-Shaped Stub Rectangular Slot Antenna with Tuning Pad for UWB Applications. *Electronics Letters*, 41, 1095-1097 (2005).

[13] Jan, J. Y., Kao, J. C. Novel Printed Wide-Band Rhombus-like Slot Antenna with an Offset Microstrip-fed Line. *IEEE Antennas & Wireless Propagation Letters*, 6, 249 – 251 (2007).

[14] Sze, J. Y., Chang, C. C. Circularly Polarized Square Slot Antenna with a Pair of Inverted-L Grounded Strips. *IEEE Antennas & Wireless Propagation Letters*, 7, 149-151 (2008).

[15] Gopikrishna, M., Krishna, D. D., Anandan, C. K., Mohanan, P., Vasudevan, K. P. Design of a Compact Semi-Elliptic Monopole Slot Antenna for UWB Systems. *IEEE Transactions on Antennas & Propagation*, 57, 1834-1837(2009).

[16] Chair, R., Kishk, A. A., Lee, K. F., Smith, C. E. Microstrip Line and CPW Fed Ultra Wideband Slot Antennas with U-Shaped Tuning Stub and Reflector. *Progress In Electromagnetics Research*, 56, 163-182 (2006).

[17] Chen, H. D. Broadband CPW-fed Square Slot Antennas with a Widened Tuning Stub. *IEEE Transactions on Antennas & Propagation*, 51, 1982 – 1986 (2003).

[18] Chiou, J. Y., Sze, J. Y., Wong, K. L. A Broad-band CPW-fed Strip-loaded Square Slot Antenna. *IEEE Transactions on Antennas & Propagation*, 51, 719-721(2003).

[19] Liu, Y. F., Lan, K. L., Xue, Q., Chan, C. H. Experimental Studies of Printed Wide-slot Antenna for Wide-band Applications. *IEEE Antennas & Wireless Propagation Letters*, 3, 273 – 275 (2004).

[20] Jan, J. Y., Su, J. W. Bandwidth Enhancement of a Printed Wide-slot Antenna with a Rotated Slot. *IEEE Transactions on Antennas & Propagation*, 53, 2111 – 2114 (2005).

[21] Dastranj, A., Imani, A., Moghaddasi, M. N. Printed Wide-slot Antenna for wideband applications. *IEEE Transactions on Antennas & Propagation*, 56, 3097 – 3102 (2008).

[22] Ma, T. G., Tseng, C. H. An Ultrawideband Coplanar Waveguide-fed Tapered Ring Slot Antenna. *IEEE Transactions on Antennas & Propagation*, 54, 1105-1110 (2006).

[23] IE3D Version 12. Zeland Software Inc;(2010).

[24] Liang, J., Guo, L., Chiau, C. C., Chen, X., Parini, C. G. Study of CPW-fed Circular Disc Monopole Antenna for Ultra-wideband Applications. *IEE Proceedings of Microwaves, Antennas & Propagation*, 152, 520–526 (2005).

[25] Liang, J., Chiau, C. C., Chen, X., Parini, C. G. Study of a Printed Circular Disc Monopole Antenna for UWB Systems. *IEEE Transactions on Antennas & Propagation*, 53, 3500–3504 (2005).

[26] Sze, J. Y., Wong, K. L. Bandwidth Enhancement of a Microstrip-line-fed Printed Wide-slot. *IEEE Transactions on Antennas & Propagation*, 49, 1020 – 1024 (2001).

[27] Foged, L. J., Giacomini, A. Wide Band Dual Polarized Probes for Near and Far Field Measurement Systems. *Proceedings of the IEEE Antennas & Propagation Society International Symposium*, 5-11 July (2008), CA, USA, 5-11.

[28] Scheers, B., Acheroy, M., Vorst, A. V. Time Domain Simulation and Characterisation of TEM Horns using Normalised Impulse Response. *IEE Proceedings Microwaves, Antennas & Propagation*, 147, 463-468 (2000).

[29] Qing, X., Chen, Z. Transfer Functions Measurements of UWB Antenna. *Proceedings of the IEEE Antennas & Propagation Society International Symposium*, 20-25 June (2004), CA, USA, 2532-2535.

[30] Guo, L., Liang, L., Chiau, C. C., Chen, X., Parini, C. G., Yu, J. Performance of Ultra-wideband Disc Monopoles in Time Domain. *IET Microwaves, Antennas & Propagation*, 1, 955-959 (2007).

Chapter 8

UWB Antennas for Wireless Applications

Osama Haraz and Abdel-Razik Sebak

Additional information is available at the end of the chapter

1. Introduction

Currently, there is an increased interest in ultra-wideband (UWB) technology for use in several present and future applications. UWB technology received a major boost especially in 2002 since the US Federal Communication Commission (FCC) permitted the authorization of using the unlicensed frequency band starting from 3.1 to 10.6 GHz for commercial communication applications [1]. Although existing third-generation (3G) communication technology can provide us with many wide services such as fast internet access, video telephony, enhanced video/music download as well as digital voice services, UWB –as a new technology– is very promising for many reasons. The FCC allocated an absolute bandwidth up to 7.5 GHz which is about 110% fractional bandwidth of the center frequency. This large bandwidth spectrum is available for high data rate communi-cations as well as radar and safety applications to operate in. The UWB technology has another advantage from the power consumption point of view. Due to spreading the ener-gy of the UWB signals over a large frequency band, the maximum power available to the antenna –as part of UWB system– will be as small as in order of 0.5mW according to the FCC spectral mask. This power is considered to be a small value and it is actually very close to the noise floor compared to what is currently used in different radio communica-tion systems [2].

1.1. Different UWB Antenna Designs

UWB antennas, key components of the UWB system, have received attention and significant research in recent years [3]-[28]. With theincreasing popularity of UWB systems, there have been breakthroughs in the design of UWB antennas. Implementation of a UWB system is facing many challenges and one of these challenges is to develop an appropriate antenna. This is because the antenna is an important part of the UWB system and it affects the overall performance of the system. Currently, there are many antenna designs that can achieve broad bandwidth to be used in UWB systems such as the Vivaldi antenna, bi-conical antenna, log periodic antenna and spiral antenna as shown in Fig. 1. A Vivaldi antenna [3]-[4] is one of the candidate antennas for UWB operation. It has a directional radiation pattern and hence it is not suitable for either indoor wireless communication or mobile/portable devices which need omni-directional radiation patternsto enable easyand efficient communication between transmitters and receivers in all directions. Mono-conical and bi-conical antennas [5] have bulky structures with large physical dimensions which limit their applications. Also, log periodic [6] and spiral antennas [7] are two different UWB antennas that can operate in the 3.1-10.6 GHz frequency band but are not recommended for indoor wireless communicationapplications or mobile/portable devices. This is because they have large physical dimensions as well as dispersive characteristics with frequency and severe ringing effect [6]. This is why we are looking for another candidate for UWB indoor wireless communications and mobile/portable devices that can overcome all these shortcomings. This candidate is the planar or printed monopole antenna [8]-[28]. Planar monopole antennas [8]-[10] with different shapes of polygonal (rectangular, trapezoidal...etc), circular, elliptical...etc have been proposed for UWB applications as shown in Fig. 2.

1.2. UWB Antennas for Wireless Communications

Due to their wide frequency impedance bandwidth, simple structure, easy fabrication on printed circuit boards (PCBs), and omni-directional radiation patterns, printed PCB versions of planar monopole antennas are considered to be promising candidates for applications in UWB communications. Recent UWB antenna designs focus on small printed antennas because of their ease of fabrication and their ability to be integrated with other components on the same PCBs [11]-[19]. Fig. 3 illustrates several realizations of planar PCB or printed antenna deigns.

1.3. UWB Antennas with Bandstop Function

However, there are several existing NB communication systems operating below 10.6 GHz in the same UWB frequency band and may cause interference with the UWB systems such as IEEE 802.11a WLAN system or HIPERLAN/2 wireless system. These systems operate at 5.15-5.825 GHz which may cause interference with a UWB system. To avoid the interference with the existing wireless systems, a filter with bandstop characteristics maybe integrated with UWB antennas to achieve a notch function at the interfering frequency band [21]-[28]. Fig. 4 shows several developed bandstop antenna designs.

This chapter focuses on the development of different novel UWB microstrip-line-fed printed disc monopole and hybrid antennas with an emphasis of their frequency domain performance. Different antenna configurations are proposed and designed in order to find a good candidate for UWB operation. The reasonable antenna candidate should satisfy UWB performance requirements including small size, constant gain, radiation pattern stability and phase linearity through the frequency band of interest. Also, the designed UWB antenna should have ease of manufacturing and integration with other mi-crowave components. We have simulated, designed, fabricated and then tested experi-mentally different printed disc monopole antenna prototypes for UWB short-range wireless communication applications. The printed disc monopole antennas are chosen because they have small a size and omni-directional radiation patterns with large bandwidth. In order to understand their operation mechanism that leads to the UWB characteristics, those antenna designs are numerically studied. Also, the important physical parameters which affect the antenna performances are investigated numerically using extensive parametric studies in order to obtain some quanti-tative guidelines for designing these types of antennas.

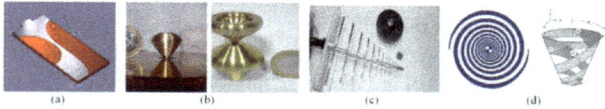

Figure 1. (a) Vivaldi antenna [4] (b) Mono-conical and bi-conical antenna [5] (c) Log-periodic antenna [6] and (d) Spiral and conical spiral antenna [7].

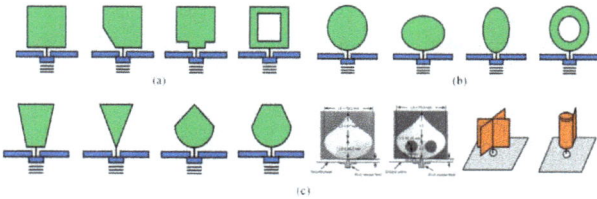

Figure 2. Modified shape planar antennas for UWB applications (a) rectangular, (b) circular and elliptical, (c) other shapes.

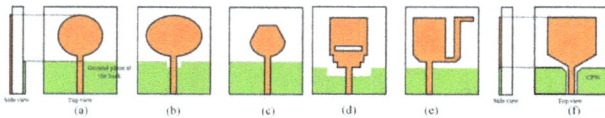

Figure 3. Planar PCB or printed antenna designs [8]-[20].

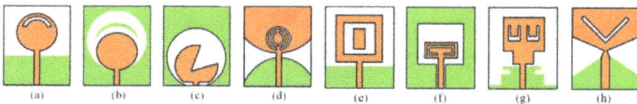

Figure 4. Printed antenna designs with single bandstop functions [21]-[28].

2. Operation Mechanism of UWB Monopole Antennas

Printed disc monopole antennas are considered to be good candidates for UWB applications because they have a simple structure, easy fabrication, wideband characteristics, and omni-directional radiation patterns [11]-[28]. The geometry of the reference printed circular disc monopole antenna is shown in Fig. 5. To determine the initial parameters of the printed circular disc monopole antenna, we should first understand their operation mechanism. It has been shown that disc monopoles with a finite ground plane are capable of supporting multiple resonant modes instead of only one resonant mode (as in a conventional circular patch antenna) over a complete ground plane [29]. Overlapping closely spaced multiple resonance modes (f1, f2, f3, ..., fN) as shown in Fig. 6 can achieve a wide bandwidth and this is the idea behind the UWB bandwidth of circular disc monopole antennas. The frequency of the first resonant mode can be determined by the size of the circular disc. At the first resonance f1, the disc antenna tends to behave like a quarter-wavelength monopole antenna, i.e. $\lambda/4$. That means the diameter of the circular disc is $2r = \lambda/4$ at the first resonant frequency.

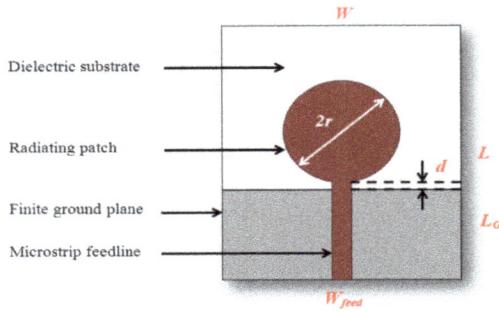

Figure 5. The configuration of the reference printed circular disc monopole antenna showing the necessary antenna parameters.

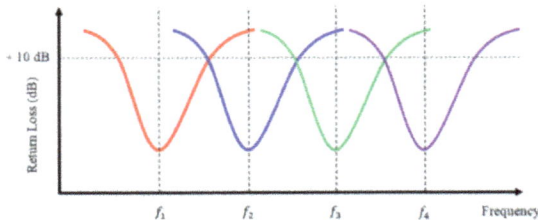

Figure 6. The concept of overlapping closely-spaced multiple resonance modes for the reference circular disc monopole antenna (reproduced from [30]).

Then the higher order modes f2, f3, ...,fN will be the harmonics of the first or fundamental mode of the disc. Unlike the conventional patch antennas with a complete ground plane, the ground plane of disc monopole antennas should be of a finite length LG to support multiple resonances and hence achieve wideband operation. The width of the ground plane W is found to be approximately twice the diameter of the disc or W=λ/2 at the first resonant [17].

The printed disc monopole antenna can be fed using different feeding techniques such as microstrip line, coplanar waveguide (CPW), aperture coupling, or proximity coupling. In the case of a microstrip line feed, the width of the microstrip feed line Wfeed is chosen to achieve a 50Ω characteristic impedance. The other antenna parameters such as the feed gap between the finite ground plane and the radiating circular disc d and the length of the finite ground plane LG can be determined using a full-wave EM numerical modeling techniques. The small feed gap between the finite ground plane and the radiating circular disc d is a very critical parameter which greatly affects the antenna impedance matching between the microstrip feedline and the radiating disc.

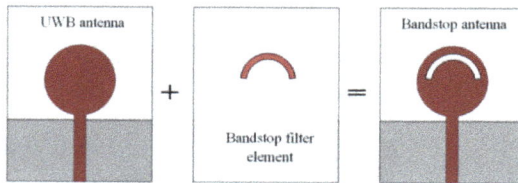

Figure 7. The idea of integrating a bandstop filtering element to the reference circular disc monopole antenna.

To avoid interference with some existing wireless systems in the 5.15-5.825 GHz frequency band, a filter with bandstop characteristics maybe integrated with UWB antennas to achieve a notch function at the interfering frequency band. The idea of integrating a bandstop filtering element to the monopole antenna is illustrated in Fig. 7. Recently, several techniques have been introduced to achieve a single band notch within this frequency band. The most popular technique is embedding a narrow slot into the radiating patch. The slot may have different shapes such as C- shaped, slit ring resonator (SRR), L- shaped,U- or V- shaped, π-shaped slot....etc. Some other techniques are based on using parasitic strips, i.e., inverted C-shaped parasitic strip. Other techniques are based on using a slot defected ground structure in the ground plane, i.e., H-shaped slot DGS.

3. UWB Disc Monopole Antennas

As mentioned in the introductory section of this chapter, there are several types of printed disc monopoles which exhibit ultra-wide impedance bandwidth. Here, different categories of disc monopoles will be investigated both numerically and experimentally.

3.1. Printed Circular Disc Monopole Antenna with Two Steps and a Circular Slot

For better understanding the antenna characteristics, the antenna reflection coefficient (S11) curves are plotted in decibel or dB scale, i.e. (S11dB = 20 log|S11| = −Return loss RL).The geometry and photograph of the proposed printed circular disc monopole antenna with two steps and a circular slot is shown in Fig. 8. The radiating element is fed by a 50Ω microstrip feed line with width of Wf = 4.4 mm. The substrate used in our design is Rogers RT/duroid 5880 high frequency laminate with thickness of h = 1.575 mm, relative permittivity of εr = 2.2 and loss tangent of tanδ = 0.0009. A finite ground plane of length LG and width W lies on the other side of the substrate. The feed gap of width d between the finite ground plane and the radiating patch is a very critical parameter for antenna matching purposes and to obtain wide bandwidth performance. This proposed antenna has a reduction in the overall antenna surface area compared to those reported in [16] and [19]. A parametric study is carried out to investigate the effect of antenna physical parameters such as the width of the substrate W, the width of the feed gap d, the radius of circular slot RS and the steps dimensions W1, W2, L1 and L2 on the performance of the proposed UWB antenna.

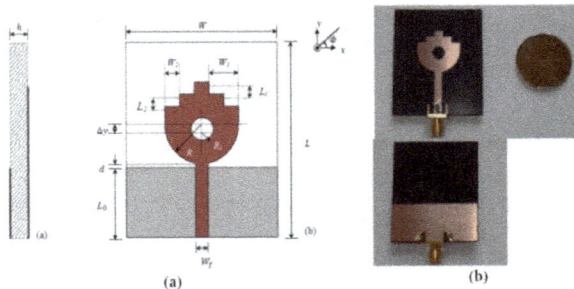

Figure 8. (a) Geometry and (b) photograph of the proposed microstrip line fed monopole antenna.

3.1.1. Design Analysis

During the parametric study, one parameter varies while all other parameters are kept fixed. The optimized antenna parameters are: W = 41 mm, L = 50 mm, LG = 18 mm, R = 10 mm, Δy = 2 mm, RS = 3 mm, W1 = 8 mm, W2 = 4 mm, L1 = 3 mm and L2 = 3 mm. Fig. 9 shows the simulated antenna reflection coefficient (20 log|S11|) curves using CST Microwave Studio TM package for different values of substrate width W, feed gap width d, slot radius RS and the steps dimensions W1, W2, L1 and L2. It can be noticed from results that the smallest substrate width for obtaining the maximum available bandwidth is W = 41 mm. It can be also seen that the reflection coefficient impedance bandwidth is greatly dependent on both the feed gap width d and the circular slot radius RS and by controlling these two parameters, the impedance matching between the radiating patch and the feed line can be easily controlled. By tuning the width of the feed gap d, the maximum achieved impedance bandwidth is determined. The circular slot inside the radiating patch acts as an impedance matching element which controls the antenna impedance matching as well as the antenna bandwidth. Also, the circular slot inside the radiating patch can be used for miniaturizing the monopole antenna. Also, it can be noticed that the rectangular steps have no remarkable effect on the overall antenna impedance bandwidth. The opti-mum values for feed gap width, slot radius and steps dimensions are d = 1 mm, RS = 3 mm and W1 (= 2W2) = 8 mm and L1 (= L2) = 3 mm, respectively.

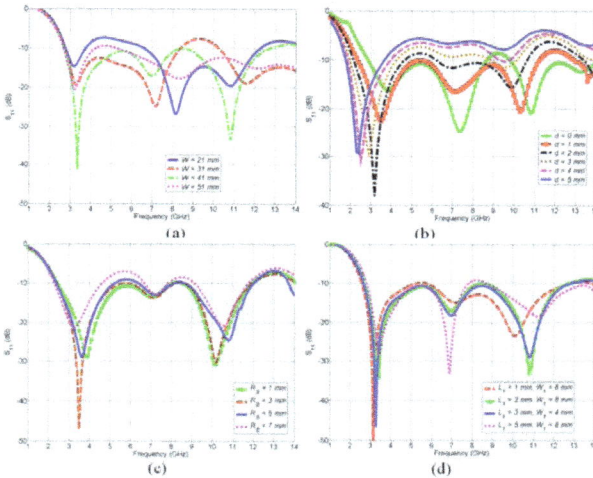

Figure 9. Parametric studies of effect of (a) substrate width W (b) feed gap width d (c) circular slot radius RS and (d) steps dimensions W1 and L1 on antenna reflection coefficient.

Cutting out two rectangular steps and a circular slot from the radiating patch to reduce the overall metallic area and hence reduce the antenna copper losses without affecting the antenna operation or disturbing the current distribution of the antenna is a challenging task. This can be done by investigating the antenna surface current distributions. Fig. 9 presents the antenna surface current and electric field distributions for the proposed disc monopole antenna. From the electric field distributions, it is noticed that the monopole antenna supports multiple resonant modes. It can be seen that the current distribution is mainly located close to the radiating patch edges rather than in the center. For increasing the maximum achieved impedance bandwidth, the lower resonant frequency should be decreased. This can be done by increasing the antenna perimeter which directly affects lower resonant frequency and then the antenna impedance bandwidth. To increase the antenna perimeter, cutting out steps from the radiating patch are used here. This is simply because the surface current will take longer path when the antenna perimeter p is larger and the new antenna with larger perimeter appears to be like a longer length monopole and then the lowest reso-nance frequency fL will be decreased according to [14]:

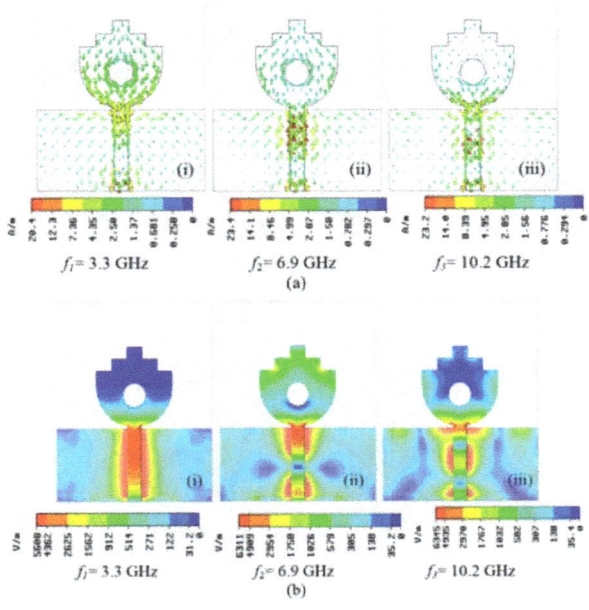

Figure 10. Simulated (a) surface current and (b) electric field distributions at the three re-sonant frequencies 3.3, 6.9 and 10.2 GHz.

$$\varepsilon_{eff} \approx (\varepsilon_r + 1)/2 \qquad (1)$$

$$f_L (\mathrm{GHz}) = 300 / \left(p\sqrt{\varepsilon_{eff}} \right) \qquad (2)$$

where εeff is the effective dielectric constant and the perimeter p units are in millimeters.

For example, in the proposed antenna design, p = 71.4 mm, εr = 2.2, then εeff = 1.6 and the calculated lower resonant frequency using Eq. (2) is found to be fL ≈ 3.3 GHz. From the si-mulated and measured reflection coefficient results shown in Fig. 10, the lower resonant fre-quency is fL ≈ 3.3 GHz which agrees well with the calculated value.

3.1.2. Experimental and Simulation Results

A prototype of the microstrip-line-fed monopole antenna with optimized dimensions was fabricated as shown in Figure8and tested experimentally in the Applied Electromagnetics

Laboratory at Concordia University. All scattering parameters measurements were carried out using Agilent E8364B programmable network analyzer (PNA). The measured and simulated reflection coefficient (S11) curves are presented in Fig. 11. It can be noticed that both measured and simulated results are in good agreement with each other and the measured 10 dB return loss bandwidth ranges from 3.0 to 11.4 GHz which covers the entire UWB frequency spectrum. Compared to the simulated results, the second resonant frequency at 7 GHz is shifted up while the third resonant frequency at 10 GHz is shifted down. This may be due to the sub-miniature version A (SMA) connector losses and/or substrate losses especially at high frequencies (7-10 GHz). Even the loss effect of the substrate is modeled correctly and taken into account in the simulations; the simulation results did not change too much and did not agree with the measured results. In general, the proposed antenna exhibits an UWB impedance bandwidth (3.1-10.6 GHz) in both simulated and measured results.

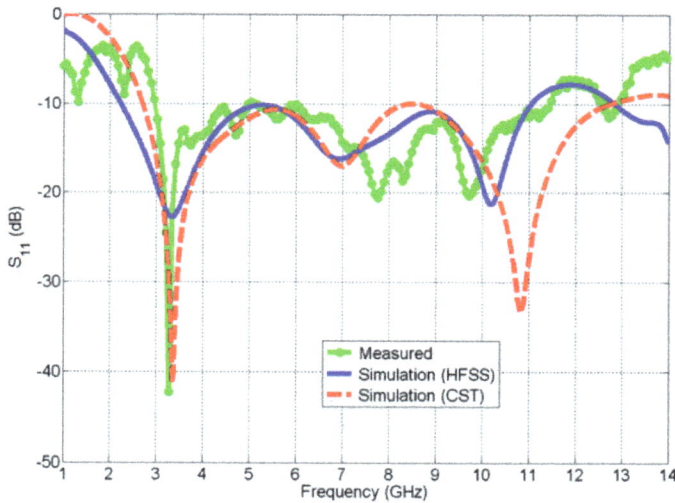

Figure 11. Measured and simulated reflection coefficient curves of the proposed antenna.

For further understanding the antenna performance, the Ansoft HFSS simulated maximum realized total directive gain in the boresight direction and the phase of reflection coefficient ∠S11 for the proposed antenna are presented in Fig. 12. The boresight of directional antenna is defined as the direction of maximum gain of the antenna. For most of antennas, the boresight is the axis of symmetry of the antenna, i.e. z-axis. It can be seen that the antenna has good gain stability across the frequency band of interest (3.1-10.6 GHz).It ranges from 3.4 dB to 5.2 dB with gain variation of about 2dB. The behavior of the phase of reflection coefficient ∠S11 versus frequency is also studied and shown in the same figure. It can be noticed that the phase seems to be linear across the whole UWB frequency range.

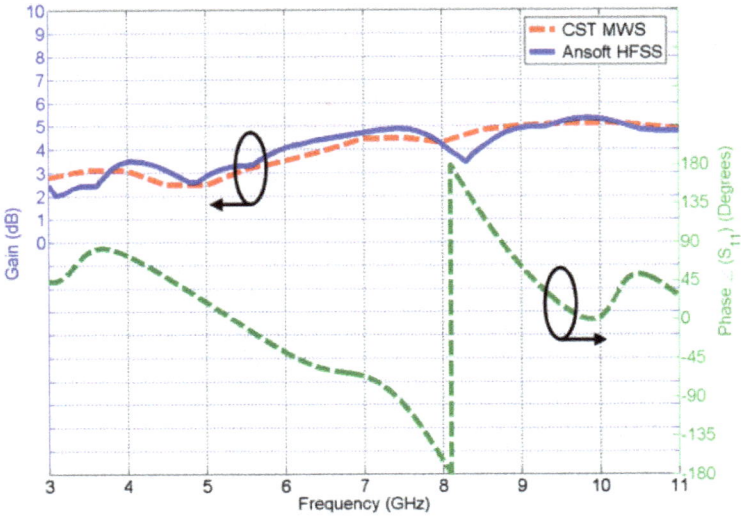

Figure 12. The simulated gain and phase of reflection coefficient ∠S11 versus frequency of the proposed microstrip-line-fed monopole antenna.

Fig. 13 shows the radiation characteristics for the proposed antenna. Both yz-cut plane (E-plane) and xz-cut plane (H-plane) radiation patterns have been simulated using Ansoft HFSS and measured in an anechoic chamber at the three resonant frequen-cies 3.3, 6.8, and 10.2 GHz. From the measured results, the proposed antenna has omni-directional radiation pattern in H-plane at lower frequency (3.3 GHz) and near omni-directional at higher fre-quencies (6.9 and 10.2 GHz) with good agreement with simula-tions. The measured E-plane radiation patterns agree with the simulations especially at lower frequency (3.3 GHz) while the agreement is not as good as the H-plane patterns at higher frequencies (6.9 and 10.2 GHz). There are some ripples and discrepancies in the measured radiation patterns especial-ly at the higher frequencies which may be due to sen-sitivity and accuracy of the measuring devices at higher frequencies in addition to the ef-fects of the SMA feed connector and the coaxial cable. The E-plane is identified by most of UWB antenna patterns which is perpen-dicular to H-plane (almost symmetric). Re-searchers in UWB antenna typically define E-plane as the plane containing the feedline and the maximum radiation of the antenna. H-plane is the plane perpendicular to E-plane.

We have investigated both simulated and measured E-plane patterns. From simu-lations, nulls in E-plane at θ = 90° depend on the size of the finite ground plane and the contact point of SMA feed connector in particular at the upper edge frequency. By searching several published UWB antennas of similar disc monopole antennas, similar behavior of measured results are reported in many papers including [31]-[34].

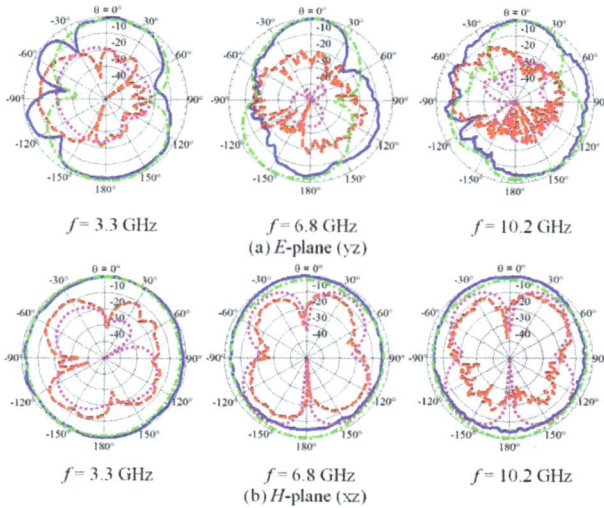

f = 3.3 GHz f = 6.8 GHz f = 10.2 GHz
(a) *E*-plane (yz)

f = 3.3 GHz f = 6.8 GHz f = 10.2 GHz
(b) *H*-plane (xz)

Figure 13. Measured co-pol (blue solid line), cross-pol (red dashed line), Ansoft HFSS simulated co-pol (green dash-dotted line) and cross-pol (magenta dotted line), (a) E-plane and (b) H-plane radiation patterns of the proposed antenna.

3.1.3. UWB Bandstop Antenna Design

A modification can be made to the above designed antenna for achieving the bandstop function to avoid possible interference to other existing WLAN systems. A very narrow arc-shaped slot is cut away from the radiating patch as shown in Fig. 14 (a) will act as a filter element to make the antenna will not respond at the bandstop frequency. For perfect band-rejection performance of UWB antenna, the return loss of the stop-band notch should be almost 0dB or the reflection coefficient is almost 1.0. However, in our first band-stop antenna design, we could achieve voltage standing wave ratio (VSWR) of about 8 (reflection coefficient is 0.78 or -2.1 dB). The arc-shaped slot filter element di-mensions will control both the bandstop frequency fnotch and the rejection bandwidth of the band-notched filter BWnotch. The arc-shaped slot filter dimensions are: the radius of the slot R1, the thickness of the slot T and the slot angle 2α. Fig. 14 (b) illustrates the simulated reflection coefficient curves using both HFSS and CST MWS for comparison. From the simulation results, it can be seen that the band-notched characteristic in the 5.0-6.0 GHz band is achieved with good agreement between them.

Parametric studies were carried out to address the effect of arc-shaped slot dimen-sions on the band-notched performance. Figures 15 shows the effect of varying the slot radius R1, slot thickness T and the slot angle 2α parameters on the simulated antenna ref-lection coefficient, respectively. From results in Fig. 15 (a) & (c), it can be seen that the notch frequency fnotch decreases by increasing both the arc-shaped slot radius R1 and the angle 2α while the notch bandwidth BWnotch is almost the same. On the other side, both the notch frequency and bandwidth increase at the same time by increasing the slot thickness T. For achieving a band-notched performance in the 5-6 GHz frequency band, the arc-shaped slot parameter dimensions are: R1 = 7.5 mm, T = 0.7 mm and $2\alpha = 160°$.

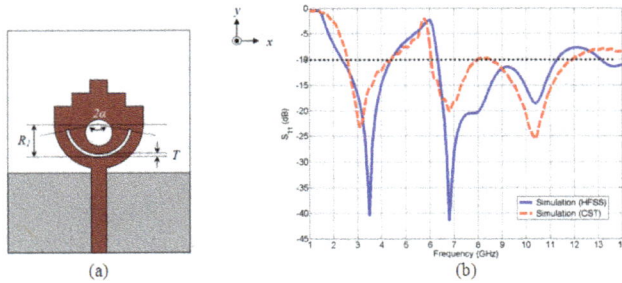
(a)   (b)

Figure 14. (a) Geometry of the band-notched antenna, R1 = 7.5 mm, T = 0.7 mm and $2\alpha = 160°$ (b) Simulated reflection coefficient curves versus frequency.

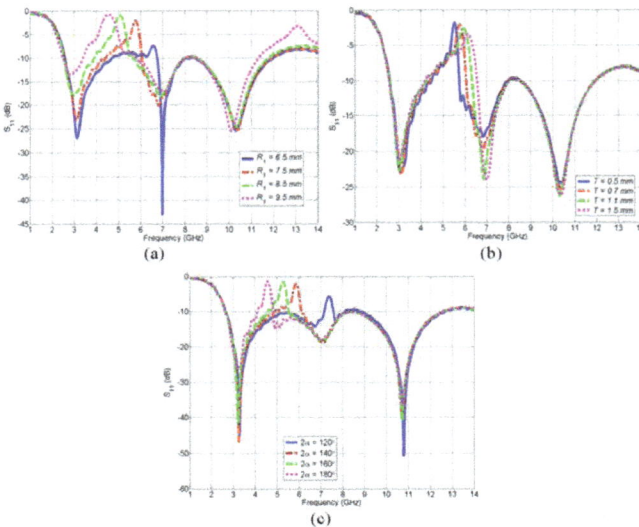
(a)   (b)

(c)

Figure 15. Simulated reflection coefficient curves versus frequency for different values of (a) arc-shaped slot radius R1, (b) thickness of the slot T and (c) the slot angle 2α.

3.2. Maple-leaf Shaped Monopole Antennas

In this section, we developed different maple-leaf shaped monopole antennas with two band-rejection techniques for the 5.0-6.0 GHz frequency band. Fig. 16 (a) & (b) show the geometrical configuration and the photograph of the proposed UWB maple-leaf-shaped monopole antenna prototype. The radiating element consists of a maple-leaf-shaped patch as a radiating element which represents the Canada flag symbol. The radiat-ing patch is fed by a microstrip line and both are etched on a Rogers RT Duroid 5880 substrate with dielectric constant εr = 2.2, dielectric loss tangent tanδ = 0.0009, and thickness h = 1.575 mm. The proposed antenna parameters L1 ~ L10 are determined using an extensive parametric study and optimization in both Ansoft HFSS and CST MWS to address the effect of those parameters on the overall performance of the antenna. Details of the optimized parameters are summarized in Table 1. Our target here is to design a compact antenna for UWB operation. So, we tried to reduce the overall antenna size by reducing the substrate dimensions from 50 × 41 mm2 as in the previous antenna design to 35.48 × 30.56 mm2 as in the present antenna design. Here, there is a reduction in the an-tenna size by almost 47% compared to our first proposed antenna prototype, i.e. circular disc monopole antenna with two steps and a circular slot.

Parameter	W	L	LG	W1	Wf	d	L1	L2
Value (mm)	30.48	35.56	12.95	5.59	4.06	0.84	2.27	7.47
Parameter	L3	L4	L5	L6	L7	L8	L9	L10
Value (mm)	2.65	4.10	4.34	3.05	5.39	7.73	4.02	5.24

Table 1. Maple-leaf Shaped Printed Monopole Antenna Dimensions (Units in mm).

The maple-leaf shaped monopole antenna is used to achieve wider impedance matching bandwidth by introducing many leaf arms into the main radiating patch. This will lead to increasing the overall perimeter of the antenna and hence the monopole an-tenna looks bigger in size than its real physical size. This is simply because the current takes paths close to the edges rather than inside the radiating patch. The proposed maple-leaf shaped monopole antenna has a wider bandwidth with smaller size compared to the first UWB antenna design (stepped monopole antenna).

Fig. 17 (a) illustrates the simulated and measured reflection coefficient curves against the frequency for the designed maple-leaf antenna. It can be noticed from the re-sults that the proposed antenna exhibits a simulated impedance bandwidth from 3 to 13 GHz with good agreement between Ansoft HFSS and CST simulation programs while the measured impedance bandwidth becomes dual-band, one in 4.1-7.0 GHz and the other one in 8.7-13.3 GHz. The explanation for the difference between the measured and simulated results can be easily understood if we mention that both simulated reflection coefficient curves are already very close or even touch the -10 dB level in the region 7.0-9.0 GHz frequency band. So, if there is any manufacturing error in the antenna parameters L1 ~ L10 during the fabrication proposes of the antenna prototype will be a big issue. This is in addition to calibration errors during S-parameters measurement and the effect of SMA connector which was not taken into account during simulations. Also, the manufacturing tolerance as well as the effect of SMA connector has been simulated in CST MWS program and simulation results are shown in Fig. 17 (b) and it is found from the obtained result that it confirms the above explanation.

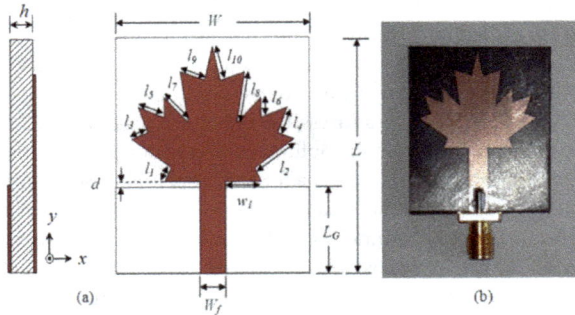

Figure 16. (a) Geometry and (b) photograph of the proposed maple-leaf shaped printed monopole antenna prototype.

The antenna radiation characteristics across the whole UWB frequency band were also investigated. Fig. 18 shows both the measured and simulated E- and H-plane radiation patterns at frequencies 3, 5, 7, and 9 GHz, respectively. The measured H-plane radiation patterns are very close to those obtained in the simulation. It can be noticed that the H-plane patterns are omni-directional at all frequencies of interest. The measured E-plane patterns follow the shapes of the simulated ones, though the agreement is not as good as the H-plane patterns. There are some fluctuations, ripples and distortions on the measured curves, which may be caused by the SMA feed connector and the coaxial cable.

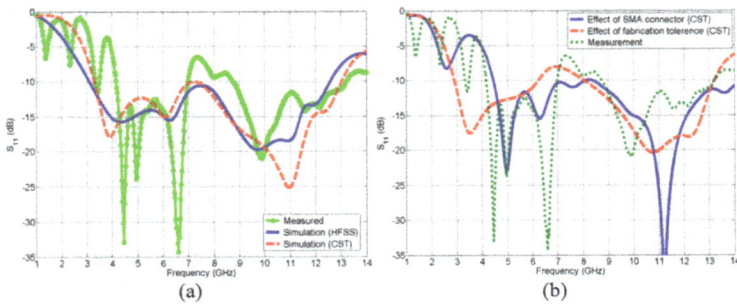

Figure 17. (a) Measured and simulated reflection coefficient curves of the maple-leaf an-tenna (b) effect of fabrication tolerance on the performance of maple-leaf antenna.

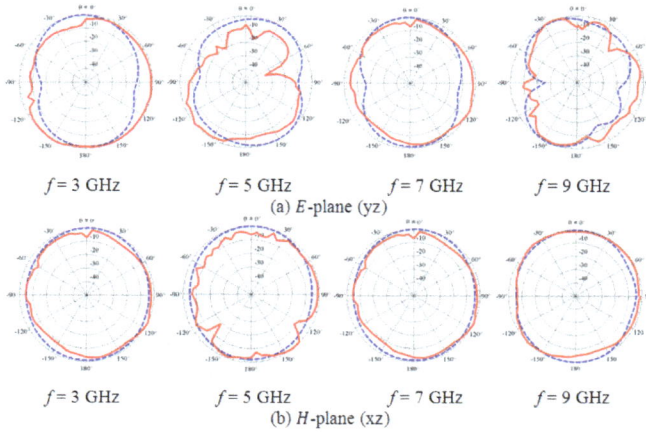

$f = 3$ GHz $f = 5$ GHz $f = 7$ GHz $f = 9$ GHz
(a) E-plane (yz)

$f = 3$ GHz $f = 5$ GHz $f = 7$ GHz $f = 9$ GHz
(b) H-plane (xz)

Figure 18. Measured (red solid line) and simulated (blue dashed line) (a) E-plane and (b) H-plane radiation patterns of the maple-leaf antenna.

3.2.1. Bandstop Antenna Design Prototypes

We developed two different band-notched antennas using two different tech-niques for band rejection. Fig. 19 (a) introduces the first proposed band-notched anten-na which is de-signed by modifying the above maple-leaf antenna by cutting a narrow H-shaped slot away from the radiating patch. The H-slot acts as a filtering element where slot dimensions con-trol the rejection band of the band-notched filter. Fig. 19 (b) presents the second proposed band-notched antenna which is designed by cutting two narrow rectangular slits in the ground plane making a DGS. In the maple-leaf band-stop antennas, we achieved VSWR of 10 (reflection coefficient is 0.82 or -1.7 dB) with H-shaped slot and VSWR of 24 (reflection coefficient is 0.92 or -0.7 dB) with two slits in the ground. It can be concluded that using two slits in the ground plane achieves better rejection characteristics compared to using narrow slots (either arc-shaped or H-shaped) in the radiating patch.

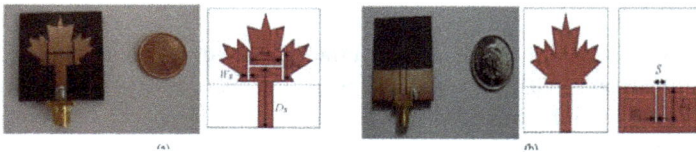

Figure 19. Photograph and geometry of the proposed bandstop antennas using (a) H-slot (b) two slits.

In both techniques, we can control both the notch center frequency fnotch and the band-width BWnotch by adjusting the H-slot and the two slits dimensions, respectively. In the first band-notched antenna, we adjust the slot length LS, thickness WS, and location from the substrate edge DS to control the bandstop characteristic. In the second band-notched antenna, we control the bandstop characteristic by adjusting the two rectangular slits length LS, thickness WS, and distance between them S. The remarkable thing here is that the notch center frequency fnotch is controlled by adjusting the mean length of the slot or the two slits to be about one half-wavelength, i.e. $\lambda/2$ at the desired notched frequency. For example, the calculated mean length of the H-shaped slot is about 26 mm and the calculated $\lambda/2$ at the notch frequency fnotch = 5.5 GHz is 27.7 mm. It is found that the notch bandwidth BWnotch can be controlled by adjusting the thickness of the slot or the two slits.

Fig. 20 (a) & (b) present the simulated and measured reflection coefficient curves of both band-notched antennas with H-slot (WS = 0.65 mm, LS = 8.6 mm and DS = 18.6 mm) and two slits (WS = 0.5 mm, LS = 10.2 mm, and S = 3 mm), respectively. It is ob-vious from the results that the bandstop function in the 5.0-6.0 GHz is successfully achieved for both antenna de-signs. The discrepancies in the 7-9 GHz frequency band come from the maple-leaf antenna itself not from the filter elements for band rejection. It can also be noticed that these discrep-ancies in the 7-9 GHz frequency band are more re-markable in the first prototype than the second one. This is may be due to the effect of using DGS in the finite ground plane en-hanced the antenna performance in the 7-9 GHz frequency band.

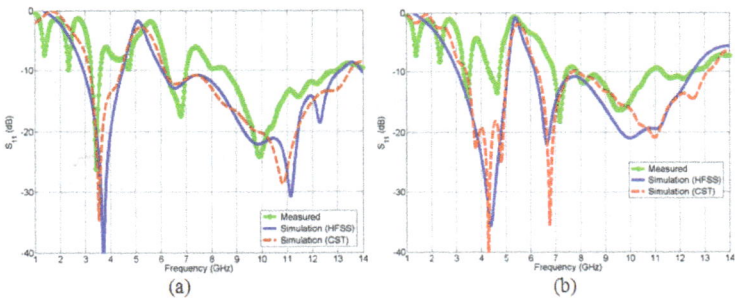

Figure 20. Measured and simulated reflection coefficient curves for bandstop antennas (a) using an H-slot and (b) using two slits.

Fig. 21 and Fig. 22 show the CST simulated surface current distributions over different fre-quencies, i.e. 3, 5.5 and 7 GHz for both band-notched antenna designs with H-slot and two slits, respectively. It can be noticed that at the bandstop frequency 5.5 GHz, nearly all the currents are trapped at the H-shaped slot or two slits which are preventing the current from radiation while at the radiation frequencies 3 and 7 GHz, the current is uniformly distribut-ed through the whole radiating patch.

The CST simulated antenna maximum realized gains in the bore-sight direction versus frequency for the maple-leaf antenna, band-notched antennas with H-slot and two slits are presented in Fig. 23. It can be seen that the maple-leaf antenna gain is almost stable over the whole frequency band and it ranges from 2 dB to 4.3 dB with gain variation about 2.3 dB through the whole frequency band of interest. For band-notched antenna designs with H-slot and two slits, a sharp gain decrease is remarkably happened in the 5.0-6.0 GHz frequency band. Gain results ensure that the band-notched antennas are not responding in the bandstop frequency range between 5.0 and 6.0 GHz.

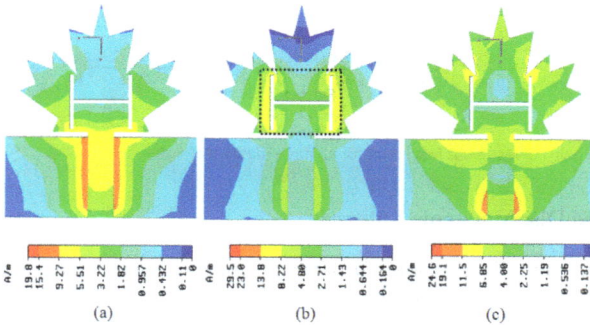

Figure 21. Current distributions for the first bandstop antenna at the (a) radiating frequency f1 = 4 GHz, (b) bandstop frequency f2 = 5.5 GHz and (c) the radiating frequency f3 = 7 GHz.

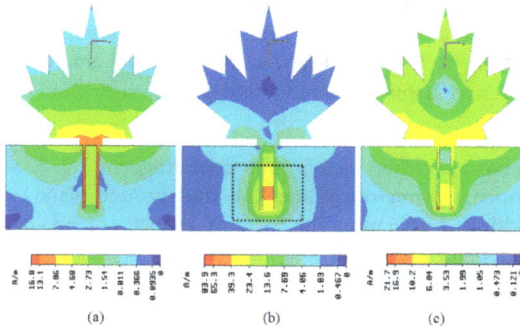

Figure 22. Current distributions for the second bandstop antenna at the (a) radiating frequency f1 = 4 GHz, (b) bandstop frequency f2 = 5.5 GHz and (c) the radiating frequency f3 = 7 GHz.

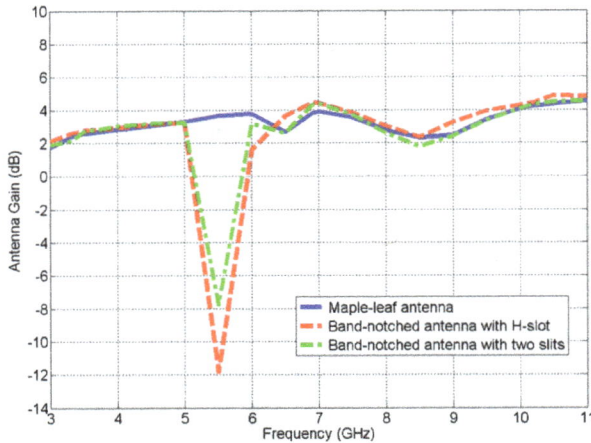

Figure 23. Simulated gain curves versus frequency for all three maple-leaf antennas.

3.3. Other Shaped Disc Monopole Antennas

In this section we continue to enhance the UWB antenna performance to obtain a compact in size antenna with maximum possible impedance bandwidth for UWB opera-tion. We are considering the design of two compact omni-directional UWB antennas with different shape of radiating patches. The first design is the butterfly-shaped monopole antenna while the second one is trapezoidal-shaped monopole antenna with a bell-shaped cut as shown in Fig. 24 (a) and (b), respectively. The butterfly-shaped monopole an-tenna size is 35 × 35 mm2 which is bigger than the previous maple-leaf-shaped antenna (35.5 × 30.5 mm2) by about 13%. The other proposed design is the trapezoidal-shaped monopole antenna of size 34 × 30 mm2 which is smaller than the maple-leaf-shaped an-tenna by about 6%. The best candidate among all printed disc monopole antennas from the antenna size point of view is the trape-zoidal antenna with bell-shaped cut. Moreover, the candidate antenna still has UWB impe-dance bandwidth with reasonable stable radia-tion characteristics and constant gain through the desired frequency range.

Both proposed antennas are etched on 1.575mm-thick Rogers RT 5880 substrate and fed by 50Ω characteristic impedance microstrip line. The finite ground plane length is LG = 10 mm and the feed gap width is d = 0.5 mm. The butterfly-shaped antenna consists of a radiating element of two overlapped elliptical discs of major radius a = 16.6 mm and a minor radius b = 10.4 mm (elliptically ratio a/b ≈ 1.6 forming the two wings of the butterfly). Two annular slot rings of an outer radius r1 = 2 mm and an inner radius r2 = 1 mm have been cut out from the radiating patch. They are located at distance c (= e) = 5.2 mm from the two ellipses' edges. These slot rings can increase the bandwidth of the proposed antenna and they are useful to reduce the overall metallic area.

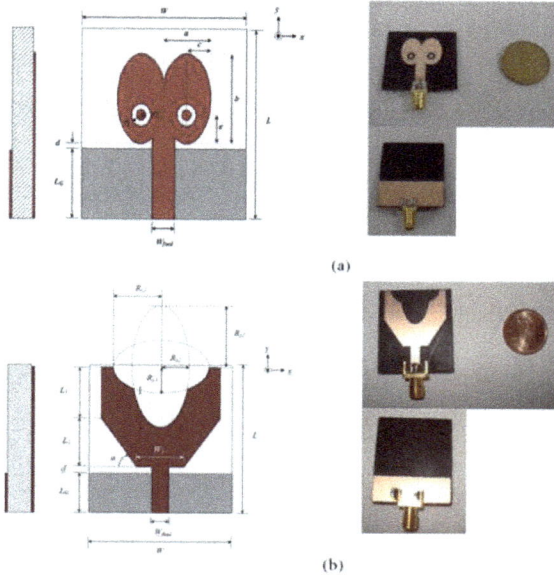

Figure 24. Geometry and photograph of the (a) butterfly-shaped (b) trapezoidal-shaped monopole antenna.

The trapezoidal-shaped antenna consists of a trapezoidal patch of dimensions L1 = 12 mm, L2 = 11 mm, W1 = 10 mm and bevel angle α = 55.7°. Two elliptical cuts have been cut out from the radiating patch forming a bell shaped cut. The first elliptical cut is of a major radius Rx1 = 10 mm and a minor radius Ry1 = 6 mm (elliptically ratio Rx1/Ry1 = 1.67). The second elliptical cut is of a minor radius Rx2 = 6 mm and a major radius Ry2 = 14 mm (elliptically ratio Ry2/Rx2 = 2.33). An antenna prototype of both structures with optimized parameters has been fabricated for experimental investigation.

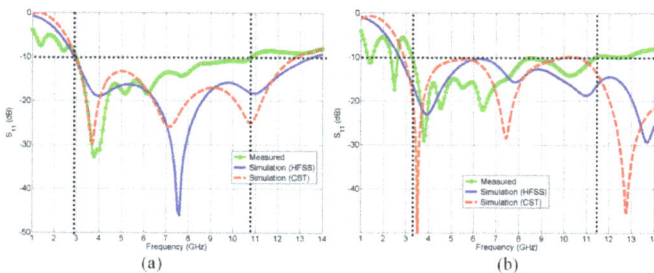

Figure 25. Measured and simulated reflection coefficient curves of the (a) butterfly antenna and (b) trapezoidal antenna.

The measured and simulated reflection coefficient curves against frequency for butterfly and trapezoidal antennas are plotted in Fig. 25, respectively. It is observed from the results that the simulated reflection coefficient with Ansoft HFSS and CST are almost in good agreement and both antennas exhibit wide impedance bandwidth from 3 GHz to beyond 12 GHz (FBW is > 110%) for both antennas. The measured results shows that the both antenna designs still have wide impedance bandwidth covering the UWB frequency range. It is shown that there are different resonances occur at different frequencies across the UWB frequency range and the overlap among these resonances achieve the wide bandwidth characteristic of those types of printed monopole antenna. The measured and simulated E- and H-plane radiation patterns at frequencies 3, 5, 7 and 9 GHz are illustrated in Fig. 26 and Fig. 27, respectively. As expected, both antennas exhibit a dipole-like radiation patterns in E-plane and good omni-directional radiation patterns in H-plane.

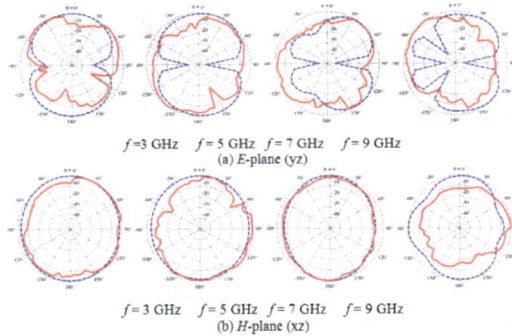

f=3 GHz f= 5 GHz f= 7 GHz f= 9 GHz
(a) *E*-plane (yz)

f= 3 GHz f= 5 GHz f= 7 GHz f= 9 GHz
(b) *H*-plane (xz)

Figure 26. Measured (red solid) and simulated (blue dashed) (a) E-plane and (b) H-plane radiation patterns for butterfly antenna.

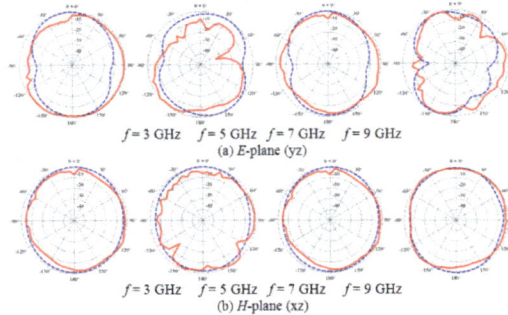

f= 3 GHz f= 5 GHz f= 7 GHz f= 9 GHz
(a) *E*-plane (yz)

f= 3 GHz f= 5 GHz f= 7 GHz f= 9 GHz
(b) *H*-plane (xz)

Figure 27. E- and H-plane radiation patterns of the trapezoidal antenna. Blue dashed lines for simulated and red solid lines for measured.

Both physical and electrical properties of different UWB disc monopole antennas for short-range wireless communications are summarized in Table 2. The comparison includes the overall antennas dimensions, 10 dB return loss bandwidth, realized gain and groub delay features. It can be seen that the Trapezoidal monopole antenna with bell-shaped cut is the good candidate among all proposed antenna designs in terms of both physical and electrical propoerties.

Parameter	Circular disc monopole with two steps and a circular slot	Maple-leaf antenna	Butterfly antenna	Trapezoidal antenna with bell-shaped cut
Dimensions (mm)	41 × 50 × 1.575	30.5 × 35.5 × 1.575	35 × 35 × 1.575	30 × 34 × 1.575
10 dB RL bandwidth (GHz)	3.0~11.5	4.1~7.0, 8.7~13.3 (Dual-band)	3.0~10.8	3.2~11.4
10 dB RL bandwidth (%)	117%	52%, 42%	113%	112%
Realized gain (dB)	3.4~5.2 ±1.8	2.0~4.3 ±2.3	2.0~4.7 ±2.7	2.7~5.3 ±2.6
Group delay (ns)	4.2	2.7	1.5	4.2

Table 2. Comparison among Different UWB Antenna Design Prototypes.

3.4. Transmission Characteristics of UWB Antennas

In this section, we investigate the transmission/reception (Tx/Rx) characteristics of different UWB antennas discussed above in both time and frequency domains. We set up various scenarios and study the communication link between two identical prototype an-tennas. The distance between the transmitting and receiving antennas is assumed to be 30 cm which is approximately 3 wavelengths at the lower frequency of the considered band of operation (antennas are in the far field of each other). Two different scenarios are established for our study. The first one is the face-to-face scenario where the two identic-al antennas are placed in vertical position facing each other at a separation distance be-tween the two antennas of d as shown in Fig. 28(a). The second case is the end-to-end scenario where the two antennas are placed in horizontal position facing each other at a separation distance d as shown in Fig. 28(b). This study is carried out calculated in the E-plane (ϕ = 90°) at different observation angles θ.

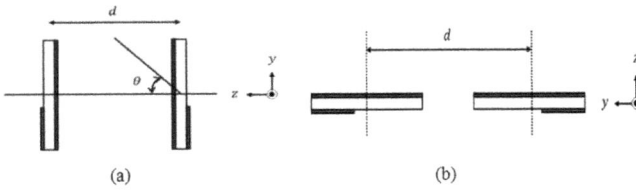

Figure 28. Configuration of UWB transmission system in case of (a) face-to-face scenario and (b) end-to-end scenario.

3.4.1. Time-Domain Characteristics

For a complete description of the antenna characteristics, the time domain behavior is calculated in the E-plane ($\phi = 90°$) at different observation angles: $\theta = 0°, 30°, 60°, 90°$. Referring to Fig. 29(a), the incident wave arriving at the receiving antenna is assumed to be the fourth derivative of a Gaussian function

$$s_i(t) = A\left(3 - 6\left(\frac{4\pi}{\tau^2}\right)t^2 + \left(\frac{4\pi}{\tau^2}\right)^2 t^4\right) \bullet e^{-2\pi\left(\frac{t}{\tau}\right)^2}\left(V/m\right) \tag{3}$$

where $A = 0.1$ and $\tau = 0.175$ ns. The normalized spectrum of this pulse is illustrated in Fig. 29(b), and proves to comply with the required FCC indoor emission mask. Further refining the pulse spectrum can be achieved by utilizing some optimization algorithms. The pulse spectrum is then multiplied by the normalized antenna transfer functions and an inverse Fourier transform (IFT) is performed to achieve the required time domain response. The output waveform at the receiving antenna terminal can therefore be expressed by where represents an ideal bandpass filter from 1 to 18 GHz.

Fig. 30 presents the CST Simulated radiation waveforms in the E-plane at different angles $\theta = 0°, 30°, 60°, 90°$ in face-to-face scenario for different UWB antenna prototypes.

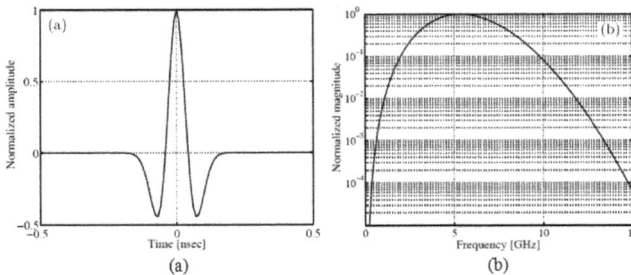

Figure 29. (a) Received UWB pulse shape and (b) spectrum of a single received UWB pulse [35].

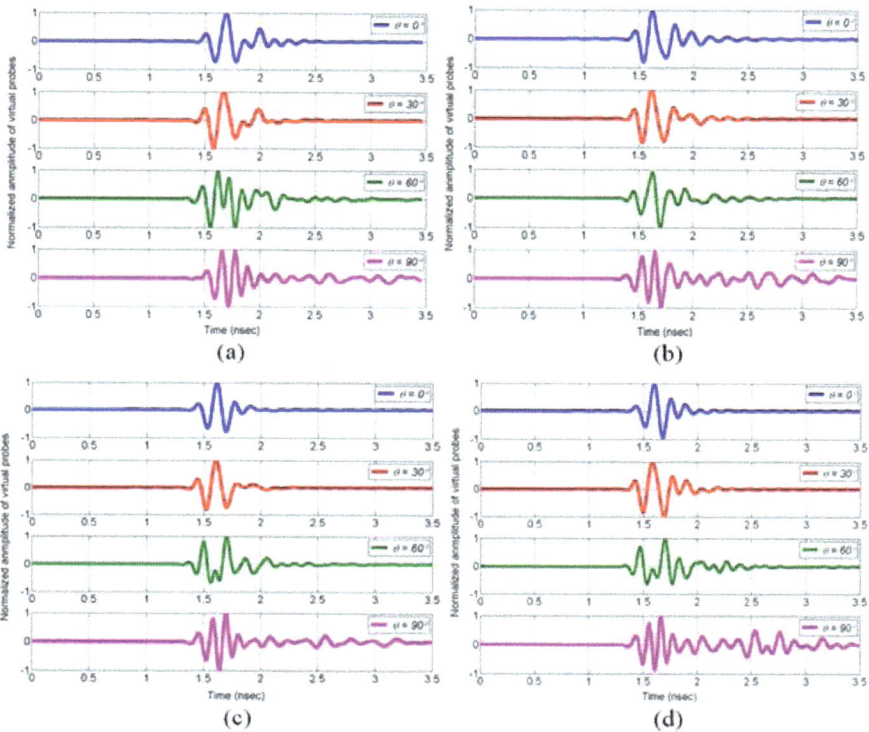

Figure 30. CST Simulated radiation waveforms in the E-plane at different angles θ = 0°, 30°, 60°, 90° in face-to-face scenario for (a) circular disc with two steps and a circular slot antenna (b) maple-leaf monopole antenna (c) butterfly monopole antenna (d) trapezoidal monopole antenna.

3.4.2. Frequency-Domain Characteristics

Since virtual probes are situated in the E-plane (ϕ = 90°), we expect the Tx/Rx system frequency-domain transfer function in face-to-face scenario to become more flat than end-to-end scenario. The separation distance between two transmit and receive antennas is set to d = 30 cm. The simulated impulse responses for both scenarios are given in Fig. 31(a) and (b), respectively. It is shown the ringing effect is slightly less in the face-to-face case compared to the end-to-end case. Fig. 32 shows the simulated transmission coefficients |S21| against frequency at different angles θ = 0°, 30°, 60°, 90° in face-to-face scenario for different UWB antenna prototypes.

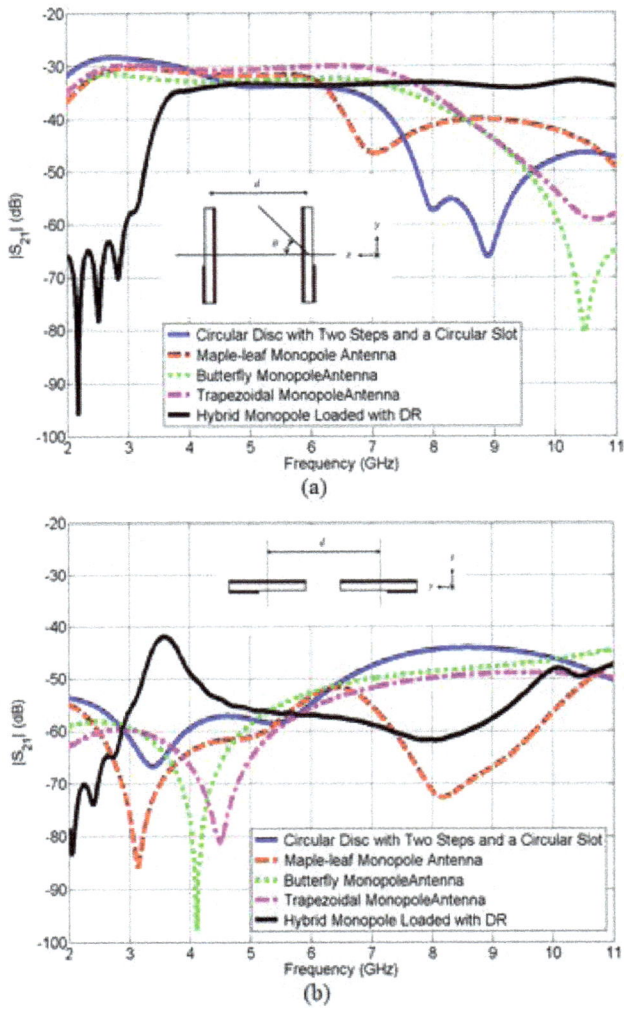

Figure 31. CST Simulated transmission coefficients |S21| as a function of frequency for different UWB antennas in case of (a) face-to-face scenario (b) end-to-end scenario.

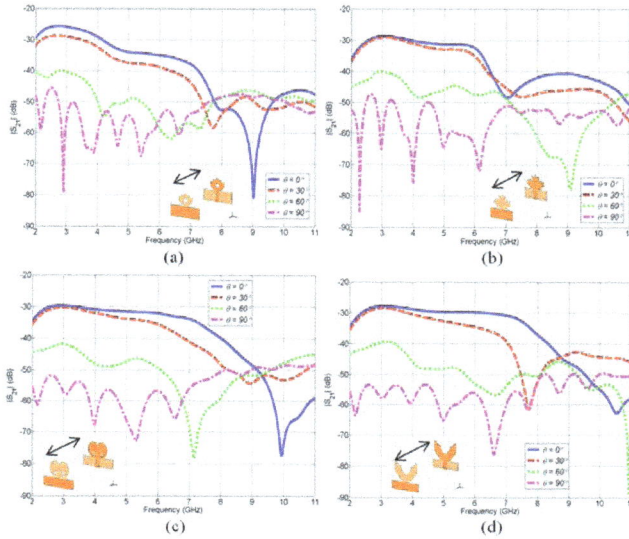

Figure 32. CST Simulated transmission coefficients |S21| as function of frequency at different angles θ = 0°, 30°, 60°, 90° in face-to-face scenario for (a) circular disc with two steps and a circular slot antenna (b) maple-leaf monopole antenna (c) butterfly monopole antenna (d) trapezoidal monopole antenna.

4. Summary

In this chapter, different UWB disc monopole antennas have been developed in microstrip PCB technology to achieve low profile and ease of integration. Parametric studies to see the effect of some antenna parameters on its performance have been numer-ically investigated. For further understanding the behavior of the proposed antennas, sur-face current distribu-tions have been simulated and presented. Different techniques for obtaining bandstop func-tion in the 5.0-6.0 GHz frequency band to avoid interference with other existing WLAN systems have been numerically and experimentally presented. The effects of band-notched parameters on the band-notch frequency and bandwidth have been studied. The chapter has investigated the frequency domain performances of different printed disc monopole anten-nas and hybrid antenna. Experimental as well as the simulated results have confirmed UWB characteristics of the proposed antennas with nearly stable omni-directional radiation prop-erties over the entire frequency band of interest. These features and their small sizes make them attractive for future UWB applications.

Acknowledgements

This research is partially supported by the King Saud University - National Plan for Sciences and Technology (NPST) through Research Grant 09ELE858-02 and by KACST Technology Innovation Center in RFTONICS hosted by King Saud University.

Author details

Osama Haraz[1*] and Abdel-Razik Sebak[1,2*]

*Address all correspondence to: osama_m_h@yahoo.com

1 Electrical and Computer Engineering Department, Concordia University, Canada

2 KACST Technology Innovation Center in RFTONICS, PSATRI, King Saud University, Saudi Arabia

References

[1] FCC. (2002). First report and order, revision of part 15 of the commission's rules regarding ultra-wideband transmission systems FCC.

[2] di Benedetto, M. G., Kaiser, T., Molisch, A. F., Oppermann, I., Politano, C., & Por-cino, D. (2006). UWB Communications Systems: A Comprehensive Overview: Hindawi.

[3] Linardou, I., Migliaccio, C., Laheurte, J. M., & Papiernik, A. (1997). Twin Vivaldi antenna fed by coplanar waveguide. *Electron. Lett.*, 33(22), 1835-1837.

[4] Kim, S. G., & Chang, K. (2004). Ultra Wideband Exponentially-Tapered Antipodal Vivaldi Antennas. *IEEE Antennas and Propagation Society Symposium*, Monterey, CA, June, 3, 2273-2276.

[5] Sibille, A. (2005). Compared Performance of UWB Antennas for Time and Frequency Domain Modulation. 28th URSI General Assembly, NewDelhi, India.

[6] Licul, S, Noronha, J. A. N., Davis, W. A., Sweeney, D. G., Anderson, C. R., & Bielawa, T. M. (2003). A parametric study of time-domain characteristics of possible UWB antenna architectures. *IEEE 58th Vehicular Technology Conference*, VTC 2003-Fall, October, 5, 6-9.

[7] Sego, D. J. (1994). Ultrawide Band Active Radar Array Antenna for Unmanned Air Vehicles. *Proc. IEEE Nat. Telesyst. Conf.*, 13-17.

[8] Su, S. W., Wong, K. L., & Tang, C. L. (2004). Ultra-wideband square planar monopole antenna for IEEE 802.16a operation in the 2-11 GHz band. *Microwave Opt. Tech-nol. Lett.*, 42(6), 463-466, Sept.

[9] Chen, Z. N., Ammann, M. J., & Chia, M. Y. W. (2003). Broadband square annular pla-nar monopoles. *Microwave Opt. Technol. Lett.*, 36(6), 449-454, Mar.

[10] Chen, Z. N., Ammann, M. J., Chia, M. Y. W., & See, T. S. P. (2002). Annular planar mono-pole antennas. *IEE Proc. Microw. Antennas Propag.*, 149(4), 200-203, Aug.

[11] Ahmed, O. M., Elboushi, A., & Sebak, A. R. (2012). Design of Half Elliptical Ring Mo-nopole Antennas with Elliptical Slot in Ground Plane for Future UWB Applications, *Microwave and Optical Technology Letters*, 54(1), 181-187, January.

[12] Ahmed, O. M. H., & Sebak, A. R. (2011). A Novel Printed Monopole Antenna for Fu-ture Ultrawideband Communication Systems. Microwave and Optical Technology Letters August , 53(8), 1837-1841.

[13] Osama, M. H. Ahmed, & Sebak, Abdel-Razik. (2010). Planar Ultrawideband Antenna Array for Short-Range Wireless Communications. *Microwave and Optical Technology Letters*, 52(5), 1061-1066, Mar.

[14] Azenui, N. C., & Yang, H. Y. D. (2007). A printed crescent patch antenna for ultrawi-deband applications. *IEEE Antennas Wireless Propag. Lett.*, 6, 113-116.

[15] Ahmed, O., & Sebak, A. R. (2008). A Printed Monopole Antenna with Two Steps and a Circular Slot for UWB Applications. *IEEE Antennas Wireless Propag. Lett.*, 7, 411-413.

[16] Hsu, C. H. (2007). Planar multilateral disc monopole antenna for UWB application. *Microwave Opt. Technol. Lett.*, 49(5), 1101-1103, May.

[17] Chen, Z. N., Ammann, M. J., Chia, M. Y. W., & See, T. S. P. (2002). Annular planar monopole antennas. *IEE Proc. Microw. Antennas Propag*, 149(4), 200-203, Aug.

[18] Huang, C. Y., & Hsia, W. C. (2005). Planar elliptical antenna for ultra-wideband com-munications. *Electron. Lett.*, 41(6), 296-297, Mar.

[19] Liang, J., Chiau, C. C., Chen, X., & Parini, C. G. (2005). Study of a printed circular disc monopole antenna for UWB systems. *IEEE Trans. Antennas Propag.*, 53(11), 3500-3504, Nov.

[20] Xuan Hui, Wu, & Ahmed, A. Kishk. (2008). Study of an Ultrawideband Omnidirec-tional Rolled Monopole Antenna With Trapezoidal Cuts. *IEEE Transactions on Anten-nas and Propagations*, 56(1), 259-263, Jan.

[21] Ahmed, O. M. H., & Sebak, A. R. (2011). Numerical and Experimental Investigation of a Novel Ultrawideband Butterfly Shaped Printed Monopole Antenna with Band-stop Function. *Progress In Electromagnetics Research PIER C*, 18, 111-121.

[22] Liu, H. W., Ku, C. H., Wang, T. S., & Yang, C. F. (2010). Compact monopole antenna with band-notched characteristic for UWB applications. *IEEE Antennas Wireless Propag. Lett.*, 9, 397-400.

[23] Elboushi, A., Ahmed, O. M., & Sebak, A. R. (2010). Study of Elliptical Slot UWB Antennas with A 5.0-6.0GHz Band-Notch Capability. *Progress In Electromagnetics Research PIER C*, 16, 207-222.

[24] Osama, M. H. Ahmed, & Abdel-Razik, Sebak. (2009). A Novel Maple-Leaf Shaped UWB Antenna with a 5.0-6.0 GHz Band-Notch Characteristic. Progress In Electromagnetics Research PIER C, , 11, 39-49.

[25] Ojaroudi, M., Ghanbari, G., Ojaroudi, N., & Ghobadi, C. (2009). Small square monopole antenna for UWB applications with Variable frequency band-notch function. *IEEE Antennas Wireless Propag. Lett.*, 8, 1061-1064.

[26] Ahamadi, B., & Dana, R. F. (2009). A miniaturized monopole antenna for ultrawideband applications with band-notch filter. *IET Microwave antennas Propagations*, 3, 1224-1231.

[27] Huang, C. Y., Huang, S. A., & Yang, C. F. (2008). Band-Notched Ultra-Wideband Circular Slot Antenna with Inverted C-Shaped Parasitic Strip. July. *Electron. Lett.*, 44(15), 891-892.

[28] Kenny, S. Ryu, & Ahmed, A. Kishk. (2009). UWB Antenna with Single or Dual Band-notches for Lower WLAN Band and Upper WLAN Band. *IEEE Transactions on Antennas and Propagations*, 57(12), 3942-3950, DEC.

[29] Constantine, A. Balanis. (2005). Antenna Theory Analysis and Design. by John Wiley & Sons, INC.

[30] Liang, J. (2006). Antenna Study and Design for Ultra Wideband Communication Applications. *PhD Thesis*, July.

[31] Yan, X. R., Zhong, S. S., & Wang, G. Y. (2007). Compact Printed Monopole Antenna with 24:1 Impedance Bandwidth. *Microwave Opt. Tech. Lett.*, 49(11), 2883-2886.

[32] Yan, X. R., Zhong, S. S., & Liang, X. L. (2007). Compact printed semi-elliptical monopole antenna for super-wideband applications. *Microwave Opt. Tech. Lett.*, 49, 2061-2063.

[33] Low, Z. N., Cheong, J. H., & Law, C. L. (2005). Low-cost PCB antenna for UWB applications. *IEEE Antennas Wireless Propag Lett*, 4, 237-239.

[34] Choi, S. H., Park, J. K., Kim, S. K., & Park, J. Y. (2004). A new ultrawideband antenna for UWB applications. *Microwave Opt. Tech. Lett.*, 40(5), Mar.

[35] Ghavami, M., Michael, L., & Kohno, R. (2004). Ultra Wideband Signals and Systems in Communication Engineering. John Wiley & Sons.

Chapter 9

Dual Port Ultra Wideband Antennas for Cognitive Radio and Diversity Applications

Gijo Augustin, Bybi P. Chacko and Tayeb A. Denidni

Additional information is available at the end of the chapter

1. Introduction

Ultra-wideband (UWB) technology has become one of the most promising technology for short- range high speed data communication due to its high data transmission rate and large bandwidth. These systems utilize the frequency band from 3.1GHz to 10.6 GHz, which is allocated to the UWB systems by the Federal Communications Commission (FCC) [1]. In this ultra-wide spectrum, several unlicensed short range communication bands are overlapping such as IEEE 802.11a WLAN and HIPERLAN/2. Therefore, one of the most effective technique to eliminate these intereferences is to integrate a narro band reject filter in the UWB antenna [2-4].

In this emerging technology, antenna plays the role of a key system element. The design of low profile, easy to construct antennas in a limited space with good radiation characteristics is a challenging task for antenna engineers. The planar antennas are very attractive mainly because of their interesting physical features such as simple structure, compactness and low manufacturing cost [5, 6]. However, the requirements such as via-hole connection in probe-fed antennas, larger ground plane size in microstrip fed designs and precise alignment between layers in multilayer configurations result in increased system complexity. One of the most commonly used feeding technique for modern antennas is the Coplanar Waveguide(CPW) which facilitates key advantages such as low dispersion, less radiation loss and ease of integration with monolithic microwave integrated circuits in uniplanar configuration [7].

In this chapter, we present a comprehensive study on the design, analysis, and characterization of two Uniplanar Ultra Wide-Band(UWB) antennas which have the potential to serve the requirements of future wireless communication systems. The studies were also extended to pulse based, time domain analysis to ensure that it will enable rich broadband services of data, voice, HD video along with high speed internet.

In the first section of the chapter, we present an integrated uniplanar UWB antenna for category A cognitive radio application. The Federal Communications Commission (FCC) defines a cognitive radio as "a radio that can change its transmitter parameters based on the interaction with the environment in which it operates" [8]. This concept has originated from the urgent need to effectively utilize the available spectrum with the explosive growth in high-data rate wireless services. In general, cognitive radio networks can utilize different spectrum sensing and allocation methods namely category A and category B, in which category A uses two antennas. In these systems one antenna is wideband, feeding a receiver for spectrum sensing task meanwhile, the second antenna feeds a front end that can be tuned to the selected transmission band. The motivation behind this work is to illustrate a new integrated antenna for category A cognitive-radio systems by utilizing the uniplanar properties of coplanar wave guide [7], time domain characteristics of vivaldi inspired antennas [9] and recent developments in antenna integration techniques [10]. Although cognitive radio may initially cover lower frequencies, the integration method is demonstrated through UWB and WLAN bands.

A uniplanar antenna for diversity application is presented in the second section of this chapter. Diversity techniques are highly desirable in modern wireless communication systems to increase the channel capacity and to combat the multipath fading problem in the environment, which usually causes larger degradation in the system performance [11, 12]. There are different types of diversities and they are categorized in a broad perspective as spatial diversity, pattern diversity and polarization diversity [13-15]. Depending on the environment, footprint specificatons and the expected interference, designers can employ one or more of these methods to achieve diversity gain. In this era of compact wireless communication systems, the pattern or polarization diversity is more suitable for portable devices than spatial diversity. In present wireless communication systems, particularly in a dense environment, a UWB system with diversity technique is a promising solution to enhance the system performance with high data rate and improved resolution [13, 16]. Such a system has potential applications in advanced instruments for microwave imaging, weapon detection radar which uses short impulses and require high speed data transfer. There have been significant efforts in recent years in various designs of dual polarized UWB antennas for future wireless communication systems [14, 17-21]. However most of them offers a large size [17] multilayer structure [14, 18], complex feeding network [20] and not equipped with band notch functionality [19].To full fill this gap, a compact, uniplanar, CPW fed, dual polarized UWB antenna with embeded notch filter is presented in this chapter.

2. Integrated wide-narrowband antenna for cognitive radio applications

2.1. Antenna geometry and design

Geometry: The evolution of integrated wide-narrowband antenna configuration along with associated parameters is shown in Figure1. The antenna lies in the XZ-plane with its normal direction being parallel to the Y-axis. Compared to the existing integrated antennas for cognitive radio application [10, 22, 23] this design has advantages of uniplanar configuration, group delay variation less than 1ns and good isolation between the ports. The major design elements of this antenna are,

i.   A coplanar wave guide (CPW) to coplanar strip line (CPS) transition (Figure 1a)

ii.   Tapered slot antenna with elliptical tapering (Figure 1b) and

iii.   A rectangular loop slot antenna (Figure 1c).

In this design each of these antenna elements were effectively integrated to form a dual port antenna with ultra wideband characteristics in first port and narrow band performance in the second port. The side and top view of the developed antenna in uniplanar configuration is shown in Figure 1d

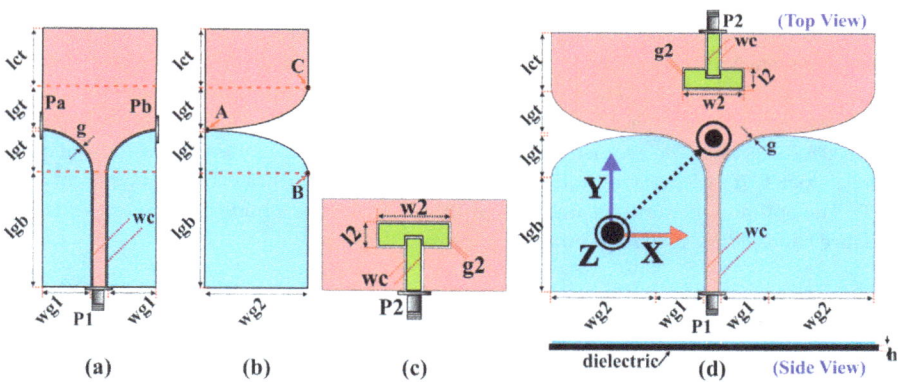

Figure 1. Geometry of the Proposed Antenna. (a) CPW to CPS transition (b) Tapered slot antenna (c) rectangular ring slot antenna (d) integrated antenna.

Design:

CPW to CPS Transition: The design is initiated by designing a coplanar waveguide with characteristic impedance 50Ω on Rogers™ TMM6 thermoset microwave laminate with dielectric permittivity (ε_r) 6, loss tangent 0.0037 and thickness (h) 0.762mm using conventional design procedure [7].The open end of the coplanar waveguide is extended with a smooth curvature to form a CPW to coplanar stripline (CPS) transition [24], which is specified by geometrical parameters lgt, wg1 and g as shown in Figure 1a. In order to maintain geometrical symmetry lgt and wg1 are maintained constant on both sides of the transmission line resulting two striplines terminated at port –Pa and port-Pb. The parameters were optimized for wide impedance bandwidth while maintaining the compact size.

Tapered Slot antenna: An elliptically tapered slot antenna characterized by two identical ellipses with X and Y radius of wg2 and lgt, respectively is shown in Figure 1b. The tapering is optimized for wideband operation specified by FCC [1], while maintaining the initial ta-

pered slot width as 'g'. The length of curvaure AB and AC are defined [25] as ¼[th]of the perimeter formed by the ellipse (1),

$$AB = AC = \frac{\pi}{2}\sqrt{\frac{(\lg t^2 + wg2^2)}{2}} \tag{1}$$

Rectangular loop slot antenna: The rectangular loop slot antenna inspired from [26], with
geometrical parameters l2, w2, g1 and g2 is shown in Figure 1c. The antenna is fed with an
inset open circuited single layer CPW stub for good impedance matching and radiation
characteristics.

The Integrated Antenna: At the first stage of integration, the elliptically tapered slot antenna is integrated to the ports Pa and Pb of CPW to CPS transition, resulting uni-planar UWB
antenna configuration. The antenna parameters wg2 and lgt were fine tuned to fix the lower
cut-off frequency to facilitate wide impedance bandwidth covering the FCC specified spectrum from 3.1GHz to 10.6GHz. In the second stage, the narrow band antenna with CPW
feed is embedded at the space between two tapered slots, without affecting the performance
of the UWB antenna. The geometrcial parameters of the antenna were optmized using commercial tool CST Microwave Studio® (CST MWS) based on finite integration technique
(FIT). The optimum parameters are listed in Table 1 which are a trade off between wide impedance bandwidth, small foot print and improved isolation.

Parameters	Value, mm	Parameters	Value, mm	Parameters	Value, mm
wc	3	lgt	10	l2	5
wg1	15	lct	12.9	g2	0.5
wg2	22	g	0.35	g1	0.25
lgb	25.8	w2	14	h	0.762

Table 1. Geometrical parameters of the integrated antenna shown in Figure 1

2.2. Simulation with experimental validation

After optimizing the integrated antenna, a prototype is fabricated using LPKF® circuit board
plotter. The entire fabrication process is relatively simple and can also be performed using
conventional low cost PCB processing technology such as photolithography. In addition, the
single layer design eliminates the requirement of alignment holes. Therefore this design facilitates accurate, efficient and cost effective fabrication. The fabricated prototype with a mechanical calibration standard is shown in Figure 2a for comparison. A perspective view of
the wide-narrowband antenna mounted for measurement in the anechoic chamber is provides in Figure 2b.

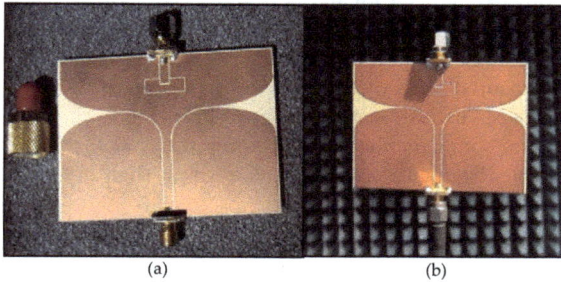

Figure 2. Photograph of the fabricated prototype (a) front view (b) perspective view

Measurement of both frequency and time domain characteristics are essential to evaluate the performance of an UWB antenna. In frequency domain the S-parameters, gain, efficiency, radiation pattern and polarization are measured and analysed in section 2.2.1. In time domain, the antenna performance is evaluated using very short pulses. The group delay, antenna transfer function, impulse response and fidelity were analysed and discussed in section 2.2.2

2.2.1. Frequency domain characteristics

The measured reflecton and transmission coefficients of the antenna along with the simulation results are plotted in Figure 3.

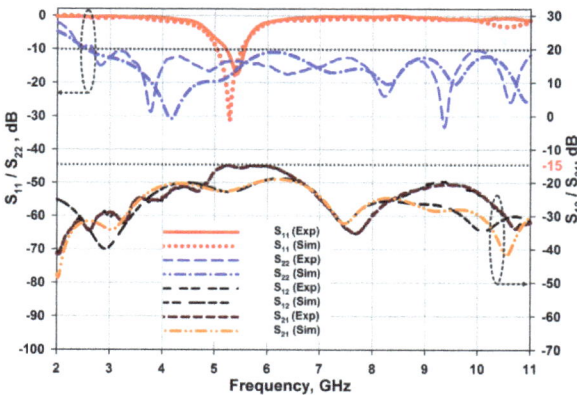

Figure 3. Simulated and measured S parameters of the Wide-Narrowband Antenna.

It is found that the UWB antenna excited through port-P1 provides a 2:1 VSWR bandwidth from 2.6GHz to 11GHz, meanwhile, the narrowband (NB) antenna excited through port-P2 provides 2:1 VSWR bandwidth from 5 GHz to 5.5 GHz. Thus the integrated antenna meets the VSWR bandwidth requirement for the FCC specified UWB band and for the WLAN spectrum. The trasmission coefficients in Figure 3 shows that, the inter-port isolation is better than 15dB which is a reasonable value to eliminate the cross talk between the antennas. The simulated and measured s-parameters are in reasonable agreement. However, there is a slight discrepancy between the theorectical and experimental results. This is mainly because of the approximation of boundary conditions of computational domain. In addition, as explained in [27], RF cables from the vector network analyzer slightly influences the measurement of small antennas.

The isolation mechanism and polarization of the electromagnetic radiation is investigated through simulated surface current analysis. The magnitude and vector plot of surface current density at 5.2 GHz is illustrated in Figure 4.

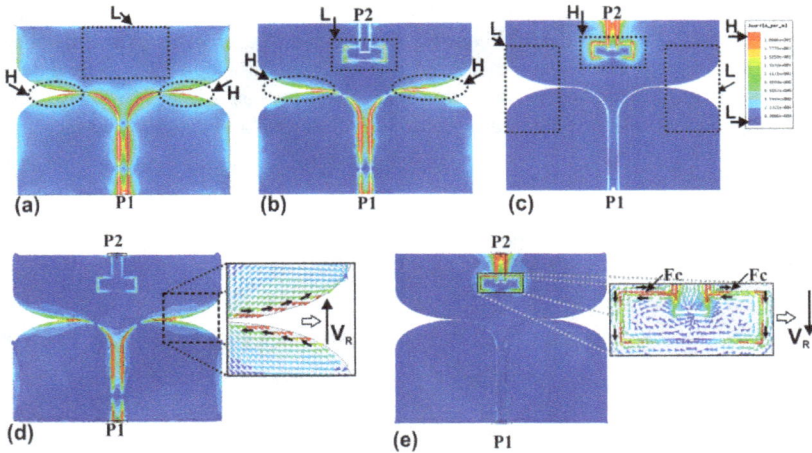

Figure 4. Simulated surface current distribution at 5.2GHz. (a) Magnitude of Jsurf of UWB antenna without integrating NB antenna (b) Magnitude of Jsurf with P1-excited, P2=50Ω (c) Magnitude of Jsurf with P2-excited, P1=50Ω (d) Vector of Jsurf with P1-excited, P2=50Ω (e) Vector of Jsurf with P2-excited, P1=50Ω

It is evident from Figure 4a that, the tapered surface regions on both sides of the integrated antenna contributes for the radiation. Meanwhile, the current excited at the top region between two tapered slot antennas, indicated by the rectangular dashed box, is almost nil. This region is effectively utilized to integrated the ring slot antenna for narrow band operation. In Figure 4b, the surface currents in the integrated configuration is provided, which shows that the current coupling from UWBA to the NBA and vice versa(figure 4c) is very low. This results in an efficient integration with good inter-port isolation. The vector analysis of surface current in the integrated antenna is shown in Figures 4 (d-e). It is clear from the plot that the dominant radiating current vector at both the tapering is vertical in direction. This shows that the polarization of the radiated electromagnetic wave from the ultra wideband antenna is vertically polarized. Similarly, the resultant current vector at the vertical slot edges of NBA are vertical in direction and in turn results in vertical polarization. However, it is worth to note that, while estimating the composite current vector, the surface regions where equal and opposite current vectors exists, indicated by Fc, are not taken into account.

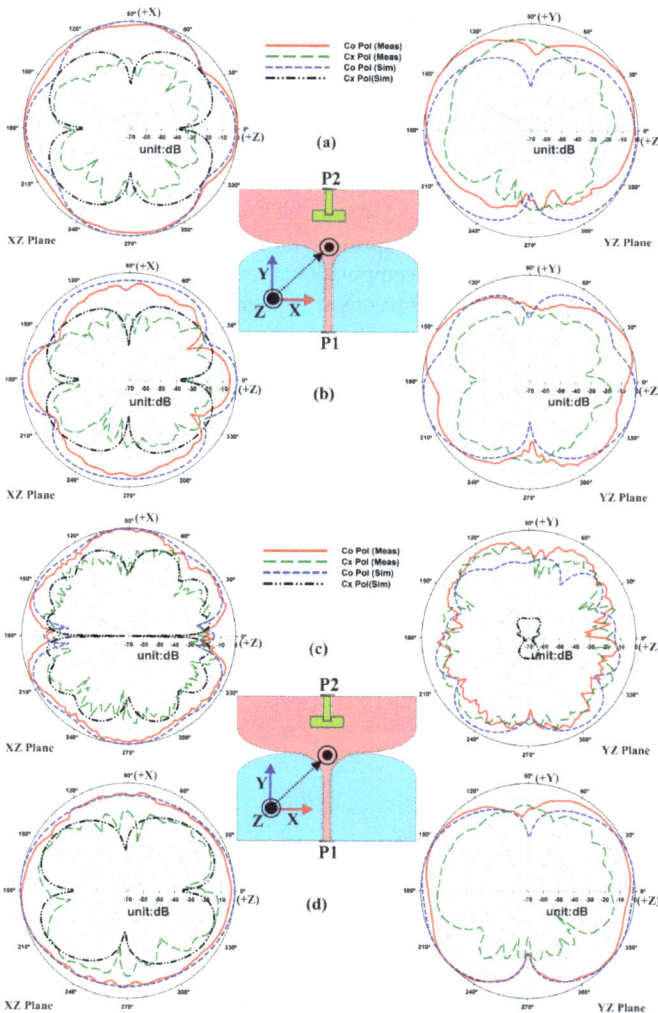

Figure 5. Measured and simulated radiation patterns (a) 3.5 GHz [P1] and (b) 5.2GHz [P1] (c) 10.5 GHz [P1] and (d) 5.2GHz [P2]

The simulated and measured radiation pattern of both the UWB and NB are presented in Figures 5(a-d). These radiation patterns were measured independently, that is, port-P2 is loaded with 50Ω termination while exciting port-P1 and vice versa. It is clear from the pattern that the UWB antenna is directional towards 0º and 180º because of the two tapered slot antennas radiating to the corresponding directions. This indicates that, the UWB and NB antenna has beam maxima at ±X and ±Z directions respectively, which are not highly attractive for applications which utilize the simultaneous utilization of both antennas.

However, the radiation patterns are suitable for cognitive radio applications such as IEEE 802.11 wireless rural area networks (WRAN) in which the spectrum sensing take plane during the interval between intra-frame and inter-frame, when the transceiver is switched off [28]. It is also visible from the radiation pattern that, at higher frequencies, the polarization purity is degraded due to the finite ground plane effect. A good agreement between simulated and measured radiation patterns are observed except for the cross polar patterns in the YZ plane. This is mainly because of the spurious reflections from the SMA connectors and cables that are not incorporated in the computational simulation. In conclusion, the patterns are similar to those observed for antennas used in cognitive radio systems [10, 22] and in wireless system terminals [29]. It is worth to note that, these patterns are also suitable for applications in indoor wireless communication systems including ad-hoc networks, where cross polar performance is not a high priority requirement and channels are dominated by rich Rayleigh fading [30]

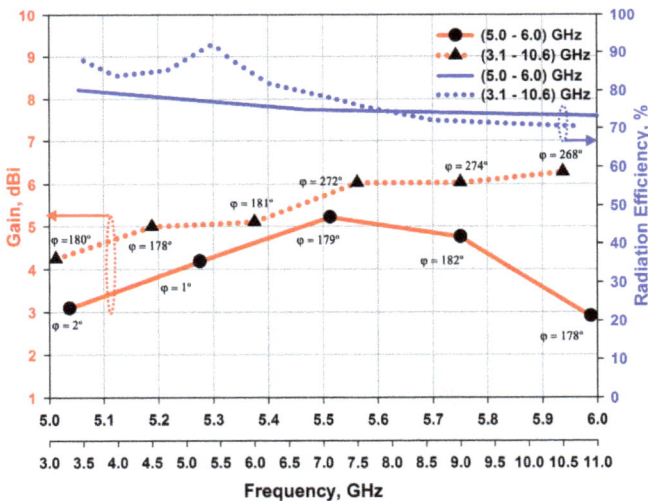

Figure 6. Separately measured gain and efficiency of UWB (port 1) and NB (port 2) antenna.

The gain of UWB and NB antenna are measured independently in the XZ-plane using the gain comparison method and shown in Figure 6. In the graph 'φ' denotes the direction of radiation corresponding to the peak gain. It is clear from the measurement results that, the gain variations are within 2.2dBi and 2.4 dBi in the ultra wide band and narrow band spectrum of the integrated antenna, respectively. The radiation efficiency is also measured using wheeler cap method [31]and incorporated in Figure 6. It is found that the average efficiency of 81% and 75% are observed for UWB antenna and NB antenna, respectively.

2.2.2. Time domain characteristics

As depicted in the introduction, the UWB antenna need to possess high level of pulse handling capabilities in order to handle high frequency impulses. In this section the time domain characteristics including group delay, antenna transfer function, implulse response and fidelity factor were measured and discussed.

In order to measure the group delay of the UWB antenna, two identical antenna prototypes were made. As illustrated in the inset of Figure 7, these two identical antennas were kept in in the anechoic chamber at far field(R=1m) with two orientations; face-to-face and side-by-side. The time domain measurement capability of the vector network analyser is utilized for this measurement. Prior to the measurement, a full two port calibration is performed to eliminate the effects of cables and connectors. The measurement is performed by exciting the identical antennas through port-P1 while port-P2 is terminated with a broadband 50Ω termination. It is clear from Figure 7 that, the group delay remains constant with variations less than a nanosecond for both orientations.

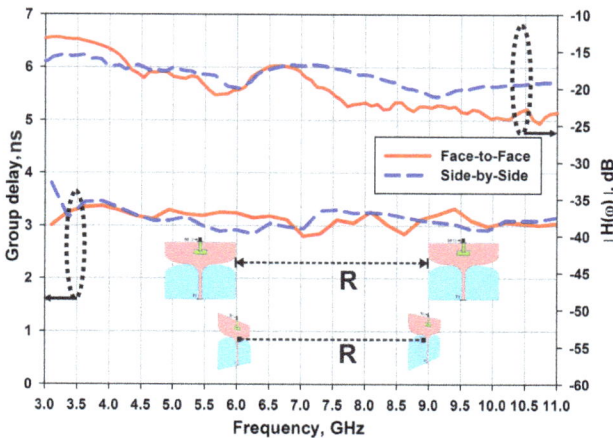

Figure 7. Measured group delay and normalized antenna transfer function in two different orientations of the UWB antenna (R = 1m)

The antenna transfer function of the antenna is also calculated using (2) and incorporated in Figure 7.

$$H(\omega) = \sqrt{\frac{2\pi R c S_{21}(\omega) e^{j\omega R/c}}{j\omega}} \qquad (2)$$

where c is the velocity of light in free space and R is the distance between two antennas [25].

The antenna transfer function remains fairly stable throughout the UWB spectrum with variations less than 10dB except for the lower end of the spectrum in the face-to-face orientation.

The transient response of the antenna is evaluated using fourth derivative of the Gaussian pulse defined by (3)

$$V_{in}(t) = A\left(3 - 6\left(\frac{4\pi}{T^2}\right)t^2 + \left(\frac{4\pi}{T^2}\right)t^4\right)e^{-2\pi\left(\frac{t}{T}\right)^2} V \Big/ m \qquad (3)$$

The impulse response of the wideband antenna is obtained by convoluting the fourth derivative of (3) with the inverse Fourier transform of antenna transfer function (2). The spectrum of this pulse fully covers the FCC band and comply with the emission standards specified when the amplitude constant A = 1.6 and pulse duration parameter T = 67ps. [32]

The input pulse and the output pulses at face-to-face orientation and side-by-side orientation are shown in Figure 8. It is evident from the plot that the UWB antenna retains the information contained in the impulse with minimum distortion.

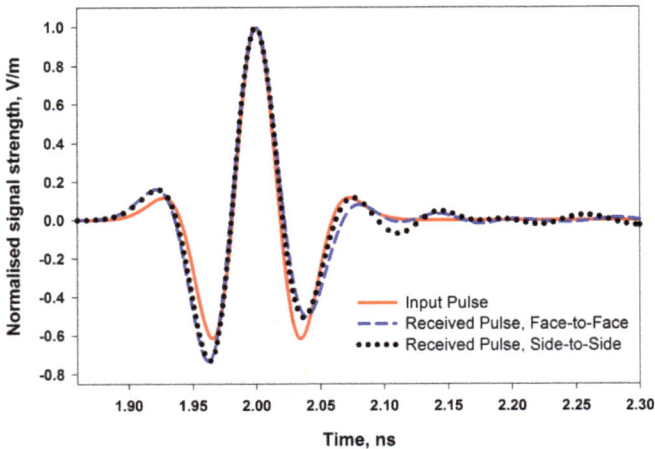

Figure 8. Input and radiated pulses of the proposed antenna.

Fidelity factor, F is an effective parameter to analyze the distortion between two pulses [33], which is defined as,

$$F = \frac{\int_{-\infty}^{\infty} S_t(t) S_r(t - \tau) d_t}{\sqrt{\int_a^\alpha |S_t(t)|^2 dt \int_a^\alpha |S_r(t - \tau)|^2 dt}} \qquad (4)$$

where τ is the delay between the input pulse St (t) and the output pulse Sr (t). The fidelity factor of the wideband antenna is also evaluated and presented in Table 2. The fidelity factor remains greater than 0.85 which shows that, the antenna imposes negligible effects on the radiated pulses.

Orientation	Fidelity Factor,F
Face-to-Face	0.86
Side – to –Side	0.88

Table 2. Fidelity Factor of the wide band antenna

2.3. Parametric analysis

A parametric analysis of the key antenna prameters which influence the lower cut-off frequency of the UWB antenna and the resonant frequency of the NB antenna is studied in this section. This will help the antenna engineers to pay more attention to those parameters during the design, optimizaton and prototyping.

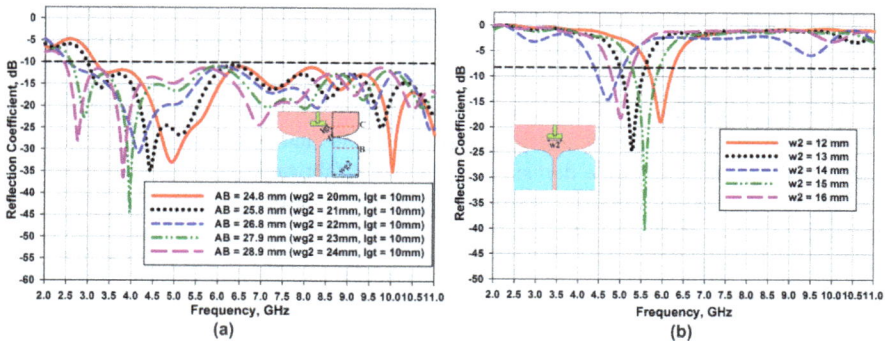

Figure 9. Influence of key antenna parameters on reflection coefficient (a) Tapering aprature, AB (b) loop width, w2

The variation of relfection coefficient with tapering aprature AB is the most sensitive parameter which determins the lower cut-off frequency of the UWB antenna. Figure 9a shows the variation of reflection coefficient with wg2 (and in turn AB). It is clear from the plot that the lower resonance shifts drastically for small variations of wg2. As wg2 increases from 20mm to 24mm the lower cut-off frequency of the UWB antenna moves from 3.2GHz to 2.4 GHz. It is also worth to note that the impedance matching throughout the wide band remains within the FCC specifications when the tapering aprature varies from 24.8mm to 28.9mm. In narrow band antenna, the variation of resonant frequency with the loop width, w2 is depicted in Figure 9(b). It is found that when the loop width varies from 12mm to 16mm the resonant frequency of the narrow band antenna drifts from 6GHz to 4.6GHz.

2.4. Conclusion

An integrated dual port antenna with good inter-port isolation in uniplanar configuration for congnitive radio systems is presented in this section. The space between two tapered slot antennas which forms the ultra wideband antenna, is effectively utilized to integrate a narrow band sqare loop slot antenna. The measurement results indicate that, the UWB and NB antenna provides a 2:1 VSWR bandwidth from 2.7 GHz to 11 GHz and 5 GHz to 5.5GHz, respectively. The antenna also provides inter-port isolation better than 15 dB throught the resonant band. Measured readiaton pattern reveals that the wide-band antenna can fulfill the needs of spectrum searching task, and the narrow-band antenna can be used for trasmission in cognitive radio systems. The time domain characteristics of the UWB antenna in the integrated configuration is also studied and the results reveals that it facilitate transmission and reception of pulses with minimum distortion. Therefore, the antenna can also be a good candidate for future applications, such as medical imaging / weapon detection systems, which are connected to the host system through high speed WLAN link.

3. Ultra-wideband slot antenna for polarization diversity applications

In this section an Ultra wideband Antenna for polarization diversity application is presented. The development, analysis and characterization of this dual port antenna is discussed in detail. Both frequency and time domain analysis of the fabricated prototype reaveals that, this antenna is an attractive element in future wireless communication systems where the challenges such as multipatch fading is a major concern.

3.1. Antenna geometry and design

The proposed antenna is inspired from the design proposed in [34], where wideband characteristics is obtained by exciting a compact annulus ground plane with a circular patch. In this work, a polarization diversity antenna is devoloped by feeding an annulus ground plane with dual orthogonal ports. This feeding mechanism is on of the effective technique for diversity antennas [14, 17].

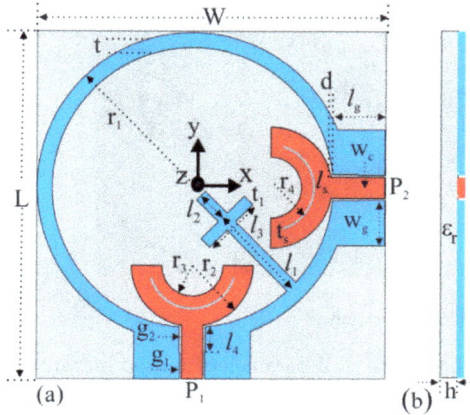

Figure 10. Geometry of the proposed antenna (a) Top view, (b) Side view

The proposed antenna geometry in cartesian coordinate system are shown in Figure 10. The basic antenna structure consists of an annulus slot and two orthogonal, identical CPW signal strips at same distance from the annulus center. Compared to dual polarized UWB slot antennas recently reported in [14, 17], the ground plane of the proposed antenna is modified as an annulus slot with radius r1 and thickness t, which creatively reduces the antenna footprint. The CPW feedline is exciting two U-shaped elements with geometrical parameter r2, r3. In addition to the the broad impedance bandwidth of this unique design, an impedance transformer with length l4 and gap g2 is also incorporated in the CPW line for further bandwidth enhancement. Since, inter-port isolation is one of the highly desirable characteristics of a diversity antenna, a cross shaped strip with dimensions l1, l2 and l3 is embeded diagonally in the antenna. In order to avoid interferences with the overlaping unlicensed bands in the UWB spectrum, an arc shaped slot resonator with specifications r4,ts and ls is also integrated in the antenna. This slot resonator facilitates band notch functionality for the diversity antenna.

Design: The antenna is realized on a Rogers® RT/Duroid 6035HTC laminate with permittivity (r) 3.6, loss tangent 0.0013, and thickness (h) 1.524mm. The CPW line with characteristics impedance 50Ω is first designed using the conventional design procedure [7]. The annulus ground plane parameters (r1) and U-shaped stub dimensions(r2,r3,d) were selected [35]to cover the FCC specified UWB spectrum. In this design the radius r1 of the ground plane determines the first resonant frequency and the radius r2 ensures the impedance matching. In conclusion, the merging of several dominant resonances, which are produced by the annulus ground, the U shaped feeding structure, and the coupling between them provides a broad impedance band width [36]. In order to increase the inter port isolation a cross shaped stub [37] is then inserted at an optimum position diagonally in the ground plane. Finally the semicircular arc shaped resonators were designed [38] and integrated for notch functionality. The antenna provides desired notch in the IEEE 802.11a and HIPERLAN/2 bands, when the length of the slot resonator ls is approximately half wave length long at the center notch frequency.

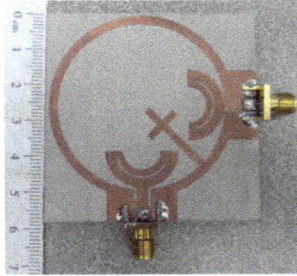

Figure 11. Photograph of the fabricated prototype

The initial analysis of the geometrical parameters and optimization of the antenna is performed using the FDTD based CST Microwave Studio®. The optimum parameters are listed in Table 3 which is a tradeoff between wide impedance bandwidth, better isolation, sharp notch and small foot print.

Parameters	Value, mm	Parameters	Value, mm	Parameters	Value, mm
W	57	g1	0.3	t	2.5
L	57	g2	0.4	t1	2
r1	23	d	0.2	ts	0.4
r2	10	lg	8.5	l1	16
r2	5	wg	7.5	l2	5
r4	7	wc	3.5	l3	10

Table 3. Geometrical parameters of the diversity antenna shown in Figure.10

3.2. Simulation and experimental results

After the initial design and optimization of the diversity antenna using 3D full wave electromagnetic solver, a prototype is fabricated using LPKF milling machine. Extreme care is taken during the milling process to ensure fabrication accuracy especially at the most sensing elements such as the width of the slot resonator. The fabricated prototype is shown in Figure 11. The measurement results in frequency and time domain are discussed in the following sections.

3.2.1. Frequency domain

The measured and simulated S-parameters of the proposed dual polarized antenna at port-1 (P_1) and port-2 (P_2) are presented in Figure 12. Due to geometrical symmetry the simulated results for both ports are identical. The slight difference in the measured S_{11} and S_{22} is owing to the fabrication inaccuracies. The antenna displays a 2:1 VSWR bandwidth from 2.80 GHz to 11GHz with an inter-port isolation better than 15dB except at the lower and higher frequency end. It also provides a notch band with high band rejection from 4.99 to 6.25GHz. The small differences in the measured and simulated results are due to the approximate boundary conditions in the computational domain. Moreover, RF cable from the vector network analyzer slightly affects the measurements of small antennas [27]

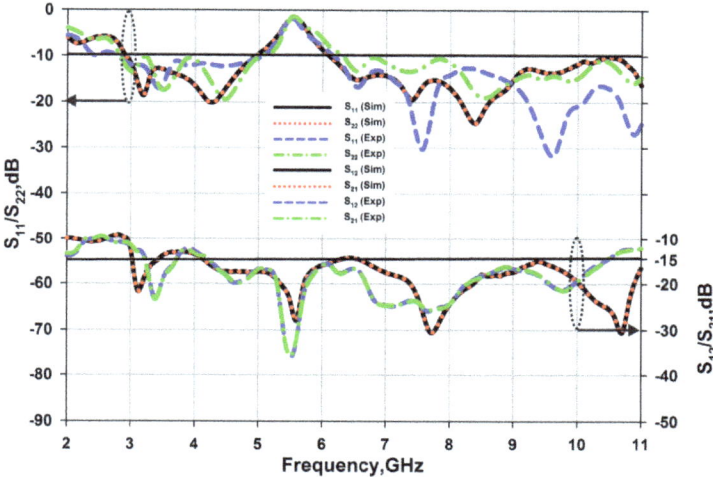

Figure 12. Simulated and Measured S-parameters of the diversity antenna.

The transmission coefficient shown in Figure 13(a) provides the influence of cross shaped stub throughout the resonant band. It is observed that, the isolation is improved by inserting the stub diagonally between the U-shaped elements. In addition the cross stub improves the impedance matching at the lower end of the spectrum.

The magnitude of surface current density at 3.5GHz and 8.5 GHz are illustrated in Figure 13(b-e) which aid better understanding about the isolation performance of the antenna. It is evident from the results that, when P1 is excited without the isolation strip, the current from P1 has a tendency to couple to P2 through the common ground plane. However the integration of the isolation strip drastically reduces the current on the ground plane around P2 (vice-versa when P2 is excited) with a strong current excitation on the strip. This results in better inter-port isolation and thereby significantly improves the diversity performance. It is also worth to note that the isolation strip has negligible influence over the surface current on the antenna elements connected through P1 at both lower and higher end of the UWB spectrum.

The far-field (2D) radiation pattern of the proposed antenna is also measured and compared with the simulation results at three different frequencies in the UWB band. The patterns are measured in a fully automated anechoic chamber, which is connected to Agilent® E8362b Performance Network Analyzer. A standard horn antenna is connected to the first port of the PNA while the second port is connected to the antenna under test. The radiation patterns in the XZ and YZ planes at 3.1 GHz, 7.5 GHz and 10.5 GHz seperately measured for both ports in orthogonal planes are illustrated in Figure 14. Nearly omnidirectional patterns are observed in the lower frequency region of the UWB spectrum meanwhile slight distortions exists at the higher frequency region. This is partially due to the effect of connecters and cables [39, 40]and magnetic current variations along the circumference of the slot [14]. It is also worth to note that the patterns at P1 and P2 are almost similar with a 90º rotation, which in turn confirms dual polarization. In general, the patters are similar to those observed for diversity applications [14] and in wireless system terminals [29]

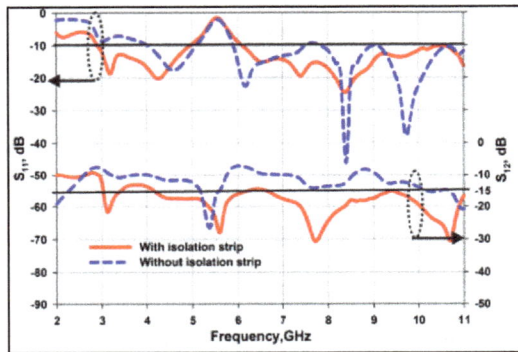

(a)

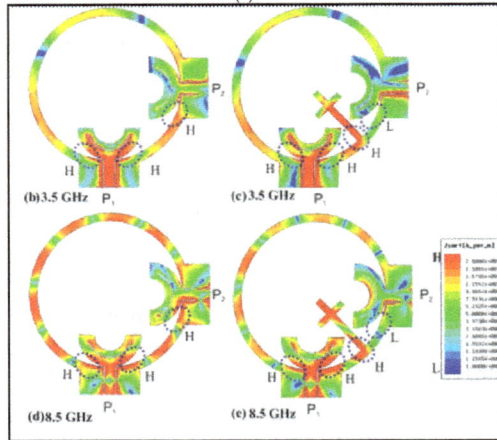

(b)

Figure 13. a) Simulated transmission coefficient, Magnitude of Jsurf (b) at 3.5GHzwithout isolation strip (c) at 3.5GHz with isolation strip (d) at 8.5GHzwithout isolation strip (e) at 8.5 GHz with isolation strip

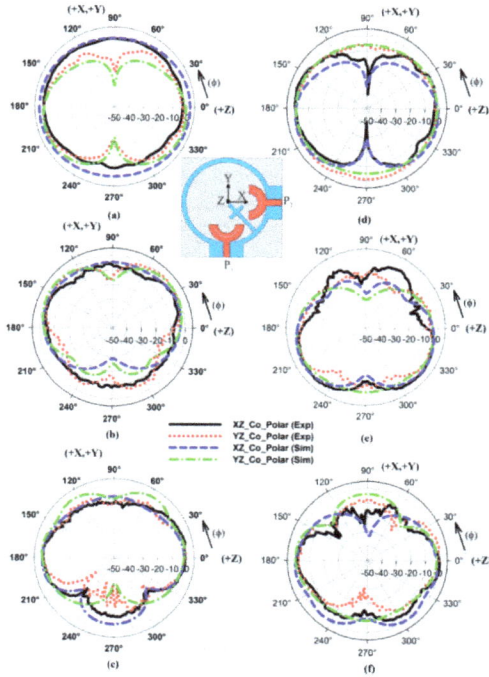

Figure 14. Measured and simulated radiation patterns (a) 3.5 GHz [P1] and (b) 7.5GHz [P1] (c) 10.5 GHz [P1] (d) 3.5 GHz [P2] and (e) 7.5GHz [P2] and (f) 10.5 GHz [P2]

In a diversity system the envelope correlation coefficient (ECC) is an important measure of diversity performance. ECC with a value of greater than 0.5 can typically degrade the diversity performance. The envelope correction coefficient of the proposed antenna is also calculated from the simulated and measured S-parameters as described in [41] using (5) and shown in Figure 15.

$$\rho_c = \frac{|S^*_{11} S_{22} + S^*_{21} S_{22}|^2}{(1 - [|S_{11}|^2 + |S_{21}|^2])(1 - [|S_{22}|^2 + |S_{12}|^2])} \tag{5}$$

It is evident from Figure 15 that, the proposed antenna has a very low value of ECC throughout the operating band which clarifies that the antenna is a good candidate for modern wireless communication systems employing polarization diversity.

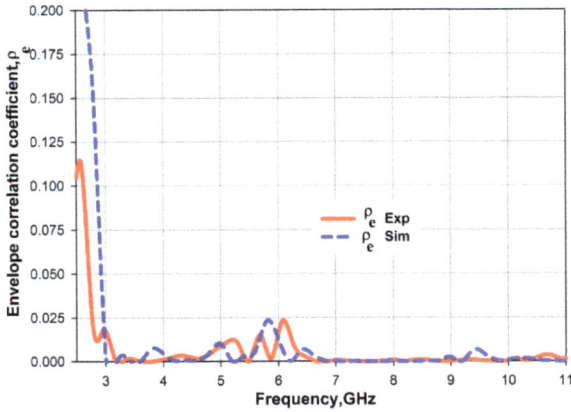

Figure 15. Measured and simulated envelope correlation coefficients from S parameters

The gain of the antenna for both ports are measured independently (when P1 is excited, P2 is terminated with 50Ω load and vice-versa) using gain comparison method. In this the gain is measured in both the planes of the radiation pattern and the peak gain is selected from either plane which gives the larger value. It is clear from the Figure 16 that, the antenna has moderate gain with variations less than 2.23dBi throughout the operating band while the gain drops up to -9.3dBi in the notch frequency.

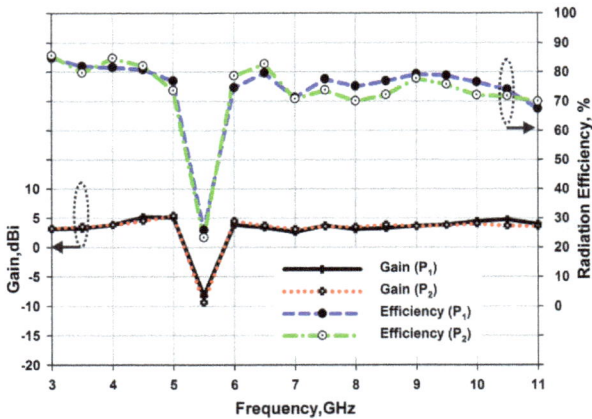

Figure 16. Measured peak gain and radiation efficiency

The efficiency of the antenna for both ports is measured using Wheeler cap method [31] and is also incorporated in Figure 16. The antenna provides efficiency better than 70% in the UWB spectrum while it drops to 25% in the notch band.

3.2.2. Time domain analysis

Advanced UWB systems are realized using an impulse-based technology in which the time domain performance are equally as important as frequency domain properties. The time domain characteristics including group delay, antenna transfer function, impulse response and fidelity are measured, analyzed and discussed in this section.

The group delay of the antenna for face to face orientation is measured using the time domain measurement facility of Anritsu Ms4647A network analyzer by exciting two identical antennas kept in the far field (when P_1 is excited P_2 is terminated with 50 Ω load and vice-versa).The antenna provides a group delay (Figure 17) which remains almost constant with variations less than 1ns except at the notch band. The antenna transfer function defined by (2) is also calculated and incorporated in Figure 17.It shows fairly flat magnitude variations for each port of the antenna, which is less than 10dB throughout the band. The impulse response of the antenna is evaluated by convoluting the modulated Gaussian monocycle defined in (3) with h(t), the inverse Fourier transform of antenna transfer function. The spectrum of this impulse fully covers the FCC band and comply with the emission standards specified when, the amplitude constant A = 1.6 and pulse duration parameter T = 67ps. The input and output waveforms for both ports are shown in Figure 18. It can be seen that the radiated pulse through two ports of the proposed antenna retain the information with minimum dispersion.

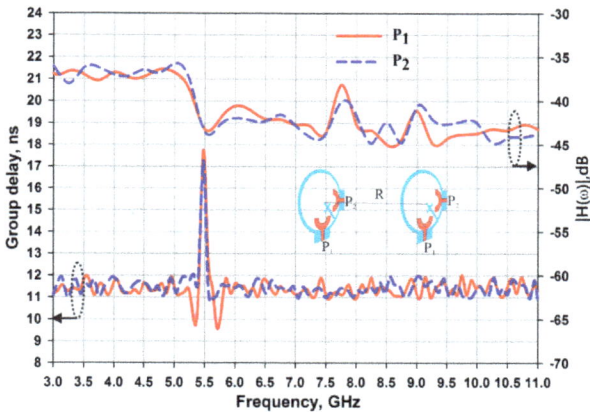

Figure 17. Measured group delay and antenna transfer function between identical antennas

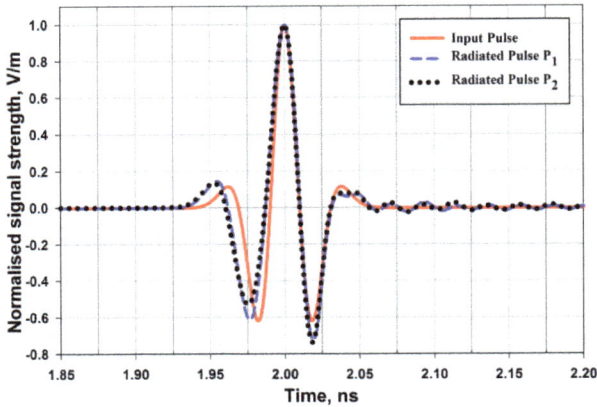

Figure 18. Input and output impulses through P_1 and P_2 of the antenna

The cross correlation between the source pulse St (t) and the radiated pulse Sr(t) is then evaluated by the fidelity factor, F using (4). As shown in Table 4, high value of Fidelity reveals that the antenna imposes negligible effects on the transmitted pulses [42].

Orientation	Fidelity Factor
P1	0.88
P2	0.85

Table 4. Fidelity Factor of the proposed antenna for both ports.

3.3. Parametric analysis

In order to provide more information to the antenna engineers during the design and optimization process, a parametric analysis of important antenna parameters which influence the lower cutoff frequency (r_1) and notch band (*ls*) are conducted and presented.

Figure 19a shows that the first resonant frequency of the antenna drifts down when the ground radius r1 is increased from 22 to 25 mm. This clarifies the initial assumption that, the first resonance frequency is determined by the radius r1. It is also clear that the ground strip length has a slight influence on the isolation characteristics. An optimum value r1=23mm is selected for required performance. The tuning of notch band with slot length ls is shown through the parametric analysis in Figure 19b. As the ls varies from 16mm to 20mm, the peak notch frequency shifts from 6.1 GHz to 5GHz. These parameters are very sensitive to the overall performance of the antenna and therefore it is required to provide extreme care during the fabrication process.

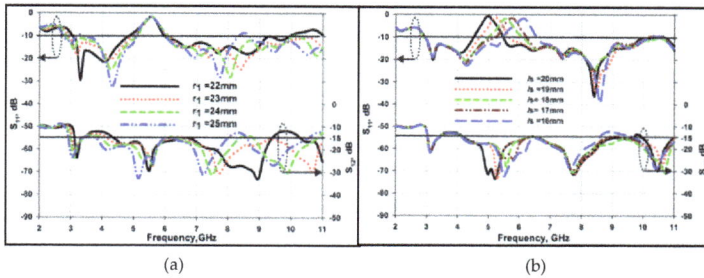

Figure 19. Effect of major antenna parameters on antenna characteristics (a) r1 (b) ls

3.4. Conclusions

A compact uniplanar dual polarized UWB antenna with notch functionality is developed for diversity applications. The antenna features a 2:1 VSWR band from 2.8-11 GHz while show-ing the rejection performance in the frequency band 4.99-6.25 GHz along with a reasonable isolation better than 15dB. The measured radiation pattern and the envelop correlation coef-ficient indicate that the antenna provides good polarization diversity performance. Time do-main analysis of the antenna shows faithful reproduction of the transmitted pulse even with a notch band.

Acknowledgements

The authors would like to acknowledge Rogers Corporation for providing high frequency laminates through university program.

Author details

Gijo Augustin , Bybi P. Chacko and Tayeb A. Denidni

National Institute of Scientific Research (INRS), Montreal QC, Canada

References

[1] FCC, "First Report and Order: Revision of Part 15 of the Commissions Rules Regard-ing Ultra-Wideband Transmission Systems," ET Docket 98-153, Apr. 2002.

[2] Chung, K., Kim, J., & Choi, J. "Wideband microstrip-fed monopole antenna having frequency band-notch function," Microwave and Wireless Components Letters, IEEE, vol. 15, no.11 pp. 766-768, (2005). .

[3] K. S. Ryu and A. A. Kishk, "UWB antenna with single or dual band-notches for lower WLAN band and upper WLAN band," Antennas and Propagation, IEEE Transactions on, vol. 57,no.12 pp. 3942-3950, 2009.

[4] A. Kishk, K. F. Lee, C. E. Smith, and D. Kajfez, "Microstrip Line and CPW FED Ultra Wideband Slot Antennas with U-Shaped Tuning Stub and Reflector," Progr. Electromagn. Res., vol. 56,pp. 163-182, 2006.

[5] X. H. Wu and A. A. Kishk, "Study of an ultrawideband omnidirectional rolled monopole antenna with trapezoidal cuts," Antennas and Propagation, IEEE Transactions on, vol. 56,no.1 pp. 259-263, 2008.

[6] S. L. S. Yang, A. A. Kishk, and K. F. Lee, "Wideband circularly polarized antenna with L-shaped slot," Antennas and Propagation, IEEE Transactions on, vol. 56,no.6 pp. 1780-1783, 2008.

[7] Coplanar waveguide circuits, components, and systems. Wiley-interscience, 2007 2001.

[8] Federal Communications Commission, "Spectrum Policy Task Force," Rep.ET Docket, Nov. (2002). (02-135), 02-135.

[9] P. J. Gibson, "The Vivaldi Aerial," in Microwave Conference, 1979. 9th European, 1979, pp. 101-105.

[10] E. Ebrahimi, J. R. Kelly, and P. S. Hall, "Integrated Wide-Narrowband Antenna for Multi-Standard Radio," IEEE Trans. Antennas Propag., vol. 59,no.7 pp. 2628-2635, Jul 2011.

[11] M. Sonkki, E. Antonino-Daviu, M. Ferrando-Bataller, and E. T. Salonen, "Planar Wideband Polarization Diversity Antenna for Mobile Terminals," Antennas and Wireless Propagation Letters, IEEE, no.99 pp. 1-1, 2011.

[12] R. G. Vaughan and J. B. Andersen, "Antenna diversity in mobile communications," Vehicular Technology, IEEE Transactions on, vol. 36,no.4 pp. 149-172, 1987.

[13] W. K. Toh, Z. N. Chen, X. Qing, and T. See, "A planar UWB diversity antenna," Antennas and Propagation, IEEE Transactions on, vol. 57,no.11 pp. 3467-3473, 2009.

[14] M. Gallo, E. Antonino-Daviu, M. Ferrando, M. Bozzetti, J. Molina-Garcia-Pardo, and L. Juan-Llacer, "A Broadband Pattern Diversity Annular Slot Antenna," Antennas and Propagation, IEEE Transactions on, no.99 pp. 1-1, 2011.

[15] C. B. Dietrich Jr, K. Dietze, J. R. Nealy, and W. L. Stutzman, "Spatial, polarization, and pattern diversity for wireless handheld terminals," Antennas and Propagation, IEEE Transactions on, vol. 49,no.9 pp. 1271-1281, 2001.

[16] Z. N. Chen and T. See, "Diversity and its applications in ultra-wideband antennas," 2009, pp. 1-4.

[17] R. Yahya and T. A. Denidni, "Design of a new dual-polarized ultra-wideband planar CPW fed antenna," 2011, pp. 1770-1772.

[18] X. Bao and M. Ammann, "Wideband Dual-Frequency Dual-Polarized Dipole-Like Antenna," Antennas and Wireless Propagation Letters, IEEE, vol. 10,pp. 831-834, 2011.

[19] G. Adamiuk, T. Zwick, and W. Wiesbeck, "Compact, dual-polarized UWB-antenna, embedded in a dielectric," Antennas and Propagation, IEEE Transactions on, vol. 58,no.2 pp. 279-286, 2010.

[20] G. Adamiuk, L. Zwirello, S. Beer, and T. Zwick, "Omnidirectional, dual-orthogonal, linearly polarized UWB Antenna," 2010, pp. 854-857.

[21] H. Yoon, Y. Yoon, H. Kim, and C. H. Lee, "Flexible ultra-wideband polarisation diversity antenna with band-notch function," Microwaves, Antennas & Propagation, IET, vol. 5,no.12 pp. 1463-1470, 2011.

[22] F. Ghanem, P. S. Hall, and J. R. Kelly, "Two port frequency reconfigurable antenna for cognitive radios," Electron. Lett., vol. 45,no.11 pp. 534-535, May 21 2009.

[23] J. R. Kelly, P. Song, P. S. Hall, and A. L. Borja, "Reconfigurable 460 MHz to 12 GHz antenna with integrated narrowband slot," Progr. Electromagn. Res.Lett., vol. 24,pp. 137-145, Sept. 2011.

[24] B. C. Wadell, "Trasmission line design handbook," Artech House, Inc., 1991.

[25] M. Gopikrishna, D. D. Krishna, C. K. Anandan, P. Mohanan, and K. Vasudevan, "Design of a Compact Semi-Elliptic Monopole Slot Antenna for UWB Systems," Antennas and Propagation, IEEE Transactions on, vol. 57,no.6 pp. 1834-1837, 2009.

[26] Carrasquillo-Rivera, R. A. R. Solis, and J. G. Colom-Ustariz, "Tunable and dual-band rectangular slot-ring antenna," in Antennas and Propagation Society International Symposium, 2004. IEEE, 2004, pp. 4308-4311 Vol.4.

[27] C. Zhi Ning, Y. Ning, G. Yong-Xin, and M. Y. W. Chia, "An investigation into measurement of handset antennas," Instrumentation and Measurement, IEEE Transactions on, vol. 54,no.3 pp. 1100-1110, 2005.

[28] P. S. Hall, P. Gardner, and A. Faraone, "Antenna Requirements for Software Defined and Cognitive Radios," Proceedings of the IEEE, vol. PP,no.99 pp. 1-9, 2012.

[29] K.L.Wong, "Planar Antennas for wireless communication," Wiley Interscience, New York, 2003.

[30] Y. J. Guo, A. Paez, R. A. Sadeghzadeh, and S. K. Barton, "A circular patch antenna for radio LAN's," Antennas and Propagation, IEEE Transactions on, vol. 45,no.1 pp. 177-178, 1997.

[31] H. G. Schantz, "Measurement of UWB antenna efficiency," in Vehicular Technology Conference, 2001. VTC 2001 Spring. IEEE VTS 53rd, 2001, pp. 1189-1191 vol.2.

[32] W. Sörgel and W. Wiesbeck, "Influence of the antennas on the ultra-wideband transmission," Eurasip journal on applied signal processing, vol. 2005,pp. 296-305, 2005.

[33] N. Telzhensky and Y. Leviatan, "Novel method of UWB antenna optimization for specified input signal forms by means of genetic algorithm," Antennas and Propagation, IEEE Transactions on, vol. 54,no.8 pp. 2216-2225, 2006.

[34] M. Chen and J. Wang, "Compact CPW-fed circular slot antenna for ultra-wideband applications," 2008, pp. 78-81.

[35] T. Denidni and M. Habib, "Broadband printed CPW-fed circular slot antenna," Electron. Lett., vol. 42,no.3 pp. 135-136, 2006.

[36] X. Qing and Z. Chen, "Compact coplanar waveguide-fed ultra-wideband monopole-like slot antenna," Microwaves, Antennas & Propagation, IET, vol. 3,no.5 pp. 889-898, 2009.

[37] Y. Cheng, W. Lu, C. Cheng, W. Cao, and Y. Li, "Printed diversity antenna with cross shape stub for ultra-wideband applications," 2008, pp. 813-816.

[38] N. Trang, D. Lee, and H. Park, "Compact printed CPW-fed monopole ultra-wideband antenna with triple subband notched characteristics," Electron. Lett., vol. 46,no. 17 pp. 1177-1179, 2010.

[39] Liang, C. C. Chiau, X. Chen, and C. G. Parini, "Study of a printed circular disc monopole antenna for UWB systems," Antennas and Propagation, IEEE Transactions on, vol. 53,no.11 pp. 3500-3504, 2005.

[40] Y. J. Cho, K. H. Kim, D. H. Choi, S. S. Lee, and S. O. Park, "A miniature UWB planar monopole antenna with 5-GHz band-rejection filter and the time-domain characteristics," Antennas and Propagation, IEEE Transactions on, vol. 54,no.5 pp. 1453-1460, 2006.

[41] S. Blanch, J. Romeu, and I. Corbella, "Exact representation of antenna system diversity performance from input parameter description," Electron. Lett., vol. 39,no.9 pp. 705-707, 2003.

[42] M. Tzyh-Ghuang and J. Shyh-Kang, "Planar miniature tapered-slot-fed annular slot antennas for ultrawide-band radios," Antennas and Propagation, IEEE Transactions on, vol. 53,no.3 pp. 1194-1202, 2005.

Circular Polarization

Chapter 10

Axial Ratio Bandwidth of a Circularly Polarized Microstrip Antenna

Li Sun, Gang Ou, Yilong Lu and Shusen Tan

Additional information is available at the end of the chapter

1. Introduction

Microstrip antenna has been widely used due to its many advantages, such as, small volume, light weight, easy to get various polarization and easy to be integrated (Dang & Liu, 1999). Microstrip antenna can adopt many methods to obtain circular polarization (Xue and Zhong, 2002). And some technologies can achieve the miniaturization of the microstrip antenna (Xue and Zhong, 2002). Also there are some methods to enhance the impedance bandwidth of the miniaturized microstrip antenna (Liu et al., 2002) ; (Wang & Gao, 2003).

In this chapter, we focus on the axial ratio bandwidth of a circularly polarized microstrip antenna. The previous reference books discussed the axial ratio bandwidth less, always said that the axial ratio bandwidth of a circularly polarized microstrip antenna was limited, and it was less than the impedance bandwidth of a linearly polarized microstrip antenna (Lin & Nie, 2002). The group of Professor Ahmed A. Kishk has done a lot of research work on the circularly polarized microtrip antenna recently (Yang et al., 2008); (Yang et al., 2007); (Yang et al., 2006); (Chair et al., 2006); (Kishk et al., 2006). We adopt theoretical analysis and simulation by CST Microwave Studio to give out the method of improving the axial ratio bandwidth of the circularly polarized microstrip antenna.

First, we briefly introduce the basic methods which can form the circular polarization for a microstrip antenna, including the single-feed and the multiple-feed. When using multiple-feed for one patch, the sequential rotation technology (Hall et al., 1989) can be adopted. Starting from the mechanism of circular polarization obtaining from multiple-feed method, the multiple-feed can improve the axial ratio bandwidth of a microstrip antenna effectively than the single-feed microstrip antenna is demonstrated by theoretical analysis and simulation. The more feeds, the better the axial ratio bandwidth is.

Then, the detail analysis of the axial ratio bandwidth including when the amplitudes have some difference and the phase excitation of the feed point has an offset according to the designed central frequency in manufacture are described.

At last, the example of circularly polarized microstrip antenna design and test are in the section 5. Due to the volume limited in the project, we choose two feeds for the microstrip antenna.

2. Circularly polarized method

2.1. Simple microstrip antennas

Generally, the configuration of the simple microstrip antenna (Ung, 2007) is showed as in Fig. 1. It can be simply formed by a dielectric substrate through photoetching technology or etching process. In the configuration, there are the metallic patch of certain shape on the top, the substrate layer of certain thickness and the ground plane on the bottom. The dielectric constant and the thickness of the dielectric substrate material, the shape and size of the top patch and the feeding method determine the performance of the microstrip antenna.

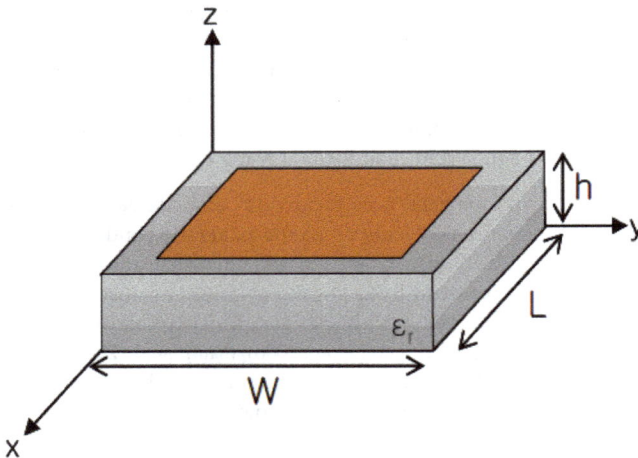

Figure 1. Configuration of the microstrip antenna

The shape of the top metallic patch can be various. Such as square, rectangle, circle, triangle, ellipse and unconventional shape, etc. The feed methods include coaxial probe feed, microstrip line feed, aperture couple feed, etc (Ung, 2007); (stutzman & Thiele, 1997). The simple microstrip antenna is usually linearly polarized. The bandwidth of the linearly polarized microstrip antenna is described by the impedance bandwidth.

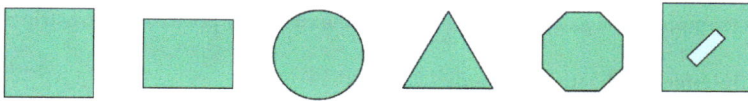

Figure 2. Patch shape

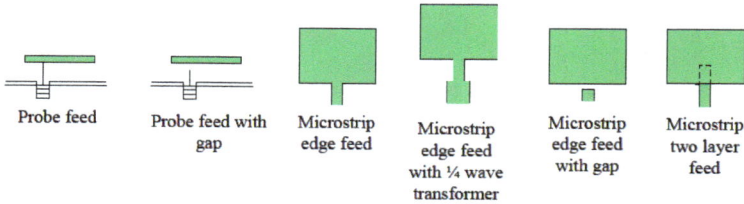

Probe feed | Probe feed with gap | Microstrip edge feed | Microstrip edge feed with ¼ wave transformer | Microstrip edge feed with gap | Microstrip two layer feed

Figure 3. Feed methods

2.2. Single-feed realization method

Single-feed for the patch to form circular polarization is based on the cavity model of microstrip antenna. The two orthogonal polarized degenerate modes which can formed the circular polarization can be obtained by corner cut, quasi-square, slot, etc, and the patch shape (Lin & Nie, 2002) can be seen in Fig.4. The feed methods can adopt coaxial probe feed, aperture couple feed, etc.

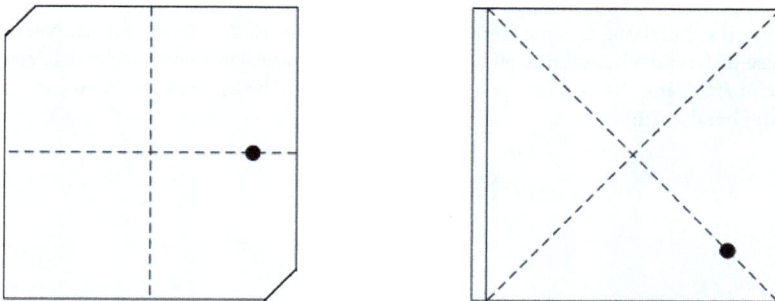

Figure 4. Patch shape of single-feed circularly polarized microstrip antenna

The axial ratio 3dB bandwidth of the circularly polarized microstrip antenna is much less than the impedance bandwidth of the linearly polarized microstrip antenna. Via application, the axial ratio 3dB bandwidth the single-feed circularly polarized microstrip antenna is limited at about 35%of the difference of the two resonant frequencies (Lin & Nie, 2002). So we must find methods to improve the axial ratio bandwidth of the circularly polarized microstrip antenna.

2.3. Multiple-feed realization method

A circularly polarized electromagnetic wave can be divided into two equal amplitudes linearly polarized components both in space and in time. Suppose that the two orthogonal polarized components are

$$\vec{E}_x = E, \quad \vec{E}_y = Ee^{j\frac{\pi}{2}},$$

then we have

$$\vec{E}_y = Ee^{j\frac{\pi}{2}} = j\vec{E}_x. \tag{1}$$

Multiple-feed for one patch can adopt the sequential rotation technology. The technology of sequential rotation is successfully used in circularly polarized antenna array design (Hall et al., 1989). Multiple-feed has an appropriate phase difference between excitations, and this can improve the axial ratio bandwidth and reduce the cross-polarization. The mode exited by each feed for one patch can be regarded as the mode exited by each element in the array. So, in the case of using M feed points, the m_{th} feed point's phase φ_{em} can be expressed as

$$\varphi_{em} = (m-1)\frac{p\pi}{M} \quad 1 \le m \le M, \tag{2}$$

where P is an integer.

Each feed point's physical position must have some symmetry, seen in fig.5. Through simulation, finding that fixing the first feed point position, other feed points rotate the corresponding phase differences between itself and the first feed point. The center is the disc center. In the case of $P<M$, and the last feed point does not rotate to the first feed point, it can improve axial ratio bandwidth.

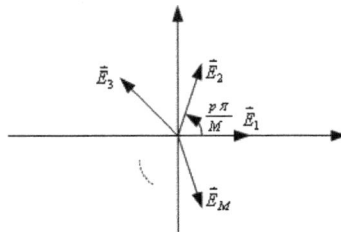

Figure 5. Feed position of multiple-feed

Suppose that

$$\vec{E}_1 = E, \ \vec{E}_2 = Ee^{j\frac{p\pi}{M}}, \ \vec{E}_3 = Ee^{j\frac{2p\pi}{M}}, \ ......, \ \vec{E}_M = Ee^{j\frac{(M-1)p\pi}{M}}$$

so the two orthogonal components are

$$E_x = \vec{E}_1 + \vec{E}_2\cos\frac{p\pi}{M} + \vec{E}_3\cos\frac{2p\pi}{M} + \cdots + \vec{E}_M\cos\frac{(M-1)p\pi}{M}$$

$$= E + Ee^{j\frac{p\pi}{M}}\cos\frac{p\pi}{M} + Ee^{j\frac{2p\pi}{M}}\cos\frac{2p\pi}{M} + \cdots + Ee^{j\frac{(M-1)p\pi}{M}}\cos\frac{(M-1)p\pi}{M}$$

$$= 1 + \cos^2\frac{p\pi}{M} + \cos^2\frac{2p\pi}{M} + \cdots + \cos^2\frac{(M-1)p\pi}{M} + \frac{1}{2}j\left[\sin\frac{2p\pi}{M} + \sin\frac{4p\pi}{M} + \cdots + \sin\frac{2(M-1)p\pi}{M}\right]$$

$$E_y = \vec{E}_2\sin\frac{P\pi}{M} + \vec{E}_3\sin\frac{2P\pi}{M} + \cdots + \vec{E}_M\sin\frac{(M-1)P\pi}{M}$$

$$= Ee^{j\frac{p\pi}{M}}\sin\frac{p\pi}{M} + Ee^{j\frac{2p\pi}{M}}\sin\frac{2p\pi}{M} + \cdots + Ee^{j\frac{(M-1)p\pi}{M}}\sin\frac{(M-1)p\pi}{M}$$

$$= \frac{1}{2}\left[\sin\frac{2p\pi}{M} + \sin\frac{4p\pi}{M} + \cdots + \sin\frac{2(M-1)p\pi}{M}\right] + j\left[\sin^2\frac{p\pi}{M} + \sin^2\frac{2p\pi}{M} + \cdots + \sin^2\frac{(M-1)p\pi}{M}\right]$$

According to the following formula,

$$\sum_{k=1}^{n}\sin(x+k\alpha) = \frac{\sin\frac{1}{2}n\alpha}{\sin\frac{1}{2}\alpha}\sin\left(x + \frac{1}{2}(n+1)\alpha\right),$$

we can get

$$\sin\frac{2p\pi}{M} + \sin\frac{4p\pi}{M} + \cdots + \sin\frac{2(M-1)p\pi}{M}$$

$$= \frac{\sin\frac{(M-1)p\pi}{M}}{\sin\frac{p\pi}{M}}\sin\left[\frac{1}{2}(M-1+1)\frac{2p\pi}{M}\right] = \frac{\sin\frac{(M-1)p\pi}{M}}{\sin\frac{p\pi}{M}}\sin(p\pi) = 0,$$

and according to

$$\sum_{k=1}^{n}\cos(x+k\alpha) = \frac{\sin\frac{1}{2}n\alpha}{\sin\frac{1}{2}\alpha}\cos\left(x + \frac{1}{2}(n+1)\alpha\right),$$

we can get

$$\cos\frac{2p\pi}{M} + \cos\frac{4p\pi}{M} + \cdots + \cos\frac{2(M-1)p\pi}{M}$$

$$= \frac{\sin\frac{(M-1)p\pi}{M}}{\sin\frac{p\pi}{M}}\cos\left[\frac{1}{2}(M-1+1)\frac{2p\pi}{M}\right] = \frac{\sin\left(p\pi - \frac{p\pi}{M}\right)}{\sin\frac{p\pi}{M}}\cos(p\pi) = -1.$$

$$1 + \cos^2\frac{p\pi}{M} + \cos^2\frac{2p\pi}{M} + \cdots + \cos^2\frac{(M-1)p\pi}{M}$$

$$= \frac{M}{2} + \frac{1}{2} + \frac{1}{2}\left(\cos\frac{2p\pi}{M} + \cos\frac{4p\pi}{M} + \cdots + \cos\frac{2(M-1)p\pi}{M}\right) = \frac{M}{2},$$

$$\sin^2\frac{p\pi}{M} + \sin^2\frac{2p\pi}{M} + \cdots + \sin^2\frac{(M-1)p\pi}{M}$$

$$= \frac{M}{2} - \frac{1}{2} - \frac{1}{2}\left(\cos\frac{2p\pi}{M} + \cos\frac{4p\pi}{M} + \cdots + \cos\frac{2(M-1)p\pi}{M}\right) = \frac{M}{2},$$

so

$$1 + \cos^2\frac{p\pi}{M} + \cos^2\frac{2p\pi}{M} + \cdots + \cos^2\frac{(M-1)p\pi}{M} = \sin^2\frac{p\pi}{M} + \sin^2\frac{2p\pi}{M} + \cdots + \sin^2\frac{(M-1)p\pi}{M} = \frac{M}{2}$$

Therefore we can get

$$E_y = jE_x. \tag{3}$$

That is (1), so the multiple-feed method above has realized the circular polarization.

3. Theoretical analysis of the axial ratio bandwidth

3.1. Axial ratio

We can use the polarization ellipse to describe the elliptical polarization. The instantaneous electric field orientation can figure out an ellipse in the space, seen in Fig.6.

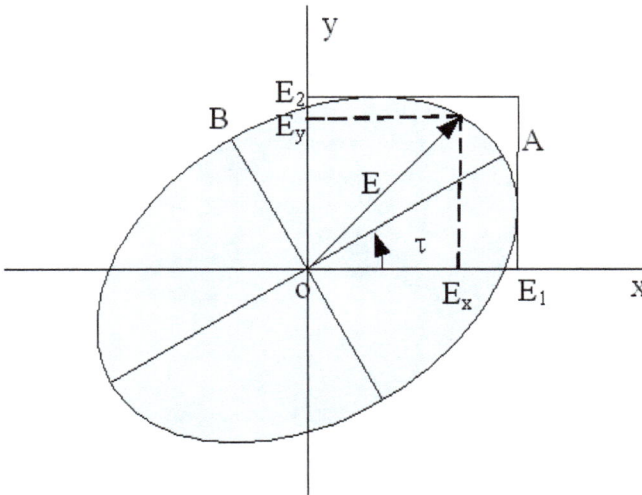

Figure 6. Polarization ellipse

The axial ratio is defined as

$$AR = OA\big/OB \quad (1 \le AR \le \infty) \tag{4}$$

where *OA* is the half major axis of the polarization ellipse, and the *OB* is the half minor axis of the polarization ellipse.

The elliptical polarization of electromagnetic wave can be divided into two linearly polarized components. One's orientation is along x-axis, and the other is along y-axis. Suppose that the two linearly polarized components are $E_x = E_1 \sin(\omega t - \beta z)$, $E_y = E_2 \sin(\omega t - \beta z + \delta)$,

where E_1 is the amplitude of the linear polarization along x-axis E_x, and E_2 is the amplitude of the linear polarization along y-axis E_y. δ is the phase difference between E_x and E_y. Based on the above, we will analyze the axial ratio bandwidth of the multiple-feed microstrip antenna in the next section.

3.2. Axial ratio bandwidth of two feeds

Assume that the amplitudes excitation of each feed are equal, mutual coupling is small, and it can be neglected. Only the frequency changes the phase excitation relationship between the feed points. In the real case, usually using power splitter with separation to realize the equal amplitude excitation, and using different microstrip line length to realize the phase excitation difference. So the assumption is reasonable.

Two feeds: *M=2, P=1*. The two orthogonal electric fields are

$$E_x = E_1 \sin(\omega t - \beta z), \tag{5}$$

$$E_y = E_2 \sin(\omega t - \beta z + \delta). \tag{6}$$

At *z=0*,

$$E_x = E_1 \sin \omega t, \tag{7}$$

$$E_y = E_2 (\sin \omega t \cos \delta + \cos \omega t \sin \delta), \tag{8}$$

where $\sin \omega t = \frac{E_x}{E_1}$, $\cos \omega t = \sqrt{1 - (\frac{E_x}{E_1})^2}$

Substitute (7) into (8), we can get

$$a E_x^{\,2} - b E_x E_y + c E_y^{\,2} = 1, \tag{9}$$

where

$$a = \frac{1}{E_1{}^2\sin^2\delta}, \ b = \frac{2\cos\delta}{E_1 E_2\sin^2\delta}, \ c = \frac{1}{E_2{}^2\sin^2\delta}.$$

Construct an ellipse equation

$$\frac{E_x^{'2}}{A^2} + \frac{E_y^{'2}}{B^2} = 1, \tag{10}$$

where

$$E_x' = E_x\cos\theta - E_y\sin\theta, \ E_y' = E_x\sin\theta + E_y\cos\theta.$$

Thus (10) becomes,

$$(\frac{\cos^2\theta}{A^2} + \frac{\sin^2\theta}{B^2})E_x^2 - (\frac{\sin 2\theta}{A^2} - \frac{\sin 2\theta}{B^2})E_x E_y + (\frac{\sin^2\theta}{A^2} + \frac{\cos^2\theta}{B^2})E_y^2 = 1. \tag{11}$$

Through (9) and (11), we can get

$$A = \sqrt{\frac{2}{a + c + \sqrt{(a-c)^2 + b^2}}},$$

$$B = \sqrt{\frac{2}{a + c - \sqrt{(a-c)^2 + b^2}}}.$$

So

$$AR = \frac{A}{B} = \sqrt{\frac{(E_1/E_2)^2 + 1 - \sqrt{(E_1/E_2)^4 + 1 + 2\cos 2\delta(E_1/E_2)^2}}{(E_1/E_2)^2 + 1 + \sqrt{(E_1/E_2)^4 + 1 + 2\cos 2\delta(E_1/E_2)^2}}}. \tag{12}$$

Two feeds, when E1/E2=1, we can get (13) from (12).

$$AR = tg\frac{\delta}{2}. \tag{13}$$

3.3. Axial ratio bandwidth of four feeds

We analyze the axial ratio bandwidth of multiple-feed antenna, in the case of amplitude excitations are equal, and mutual coupling is neglected. Four feeds, when $M=4$, $P=2$.

In other words, the phase excitation difference is 90°. At $z=0$, the two orthogonal electric fields are

$$E_x = E_1 \sin \omega t - E_3 \sin(\omega t + 2\delta),$$ (14)

$$E_y = E_2 \sin(\omega t + \delta) - E_4 \sin(\omega t + 3\delta),$$ (15)

where

$$E_y = (E_2 \cos \delta - E_4 \cos 3\delta)\sin \omega t + (E_2 \sin \delta - E_4 \sin 3\delta)\cos \omega t.$$ (16)

In the case of $E_1=E_2=E_3=E_4$,

$$\cos\omega t = \frac{2E_x \cos\delta - E_y}{-2E_1 \sin\delta}$$

$$\sin\omega t = \sqrt{1-(\frac{2E_x \cos\delta - E_y}{-2E_1 \sin\delta})^2}$$

Substitute into (16), we can get

$$aE_x^2 - bE_x E_y + cE_y^2 = 1,$$ (17)

where

$$a = \frac{1}{4E_1^2 \sin^4\delta},$$

$$b = \frac{\cos\delta \cos 2\delta}{E_1^2 \sin^4\delta} + \frac{\cos\delta}{E_1^2 \sin^2\delta},$$

$$c = \frac{\cos^2\delta}{4E_1^2 \sin^4\delta} + \frac{1}{4E_1^2 \sin^2\delta}.$$

So

$$AR = \frac{A}{B} = \sqrt{\frac{a+c-\sqrt{(a-c)^2 + b^2}}{a+c+\sqrt{(a-c)^2 + b^2}}} = \sqrt{\frac{1-2\cos^3\delta}{1+2\cos^3\delta}}.$$ (18)

That is

$$AR = \sqrt{\frac{1-2\cos^3\delta}{1+2\cos^3\delta}}. \tag{19}$$

3.4. Comparison of the two feeds and the four feeds

Next we give out the expression for phase excitation difference δ between the two feeds. The feed network substrate's relative dielectric constant is ε_r, the substrate thickness is h, and the width of the microstrip line is W. With the theory of the microstrip line, the effective dielectric constant ε_{re} is (Lin & Nie, 2002)

$$\varepsilon_{re} = \frac{\varepsilon_r+1}{2} + \frac{\varepsilon_r-1}{2}(1+\frac{12h}{W})^{-\frac{1}{2}}. \tag{20}$$

The phase velocity's wavelength λ_p of the quasi-TEM wave propagated in the microstrip line is

$$\lambda_p = \frac{c}{f\sqrt{\varepsilon_{re}}}, \tag{21}$$

where c is the velocity of light in the vacuum, and f is frequency.

Assume that the microstrip line length x which providing $90°$ phase excitation according to the central frequency, provide δ phase excitation in fact due to the changing of the frequency, $\delta/x = 360/\lambda_p$, then

$$\delta = \frac{360xf\sqrt{\varepsilon_{re}}}{c}. \tag{22}$$

The phase excitation difference of each feed in the feed network is designed according to the central frequency. The phase excitation difference which provided by the microstrip line is changing according to the changing frequency. This will affect the circular polarization out side the central frequency.

We use the CST microwave studio to simulate the multiple-feed microstrip antenna. The simulation files are showed in Fig.7. Thorough simulation and calculation, we give out the axial ratio bandwidth comparison between two feeds and four feeds in Fig.8. Through the theoretical computation, we demonstrate that multiple-feed for one patch can effectively improve the axial ratio bandwidth. The axial ratio 3dB bandwidth of two feeds can achieve 42.6%, and four feeds can achieve 74%.

Figure 7. Simulation files of two feeds and four feeds

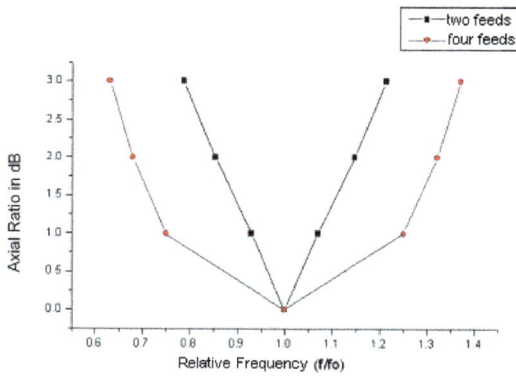

Figure 8. Axial ratio bandwidth comparison between two feeds and four feeds

4. Axial ratio bandwidth analysis when manufacture error exist

When two feeds, assume that the amplitudes excitation are equal at every frequency. But if we substitute (22) into (12), we can get the changing of the axial ratio bandwidth according to the different ratio of E1 and E2, showing in Fig.9.

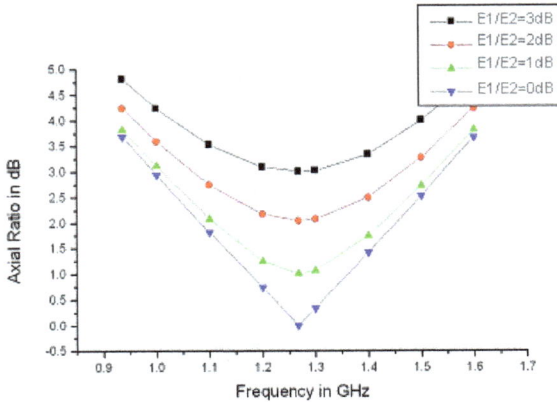

Figure 9. Axial ratio bandwidth of different amplitudes excitation of the two feeds

We can get the conclusion that the amplitude difference between the two feeds affects the axial ratio badly. When the amplitude ratio of the two feeds is 3dB, the axial ratio 3dB bandwidth has already disappeared.

Next we have a look at the axial ratio bandwidth changing when the phase excitation designed at the central frequency has an offset. In the feed network, change the microstrip line length x which provides 90° phase excitation to the length which provides 85.8° phase excitation. Using the same process, we can give out the changing of the axial ratio bandwidth when two feeds amplitudes are equal in Fig.10. When two feeds amplitudes ratio is 2dB in Fig.11.

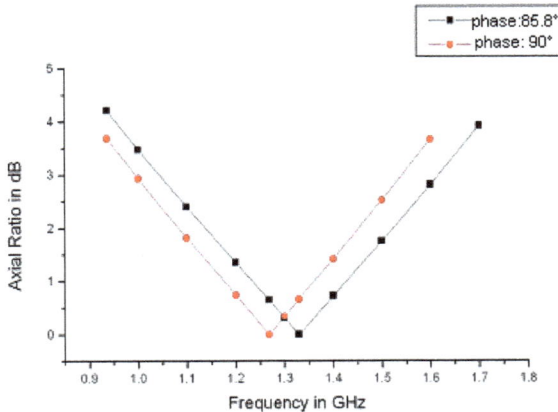

Figure 10. Axial ratio bandwidth of phase excitation has an offset at the central frequency in case of E1/E2=0dB

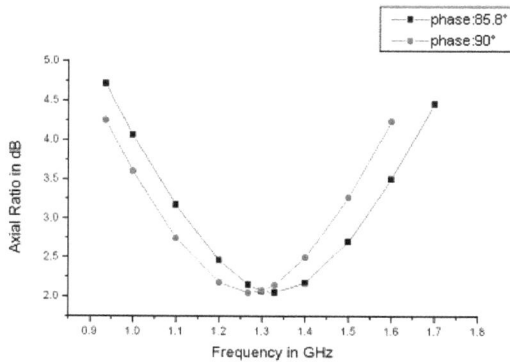

Figure 11. Axial ratio bandwidth of phase excitation has an offset at the central frequency in case of E1/E2=2dB

We can see that there is an offset on the axial ratio bandwidth when the phase excitation designed at the central frequency has an offset. From our theoretical analysis, we can get the conclusion that the multiple-feed technology can improve the axial ratio bandwidth of the microstrip antenna effectively. To get a wide band circularly polarized microstrip antenna, first, we must determine the most feed points we can use in the design according to the size limited in the project.

5. Antenna design example

5.1. Design

The more feeds, the better the axial ratio bandwidth of the circularly polarized microstrip antenna. But the feed network is more complicated and the feed network needs more space to realize.

We design a small antenna, using two feeds. Two linearly polarized components which are equal amplitude and 90°phase difference form the circular polarization. The patch shape is in Fig.12 (Hall et al., 1989), and the stubs on the patch are used to debug the resonant frequency in antenna manufacture. The feed network is in Fig.13.

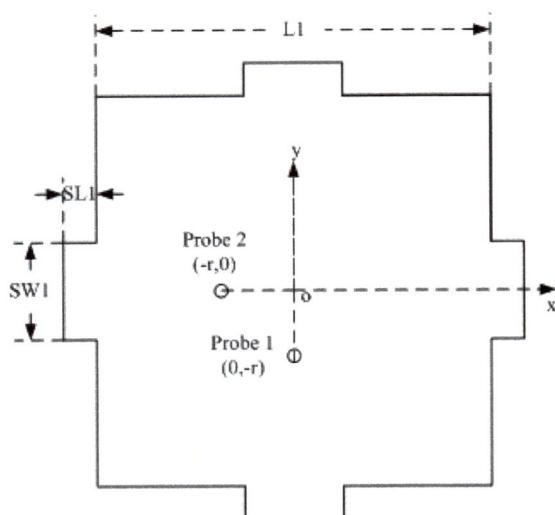

Figure 12. Patch shape

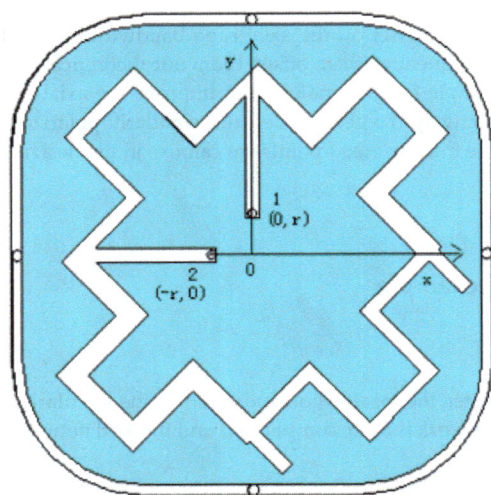

Figure 13. Feed network

5.2. Simulation analysis

Simulate the two feeds microstrip antenna we design in the above section using the CST microwave studio. We compare the difference in the axial ratio bandwidth between the single-feed and the two feeds through simulation. The configurations of the single-feed and the two feeds microstrip antenna are showed in Fig.14.

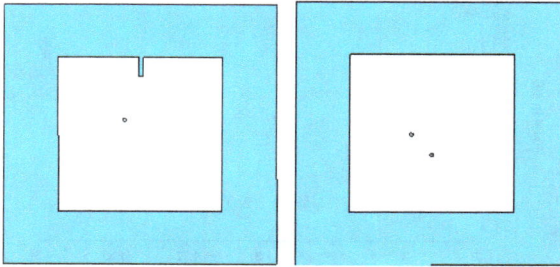

Figure 14. Simulation configuration of the single-feed and the two feeds

The simulation results of the axial ratio of the single-feed and the two feeds at zenith are showed in Fig.15. We can see that the axial ratio bandwidth of the single-feed is very limited. For the two feeds, the phase difference of the two equal amplitudes and 90° phase difference linearly polarized components according to the centre frequency change slowly and smoothly with the frequency band. This can improve the axial ratio bandwidth of a circularly polarized microstrip antenna.

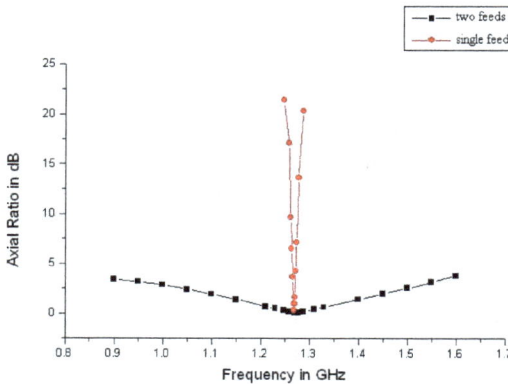

Figure 15. Axial ratio simulation results of the single-feed and the two feeds

5.3. Test result

The manufactured two feeds microstrip antenna is tested in the anechoic chamber. The test result of the axial ratio is showed in Fig.16.

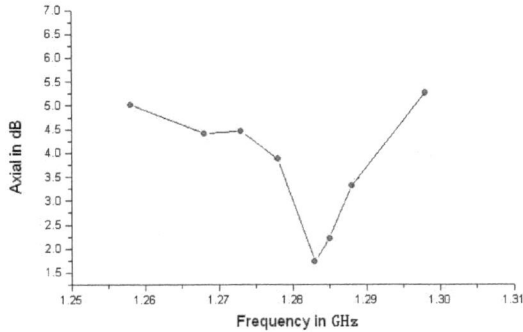

Figure 16. Axial ratio test result

In simulation, the two feeds are ideal equal amplitudes and 90° phase difference. In the manufacture, the microstrip line feed network provides the two equal amplitudes and 90° phase difference excitations. Due to the dielectric constant error of the substrate material and error of manufacture, the axial ratio bandwidth of the microstrip antenna get worse compared to the simulation result. The axial ratio 3dB bandwidth tested of the microstrip antenna is about 10MHz.

6. Conclusion

Microstrip antenna has been used in every field, due to its many advantages. Our main research topic in this chapter was how to improve the axial ratio bandwidth of a circularly polarized microstrip antenna. Multiple-feed method can realize the circular polarization for a microstrip antenna. Circularly polarized microstrip patch antenna designed by the multiple-feed method adopting the sequential rotation technology can improve the axial ratio bandwidth effectively. In this chapter, we demonstrate it by theoretical analysis.

Through simulation by CST Microwave Studio and theoretical computation, the axial ratio 3dB bandwidth of two feeds can achieve 42.6%, and four feeds can achieve 74%.

In engineering, choosing the most feed points according to the feed network space limited in the project can improve the axial ratio bandwidth of a circularly polarized microstrip antenna. And it is at the price of a complicated feed network compared to the few feed points design.

Author details

Li Sun[1], Gang Ou[2], Yilong Lu[3] and Shusen Tan[1]

1 Beijing Satellite Navigation Center, China

2 College of Electronic Science and Engineering, National University of Defense Technology, China

3 School of Electrical and Electronic Engineering, Nanyang Technological University, Singapore

References

[1] Dang, H. S, & Liu, Y. G. (1999). The analysis and design of a microstrip antenna. (in Chinese). *Journal of Detection and Control*, March. 1999) page numbers (35-39), 21(1)

[2] Hall, P. S, Dahele, J. S, & James, J. R. (1989). Design principles of sequentially fed, wide bandwidth, circularly polarized microstrip antennas, *Proceedings of IEE Microwaves. Antennas and Propagation*, 0095-0107X, October, 1989, IEE, 136(5), 381-389.

[3] Lin, C. L, & Nie, Z. P. (2002). *Antenna engineering handbook*, Publishing house of electronics industry, 7-50537-495-8

[4] Liu, Z. F, Lu, S. W, & Li, S. Z. (2002). Improved Method for Designing Wideband Microstrip Antennas. (in Chinese). *Journal of Beijing University of Aeronautics and Astronautics*, February. 2000) page numbers (15-18), 26(1)

[5] Stutzman, W, & Thiele, G. (1997). *Antenna theory and design*, Wiley, 978-0-47102-590-0US

[6] Ung Suok Kim(2007). Mitigation of signal biases introduced by controlled reception pattern antennas in a high integrity carrier phase differential GPS system, *dissertation*. Stanford University. March, 2007

[7] Wang, C. M, & Gao, X. J. (2003). Technologies of broadband microstrip antenna. (in Chinese). *Electronic Warfare Technology*, September. 2003) page numbers (23-26), 18(5)

[8] Xue, R. F, & Zhong, S. S. (2002). Survey and progress in circular polarization technology of microstrip antennas. (in Chinese). *Chinese Journal of Radio Science*, August. 2002) page numbers (331-336), 17(4)

[9] Yang, S. S, Lee, K. F, Kishk, A. A, & Luk, K. M. (2008). Design and study of wideband single feed circularly polarized microstrip antennas. *Progress In Electromagnetics Research*, PIER 80, page numbers , 45-61.

[10] Yang, S. L. S, Kishk, A. A, & Lee, K. F. (2008). Wideband Circularly Polarized Antenna with L-shaped Slot. *IEEE Transactions on Antennas and Propagations*, June. 2008) page numbers 1780-1783, 56(6)

[11] Yang, S. L. S, Chair, R, Kishk, A. A, Lee, K. F, & Luk, K. M. (2007). Study on Sequential Feeding Networks for Sub-Arrays of Circularly Polarized Elliptical Dielectric Resonator Antenna. *IEEE Transactions on Antennas and Propagation*, February. 2007) page numbers 321-333, 55(2)

[12] Yang, S. L. S, Chair, R, Kishk, A. A, Lee, K. F, & Luk, K. M. (2006). Single Feed Elliptical Dielectric Resonator Antennas for Circularly Polarized Applications. *Microwave and Optical Technology Letters*, November. 2006) page numbers 2340-2345, 48(11)

[13] Chair, R, Kishk, A. A, & Lee, K. F. (2006). Aperture Fed Wideband Circularly Polarized Rectangular Stair Shaped Dielectric Resonator Antenna. *IEEE Transactions on Antennas and Propagations*, April 2006) page numbers 1350-1352, 54(4)

[14] Kishk, A. A. (2003). Performance of planar four elements array of single-fed circularly polarized dielectric resonator antenna. *Microwave and Optical Technology Letters*, page numbers 381-384, 38(5)

Recent Advanced Applications

Chapter 11

Drooped Microstrip Antennas for GPS Marine and Aerospace Navigation

Ken G. Clark, Hussain M. Al-Rizzo,
James M. Tranquilla, Haider Khaleel and
Ayman Abbosh

Additional information is available at the end of the chapter

1. Introduction

The Navigation Satellite Timing and Ranging (NAVSTAR) GPS is a space-based system designed primarily for global real-time, all-weather navigation. There are 30 GPS satellites in six nearly circular, approximately 20,000 kilometer orbital planes, with an inclination of 55^0 relative to the equator [1]. Each satellite transmits two unique, Right Hand Circularly Polarized (RHCP) L band signals. The L1 (1.57542 GHz) carrier is bi-phase modulated with two pseudo-random noise sequences; the P and C/A codes. The L2 (1.2276 GHz) carrier is modulated only with the P code and is used mainly to determine and correct phase advance caused by the ionosphere. Superimposed on the P and C/A codes is the navigation message which contains, among other things, satellite ephemerides, clock biases, and ionosphere correction data [1].

Due to their light weight, reduced size, low cost, conformability, robustness, and ease of integration with MMIC, tremendous research has been reported over the last three decades into the use of microstrip antennas in GPS navigation [2]-[20]. Antenna designers are often faced with interrelated, strict, and conflicting performance requirements in order to meet the accuracy, continuity, and integrity of differential GPS, relative geodetic and hydrographic surveying, ship-borne and aerospace navigation [21]-[27].

The design specifications of a GPS antenna depend on the performance requirements peculiar to the application under consideration. A GPS user antenna requires RHCP and adequate co-polarized radiation pattern coverage over almost the entire upper hemisphere to track all visible satellites. Moreover, the antenna should ideally provide a uniform response in amplitude and, more critically, in phase to the full visible satellite constellation [21]. The angle cutoff and roll-off characteristics of the radiation pattern can be altered to suit the application of interest. For example, fixed ground reference stations and relative static geodetic surveying demand a rapid fall-off near the horizon, a high cross-polarization rejection, and a front-to-back gain ratio in excess of 20 dB to mitigate deleterious effects of severe multipath [28], [29].

In real-time kinematic positioning, few if any of the above constraints may be effective [30] and it may be necessary to operate the antenna under less than optimal conditions in regard to cross polarization performance if a wide beamwidth is of precedence. Precise GPS hydrographic surveying on a vessel cruising at speeds of 10 to 20 knots in open oceans is a challenging task due to the rotational disturbances from a relatively harsh sea environment. Pitch and roll amplitudes as high as 10^0 to 15^0 may be encountered in stormy weather, which presents a major obstacle to GPS derived attitude determination [21], [22]. Another envisaged application for the drooped microstrip antennas introduced in this chapter involves normal pitch or roll maneuvers of a general aviation aircraft, which may cause loss of some satellite signals for a range of flight orientation.

There is significant interest in the commercial and military sectors to develop antennas that could cover much of the upper hemisphere, including GPS satellites at elevation angles as low as 10^0, and to extend the coverage to negative elevation angles [3], [4], [15], [21]. This will lead to fewer occurrences of cycle slips and loss of lock to satellites while rising or setting, will maintain the proper Dilution of Precision (DoP) by maximizing the number of satellites in view, and will reduce the RMS error in range and velocity [1]. Notably, on the negative side, undesired multipath reflections off water and conducting bodies are also strongest at low-elevation angles. Nevertheless, whatever type of antenna is chosen, multipath reception will still have to be dealt with as a common problem [30]. It is fair to say that no single antenna design in the open literature has satisfactorily fulfilled all the above-mentioned requirements on coverage, phase center stability, and multipath rejection for real-time highly dynamic GPS marine and aerospace applications.

A pedestal ground plane is reported in [31], based on a trial-and-error experimental design approach, consisting of a cylindrical structure with a flat elevated center surrounded by sloping sides, to address beam shaping of crossed dipoles. This structure was found to be successful in improving the pattern coverage of the crossed dipole at low-elevation angles. Additional elements were also examined such as folded, serrated, rolled edges, monofilar, and quadrafiliar helices [32]-[35], although none achieved the same degree of radiation pattern control as the crossed dipole. This is attributed in part to the extent of ground plane illumination produced by the different sources and serves to highlight the importance of the ground plane as a secondary source with which to produce pattern changes. Further modifications to the ground plane using choke rings [28] were investigated primarily for multipath rejection.

It is well known that a stand-alone microstrip antenna mounted on a flat ground plane suffers from a lack of pattern control and reduced gain at low-elevation angles. This may result in a loss of contact with satellites when the antenna is mounted on a highly-dynamic platform, an attractive use for this low-profile structure. Fundamental to the design of a patch antenna is the interaction with the ground plane. In fact, the size, orientation, and shape of the ground plane are among the most important parameters that have an influence on the radiation pattern

[36]. However, a fundamental distinction exists in the relationship between a patch and the ground plane when compared with helical or dipole elements in that the ground plane of a microstrip antenna forms an integral part of the radiating structure and may not best be defined as a "secondary" source.

Building on our previous design experiences [28], [31]-[35], our research group at the University of New Brunswick, Fredericton, NB, Canada was the first to rigorously investigate the potential performance enhancements and limitations involved when these design modifications are applied to the more appealing microstrip antenna element [37]. The advantages associated with the microstrip antenna are such that one patent has been issued to a GPS manufacturer [38] based only on a downward drooped antenna structure. Neither the dimensions nor the performance of the proposed antennas were quantified in [38]. Later, a corner truncated square patch, partially enclosed within a flatly folded conducting wall, mounted on a pyramidal ground plane, was reported in [39]. However, neither the cross polarization performance nor the phase center stability were provided in [39].

In contrast to the antenna reported in [39], the drooped microstrip antennas introduced in this chapter have the ground plane and actual element deformed such that the corners or edges of the resonant cavity region fall away from the plane occupied by the element. A fundamental understanding of the operation and limitations of the drooped microstrip antenna is still lacking. A diffraction technique was attempted in [40] to model the effects of a sloping ground plane. This, however, was limited by the difficulty of implementing a realistic source term as well as the inclusion of finite lossy dielectric materials. A rigorous full-wave 3-D model, which incorporates the coaxial feed and detailed geometrical features of the drooped microstrip antenna, has not yet been reported.

For these reasons, we have performed the research reported in this chapter, which is the first to our knowledge that combines rigorous 3-D full-wave simulations and experimental measurements to provide a comprehensive characterization of downward and upward drooped microstrip antennas. A FDTD model has been developed, validated experimentally, and used to compute the input impedance and far-field radiation patterns. The FDTD model was used to examine the effects of a wide range of structural variations to gain an insight into the benefits and limitations of the proposed antennas. The parameters of interest include the location and angle of the bend, length of the ground plane, dielectric constant, and thickness of the substrate. Prototype structures were constructed, and their characteristics measured and then compared against simulated results. The authors wish to point out that the antennas described in this chapter are not intended to target multipath mitigation; on the contrary, they demonstrate the range of pattern modifications that could be accomplished by manipulating the orientation and size of the ground plane to suit GPS applications in marine and aerospace navigation.

The rest of the chapter is organized as follows: Section 2 summarizes the FDTD algorithm developed to perform the design and parametric studies and presents results from experimental tests performed to validate the implementation of the model and to demonstrate its ability to correctly predict the behavior of the drooped antennas. Section 3 addresses the design procedure, introduces parametric studies, and describes the drooped antennas constructed and tested for the control of the radiation patterns. Finally, Section 4 provides concluding remarks.

2. The contour path FDTD model: Experimental validation

The basic microstrip antenna, which consists of a conducting patch radiator rotated by 45^0 with respect to the center of the ground plane, and separated from the ground plane by a thin dielectric substrate, and the downward drooped antenna are shown in Figure1.

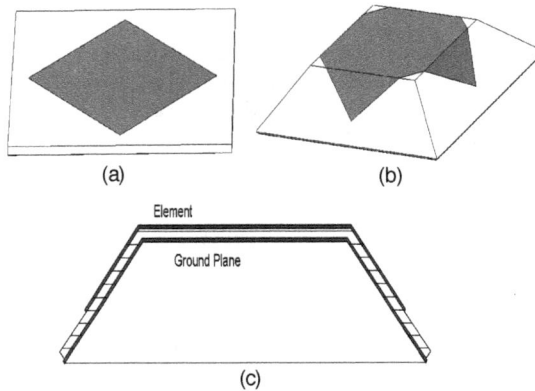

Figure 1. (a) The reference flat microstrip antenna, (b) the downward geometry, (c) cross sectional view of the resonant cavity.

The element is driven by a 50 Ω coaxial cable passing through the ground plane and the substrate. The antenna operates at the L1 GPS frequency of 1.57542 GHz since the majority of commercial GPS receivers use only the L1 frequency. Due to the relatively low cost, time savings, and repeatability, computer simulations can often complement and reduce the empiricism involved in an otherwise purely experimental approach, particularly in the initial design phase, allowing the antennas to be characterized carefully prior to their construction. Because of the complex geometries involved, the task of modeling the drooped microstrip structures is by no means a simple endeavor; it better lends itself to numerical simulation techniques. The FDTD method has been adopted in this research due to its conceptual simplicity and ease of implementation. Because it is a time-domain scheme, it is straightforward to impose a pulse excitation to perform broadband analysis using the Discrete Fourier Transform (DFT).

The FDTD algorithm is implemented in a 3-D Cartesian coordinate system with the formulation allowing for different spatial increments along each coordinate direction. Provision is made for modeling symmetrical objects by applying the Neumann boundary condition along one surface of the computational space. The antenna is excited either by a sinusoidal signal at the resonant frequency of the dominant mode or by a Gaussian pulse with a specified width and delay. The excitation is applied to either the electric or magnetic field, depending on whether a voltage or current source is desired.

In order to compute the input impedance, the instantaneous voltage and current are calculated at a fixed location in the coaxial feed by integrating the radial electric field and the magnetic field components encircling the inner conductor of the coaxial cable. A Gaussian pulse is used, and a DFT is performed to obtain broadband results. The impedance is then translated into the ground plane aperture by standard transmission line methods. The far-field radiation patterns are next determined by driving the model to steady state, using a sinusoidal wave at the fundamental resonant frequency of the antenna.

With pulse excitation, the fields must settle toward zero as energy escapes through the absorbing boundary. The condition used to judge if steady state is reached requires that the energy monitored at several observation points within the computational domain remains below 1% of the peak observed value with lower values enforced for cases continuing to display periodic oscillations in the fields. For a sinusoidal excitation, the solution must converge to an oscillation. The magnitude and phase at several observation points are extracted from the DFT at each temporal cycle. A 1% variation in steady state is permitted in phase.

The methodology we followed involves extracting the antenna characteristics for a given geometry selected from a parametric study using the FDTD code; constructing a prototype; measuring the frequency response of the input impedance, far-field radiation patterns at the measured resonant frequency; and finally comparing simulated results against measurements. To validate the operation of the FDTD model, several antenna structures were simulated and measured. The first antenna used to validate the code was a flat rectangular microstrip. The 50 mm × 47 mm patch was constructed on a square ground plane, 150 mm in side length. The substrate has a relative dielectric constant, ε_r = 4.2 and a thickness of 1.5 mm. The frequency spectrum of the real and imaginary parts of the input impedance and the far-field radiation pattern in the E plane are shown in Figs. 2 and 3, respectively.

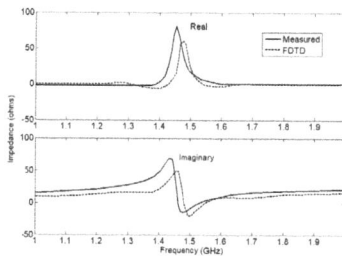

Figure 2. Measured and calculated input impedance for a 50 mm × 47 mm microstrip on a flat, 150 mm square ground plane.

The calculated impedance correctly predicts the resonant frequency measured using a network analyzer. Similarly, excellent agreement is observed between the amplitude and phase of the measured and simulated far-field radiation patterns.

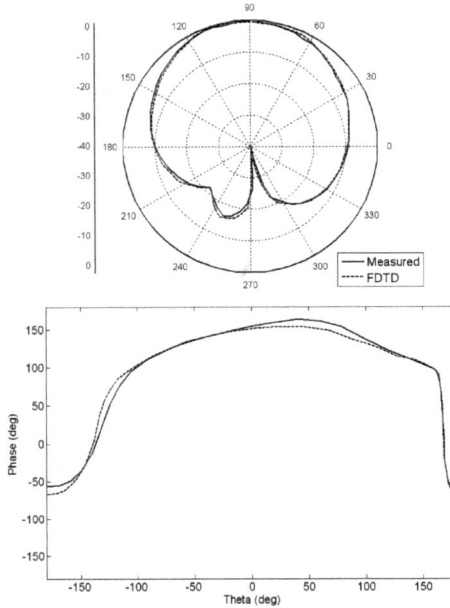

Figure 3. Normalized amplitude and phase of the far-field radiation pattern in the *E* plane for a 50 mm × 47 mm flat microstrip on a 150 mm square ground plane.

As a progression toward the drooped structure, a microstrip antenna with two edges bent down at an angle of 45^0 was modeled. This structure requires special consideration in that the boundary conditions imposed by the element and ground plane do not fall along the coordinate planes. The most straightforward approach is to discretize the sloping sides with the traditional stair-stepped approximation. However, it has been observed that this can result in a slight change in the resonant frequency of high Q structures [41]. To avoid errors associated with the conventional FDTD method, we used the contour path method introduced in [42]. In this approach, a field component adjacent to a boundary is not updated in terms of the spatial derivatives of the surrounding fields but by an integration of adjacent fields along the perimeter of the cell. This allows for partial or deformed cells, thus better approximating the drooped surfaces.

Three different step approximations, shown in Figure 4, were used to model the sloping sides, depending upon the angle of the bend. *For fine adjustments*, the ratio of the vertical to horizontal spatial increments is adjusted to yield the desired slope angle. This has an additional benefit of simplifying the implementation of the contour method when applied to field components adjacent to the sloped surfaces. Because each cell is truncated in the same way, the contour corrections proceeded without the need to calculate intercepts of the actual slope line at each cell.

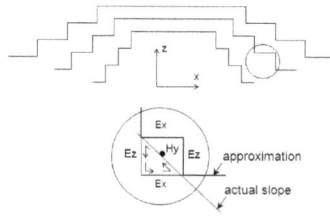

Figure 4. Stepped approximation of shallow, intermediate, and steep bend angles. Fine adjustments are made by varying the $\Delta x/\Delta z$ ratio. The contour integral correction is made to adjacent H fields to define the actual surface.

A rectangular microstrip with two 45^0 drooped edges fed by a coaxial cable was constructed and tested to compare measured and simulated input impedance, E and H plane patterns. The geometry of the antenna is shown in Figure 5 along with the phase of the E plane elevation cut. The measured phase displayed in Figure 5 showed a slight asymmetry due to the offset in antenna mount necessary to accommodate the bends and the connector. For comparison, we referenced the calculated far-field patterns to the same offset origin. The E plane and H plane patterns shown in Figs. 5 and 6 along with the impedance of Figure 7 reveal good correspondences between measured and simulated results.

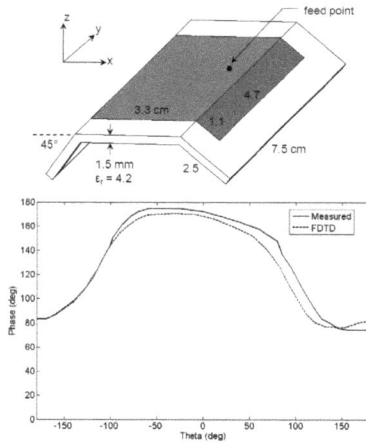

Figure 5. E plane (x-z plane) elevation phase for a 55 mm × 47 mm microstrip with a 45° bend, centered on a 83 mm × 75 mm ground plane.

Before progressing to the double-bend antenna, a test was conducted to verify that the FDTD model maintained continuity at the point where the stepped approximation changed from a 2:1 to a 1:1 ratio. To accomplish this, we modeled a 40 mm × 50 mm bent microstrip antenna near the change of an angle of 55^0 using both ratios. A comparison of the input impedance and

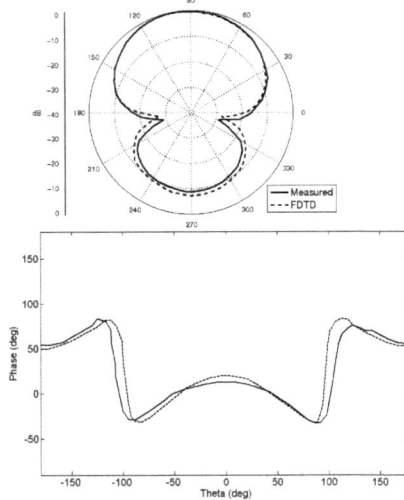

Figure 6. *H* plane elevation pattern for a 55 mm × 47 mm microstrip with a 45° bend, 10 mm from each short side, magnitude (top) and phase (bottom).

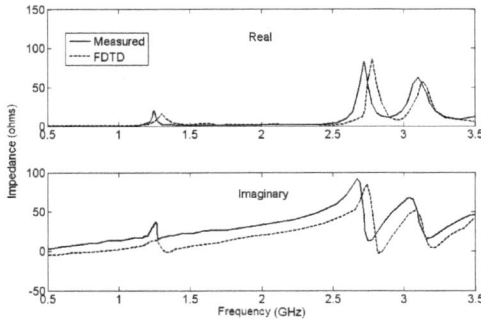

Figure 7. Measured and calculated input impedance for a 47 mm × 55 mm microstrip on a 45° bent ground plane (33 mm flat top with 11 mm bent over each angle).

elevation patterns depicted in Figure 8 show good agreement between results obtained from each approximation.

To duplicate the bend on the two remaining sides to achieve the full drooped structure, we constructed two drooped antennas and used them to verify the performance of the completed model. The first antenna, shown in Figure 9, consists of a 62 mm × 62 mm patch, a 40 mm square elevated section, printed on a 1.5 mm thick substrate, $\varepsilon_r = 4.2$, and a 60^0 droop angle. The second is a 64 mm × 64 mm patch, a 50 mm square elevated section, printed on a 3 mm substrate, $\varepsilon_r = 2.2$, and a 30^0 droop angle. Results obtained for the real and imaginary parts of the input impedance and far-field radiations patterns, as shown in Figs. 9 and 10, respectively, display good agreement between simulated and measured results. The excellent agreement demonstrated thus far between the amplitude and phase of the simulated and measured far-field radiation patterns prompted further exploration of the possibility of controlling the radiation pattern by manipulating the droop parameters.

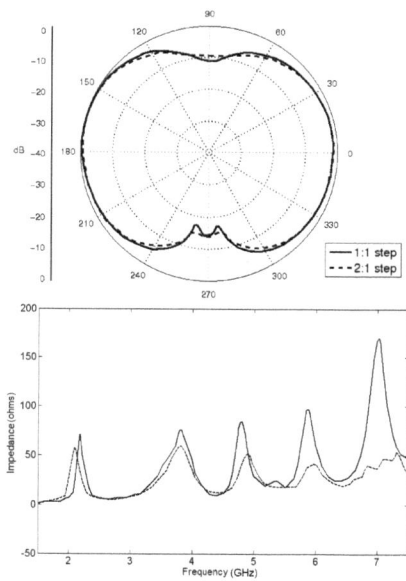

Figure 8. *E* plane pattern and impedance comparison for 50 mm × 40 mm microstrip ($\varepsilon_r = 2$) with a 55° bend, calculated for a 1:1 and 2:1 step approximation of the sloped sides.

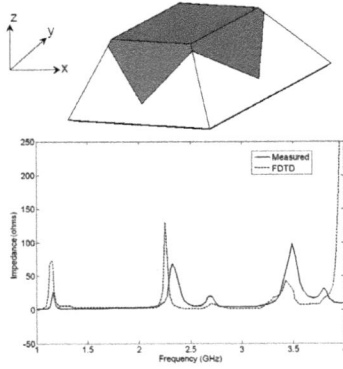

Figure 9. Measured and calculated input impedance for a 60° double bend microstrip.

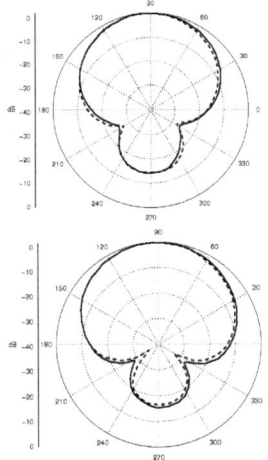

Figure 10. Measured and calculated elevation patterns for 64 mm square microstrip with a 30° bend, $\epsilon_r = 2.2$, E plane (x-z) top and H plane (y-z) bottom (1550 MHz).

3. Parametric analysis of the drooped microstrip antennas

Further parametric analyses were conducted to determine the range of structural variations that can be utilized to optimize the performance of the drooped microstrip antennas. The design process was complicated due to the several interacting parameters that must be considered in order to provide adequate low-angle coverage, uniform phase response, and polarization purity. Parameters investigated include the angle and location of the bend, length of the ground plane skirt, thickness and dielectric constant of the substrate. The results are assessed on the basis of their impact on antenna gain at bore sight, phase performance in the upper hemisphere, pattern beamwidth, cross polarization rejection, and near horizon gain reduction.

In all cases to be presented, the radiation patterns were obtained at the resonant frequency of the dominant mode. The amplitude and phase of the radiation patterns were obtained from anechoic chamber measurements in a $5^0 \times 5^0$ grid, following the procedure described in [43]. These were then analyzed to determine the 3 dB beamwidth and near horizon gain reduction with respect to bore sight (zenith). In order to determine the absolute gain, we integrated the calculated patterns over the upper and lower hemispheres with a 5^0 step in azimuth and elevation. All field calculations were then referenced to the resulting isotropic power density.

Whereas an enormous volume of literature is available on patch antennas for GPS applications [2]-[20], a close scrutiny revealed that the design objectives in the majority of these studies and the performance characterization were based entirely on the amplitude of the co- and cross-polarized radiation patterns. Only few have considered the phase response as a figure of merit in the design process and/or in the analysis or measurements [28], [31]-[35], [43]. The phase response directly weighs the arriving signals and produces a phase-shaping effect, which depends on the angle of arrival of the satellite signals. The calibration of the phase response provides invaluable information regarding the level of accuracy one can ultimately achieve for sub-centimeter static geodetic positions [43].

The measured upper hemispherical phase response of the antennas under test was matched to an ideal hemisphere in a $5^0 \times 5^0$ grid, using equal solid angle weighting. The position of the ideal hemisphere was adjusted to minimize the RMS error between the measured and ideal phase [43]. The origin of this hemisphere is defined as the "center of best fit" or "phase center," and the difference between the measured and ideal phase is defined as the "phase residual" or "phase error." The RMS value of the phase error is used as a figure of merit to describe the phase distortion introduced by the antennas considered in the rest of this chapter.

3.1. Drooped microstrip with a downward bend

For the initial test, a 40 mm × 40 mm patch was placed on three different ground planes. These, in turn, were bent at three different distances from the center, forming a flat square top that was 10, 30, or 50 mm^2. Simulations were carried out for bend angles, ranging from 0 to 90^0 in a 15^0 step. Figure 11 depicts the parameters and geometries used for the initial set of simulations.

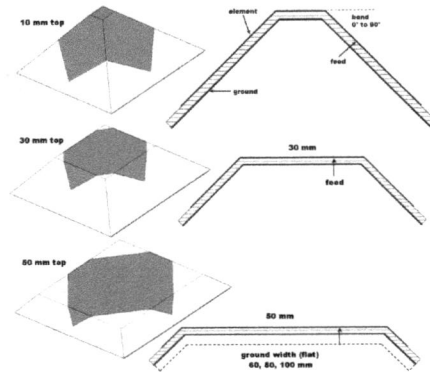

Figure 11. Initial structural variations included three bend locations at 5, 10, and 25 mm from the patch center, droop angles ranging from 0° to 90° and 3 ground planes having flat (unbent) dimensions of 60, 80, 100 mm. Three substrate materials with relative permittivity of 2.2, 4.2, and 10.

Three dielectric constants were used in the course of the simulations to examine the effects of different substrate materials; values of 2.2 and 4.2 were selected because of the availability of substrates to construct the verification cases, while an ε_r of 10 was chosen as an example of a ceramic substrate frequently used in industry. Copper tape applied to bulk Teflon, one eighth inch thick, was used to form several fixed and adjustable structures. A standard polyester-based circuit board was used to construct others.

A sequence of elevation patterns is presented in Figure 12, which displays the progression of the radiated fields with the size and bend angle of the ground plane. Not unexpectedly, smaller ground planes with larger bends allow more energy to escape off the back, until it appears that the main beam is 180^0 from the bore sight direction. This is made obvious by plotting the polar patterns for E_θ, using a linearly polarized excitation. With a circularly polarized excitation, backward radiated energy appears predominately in the cross polarized component, making the effect less noticeable. Notably, all subsequent results for gain, beamwidth, or phase are presented while exciting circular polarization as this would be the normal operating mode of the antenna.

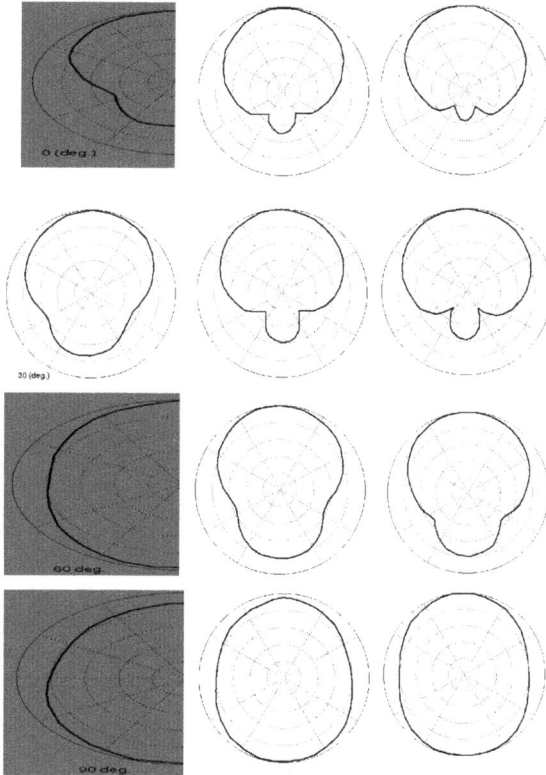

Figure 12. Elevation patterns for bend angle variations of 0° to 90° with 60 (left), 80 (center), and 100 (right) mm ground planes. E_θ component, $\varepsilon_r = 2.2$, all with 30 mm top. All patterns are normalized to 0 dB maximum, 10 dB/division.

Next, a prototype was constructed with adjustable bend plates, and the measured beamwidths were compared against simulated results. For construction simplicity, the antenna was excited using linear polarization with only two sloped sides. In Figure 13, the measured and computed beamwidths show a slight improvement with increasing bend angle.

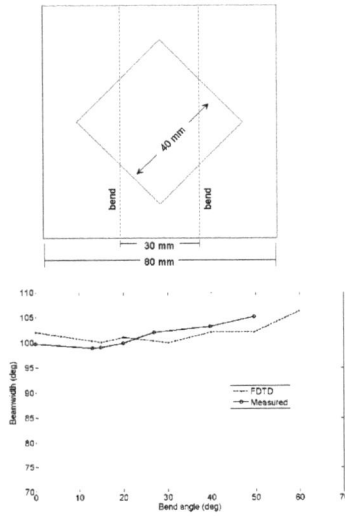

Figure 13. Measured and computed 3 dB beamwidth for an adjustable bend microstrip antenna.

Variations in bore sight gain, 3 dB beamwidth, near-horizon gain roll-off, and RMS phase error as a result of increasing the bend angle are summarized in Tables 1 to 3. The 10 mm case shows a sharp gain reduction at extreme bend angles. The results reveal a relatively minor improvement in the pattern beamwidth, even for large bend angles and varying ground plane sizes for the higher dielectric substrates. Indeed, the radiation patterns above the horizon remain virtually unchanged for bend angles up to 60⁰.

$\varepsilon_r=2.2$		Boresight gain (dBi)			3 dB Beamwidth (deg)			Near horizon gain Roll-off (dB)			RMS phase error (deg)		
Width-mm		60	80	100	60	80	100	60	80	100	60	80	100
Bend angle (10 mm top)	0	7.2	7.5	7.7	88	83	80	-11.	-12.	-13.	3.0	2.6	2.0
	15	7.1	7.5	7.7	88	84	80	-10.	-11.	-13.	2.5	3.3	3.4
	30	6.9	7.1	7.1	90	86	86	-9.7	-10.	-11.	1.5	1.8	1.7
	45	5.5	6.5	6.4	96	95	96	-9.0	-9.1	-9.1	2.2	1.6	2.0
	60	0.9	5.6	6.2	100	94	97	-7.9	-9.5	-9.0	4.3	6.5	2.6
	70	0.8	3.3	5.4	119	91	95	-5.5	-9.3	-9.0	12.	4.4	2.8
	90	-24	-25	-26	354	356	356	22	23	24	11.9	2.7	2.8

$\varepsilon_r=2.2$		Boresight gain (dBi)			3 dB Beamwidth (deg)			Near horizon gain Roll-off (dB)			RMS phase error (deg)		
Width-mm		60	80	100	60	80	100	60	80	100	60	80	100
Bend angle (30 mm top)	0	7.2	7.5	7.7	88	83	80	-11.	-12.	-13.	2.9	2.6	2.0
	15	7.1	7.5	7.8	88	83	78	-11.	-11.	-13.	2.4	3.3	3.4
	30	6.9	7.4	7.4	89	83	82	-9.6	-11.	-11.	1.5	1.8	1.7
	45	6.6	7.1	7.0	90	87	87	-10.	-10.	-11.	1.9	1.6	2.0
	60	5.5	6.7	6.5	91	90	93	-10.	-10.	-9.7	2.3	6.5	2.6
	70	5.2	6.4	6.2	92	92	96	-10.	-9.9	-9.1	1.6	4.4	2.8
	90	0.1	2.3	3.78	97	84	85	-9.6	-11	-12	8.2	2.7	2.8
Bend angle (50 mm top)	0	7.2	7.5	7.7	88	83	80	-11.	-12.	-13.	2.9	2.6	2.0
	15	7.0	7.7	7.9	89	81	77	-10.	-12.	-13.	2.4	3.3	3.4
	30	7.1	7.5	7.4	87	82	83	-9.9	-11.	-11.	1.5	1.8	1.7
	45	6.9	7.2	7.1	89	86	87	-9.6	-10.	-10.	1.9	1.6	2.0
	60	6.8	7.2	7.0	89	87	88	-9.7	-10.	-10.	2.3	6.5	2.6
	70	7.1	7.2	6.8	87	87	91	-10.	-10.	-9.4	1.6	4.4	2.8
	90	6.4	6.9	6.9	91	89	89	-9.4	-9.7	-10.	8.2	2.7	2.8

Table 1. Results for the drooped microstrip, substrate permittivity 2.2.

$\varepsilon_r=4.2$		Boresight gain (dBi)			3 dB Beamwidth (deg)			Near horizon gain Roll-off (dB)			RMS phase error (deg)		
Width-mm		60	80	100	60	80	100	60	80	100	60	80	100
Bend angle (10 mm top)	0	4.7	5.8	6.2	106	100	100	-11.6	-13.5	-14.5	1.6	1.2	1.1
	15	4.5	5.7	6.0	106	103	100	-9.9	-10.8	-10.9	2.2	3.6	4.3
	30	4.8	5.9	6.0	106	100	101	-8.6	-11.0	-10.7	1.5	1.0	1.0
	45	3.4	5.2	6.5	149	107	108	-3.7	-9.0	-9.7	2.0	1.5	1.2
	60	0.9	4.3	6.1	360	129	115	2.5	-5.7	-8.4	3.6	2.7	2.6
	70	-1.1	1.5	3.3	360	360	144	6.5	0.8	-3.8	12.8	2.9	3.3
	90	-27	-28	-29	360	360	360	-	-	-	-	-	-
Bend angle (30 mm top)	0	4.7	5.8	6.2	114	106	100	-11.6	-13.5	-14.5	1.6	1.2	1.1
	15	4.5	6.0	6.0	119	103	103	-10.6	-13.8	-13.7	1.2	1.1	0.9
	30	4.6	5.9	6.0	113	104	102	-11.1	-13.3	-13.5	0.9	0.5	0.5
	45	4.9	5.8	5.9	107	104	106	-12.2	-13.4	-13.4	1.2	0.8	0.7
	60	3.5	4.4	5.3	106	109	112	-11.3	-11.4	-12.0	1.6	1.9	1.3
	70	3.0	4.3	4.8	116	114	115	-9.2	-10.9	-11.3	0.9	0.9	0.8
	90	1.0	1.5	2.4	121	110	103	-6.7	-8.2	-10.1	1.6	1.5	1.4

ε_r=4.2		Boresight gain (dBi)			3 dB Beamwidth (deg)			Near horizon gain Roll-off (dB)			RMS phase error (deg)		
Width-mm		60	80	100	60	80	100	60	80	100	60	80	100
Bend angle (50 mm top)	0	4.8	5.8	6.2	114	106	100	-11.6	-13.5	-14.5	0.8	0.9	1.2
	15	4.9	5.9	6.3	119	103	103	-11.5	-13.6	-14.6	0.7	0.7	0.9
	30	5.3	6.0	6.2	113	104	102	-12.5	-13.6	-14.1	0.6	0.5	0.6
	45	5.0	5.8	6.0	107	104	106	-12.4	-13.4	-13.4	0.7	0.6	0.6
	60	4.7	5.6	5.8	106	109	112	-11.7	-12.8	-13.3	0.6	0.6	1.0
	70	5.2	5.5	5.4	116	114	115	-12.2	-12.5	-12.0	0.4	0.4	0.5
	90	4.4	5.2	5.4	121	110	103	-11.6	-12.6	-12.8	0.8	0.6	0.6

Table 2. Results for the drooped microstrip, substrate permittivity 4.2.

ε_r=10		Boresight gain (dBi)			3 dB Beamwidth (deg)			Near horizon gain Roll-off (dB)			RMS phase error (deg)		
Width-mm		60	80	100	60	80	100	60	80	100	60	80	100
Bend angle (10 mm top)	0	3.1	4.7	5.3	114	108	104	-9.4	-11.5	-12.5	0.80	0.49	0.53
	15	3.4	4.4	5.2	116	115	111	-9.3	-10.4	-11.4	0.89	1.0	0.87
	30	3.8	4.4	5.1	117	113	111	-9.7	-10.7	-11.3	0.59	0.46	0.42
	45	2.5	3.6	4.5	123	117	116	-7.8	-9.5	-10.6	0.41	0.41	0.38
	60	3.3	2.6	3.6	139	120	118	-8.2	-8.3	-9.5	3.0	0.80	0.86
	70	2.4	3.0	3.5	134	122	120	-7.4	-8.5	-9.3	2.0	0.72	0.97
	90	-38	-40	-43	360	360	360	-	-	-	-	-	-
Bend angle (30 mm top)	0	3.1	4.7	5.3	114	108	104	-9.4	-11.5	-12.5	0.80	0.49	0.53
	15	3.8	4.5	4.9	119	108	106	-9.6	-11.3	-11.9	0.40	0.46	0.48
	30	3.8	5.0	5.4	116	106	106	-9.8	-12.1	-12.5	0.25	0.33	0.36
	45	2.9	4.1	4.7	116	114	114	-8.9	-10.3	-10.9	0.39	0.32	0.30
	60	2.7	3.6	4.3	119	104	104	-8.6	-11.1	-12.0	0.56	1.2	1.2
	70	2.3	2.9	3.5	122	119	119	-7.8	-8.6	-9.3	0.24	0.21	0.18
	90	1.5	1.4	1.5	134	132	127	-6.0	-6.1	-6.6	0.29	0.31	0.34
Bend angle (50 mm top)	0	3.1	4.7	5.3	114	108	104	-9.4	-11.5	-12.5	0.80	0.49	.53
	15	2.9	4.6	5.4	133	114	109	-7.6	-10.7	-11.7	0.34	0.70	0.80
	30	3.2	4.9	5.5	121	110	109	-8.7	-12.4	-11.8	0.46	0.90	0.32
	45	3.4	4.5	4.8	119	112	117	-9.0	-10.7	-10.4	0.37	0.56	0.41
	60	3.0	3.7	4.9	122	124	119	-8.4	-8.9	-11.8	0.51	0.51	0.90
	70	3.7	3.6	4.5	117	123	107	-9.5	-9.1	-10.3	0.28	2.2	1.5
	90	2.0	2.6	3.2	113	110	108	-8.3	-9.2	-10.0	2.1	2.4	2.2

Table 3. Results for the drooped microstrip, substrate permittivity 10.

The phase of the elevation cuts shown in Figure 14 shows a remarkable change in the below-horizon phase. For small bend angles, the phase diminishes from the bore sight value when approaching the horizon, while at higher bends the phase increases from the bore sight value. At about 15^0 to 30^0 of bend, a region exists where the elevation phase remains relatively constant above the horizon. Bending of the structure in this manner could provide an additional beamwidth and, more importantly, the phase stability necessary to achieve a design specification, particularly if further modification of the substrate is not feasible.

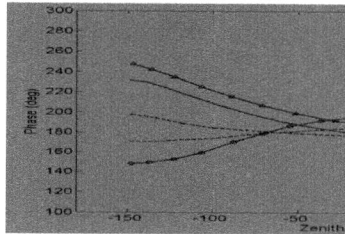

Figure 14. Elevation phase patterns for different bend angles (0^0, 15^0, 30^0, 60^0, and 70^0), with a 100 mm ground plane and a 30 mm flat ($\varepsilon_r = 4.2$).

Another consideration for GPS antennas is the cross polarization behavior. Odd reflections from nearby objects tend to be orthogonally polarized. Hence, it is important that the antenna be able to reject these, particularly near the horizon. To demonstrate the effect of bending the structure on the polarization performance, we examined the ratio of the right- to the left-hand

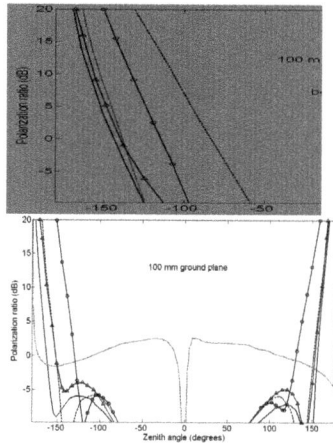

Figure 15. Cross polarization rejection (Defined as the ratio of the cross-polarized to the co-polarized components) for different bend angles (0^0, 15^0, 30^0, 60^0, and 90^0); top, $\varepsilon_r = 4.2$, bottom, $\varepsilon_r = 2.2$.

circular components for several bend angles. Almost without exception, the bend degraded the cross polarization rejection near the horizon as shown in Figure 15, making the antenna more susceptible to spurious signals.

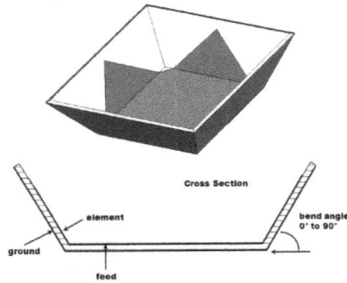

Figure 16. Geometry of the upward bend antenna.

3.2. Drooped microstrip with an upward bend

The modification of the FDTD model to accommodate upward bends was accomplished by interchanging the positions of the ground plane and the element. The structure would then be upside down in the computational space but would have an upward bend. Initially, we considered the antenna shown in Figure16, which has a 30 mm flat top on a substrate with ε_r = 2.2. Three ground plane sizes were analyzed with bend angles varied up to 90⁰. Figure 17

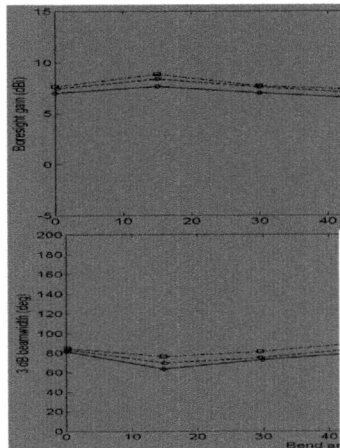

Figure 17. Gain and beam width variation with upward bend angle for three ground plane sizes, ε_r = 2.2. (Bend forming a 30 mm flat center section).

shows the gain and the 3-dB beamwidth for the upward bend cases. As seen, a noticeable beam broadening is evident at the higher bend angles with 3-dB beamwidths up to 60% greater than the equivalent flat case. One can observe, however, a distinct reduction in the beamwidth for the initial small bend angles, particularly for larger ground planes. The beamwidth did not recover to that of the equivalent downward case until the bend angle exceeded 60^0.

A prototype antenna was constructed and tested to allow comparison against experimental results. The antenna is identical to the one shown in Figure 13; only two variable upward bends were used in this case. For measurement purposes, a linearly polarized excitation was used. The initial dip in the beamwidth did not occur in the simulated and measured results shown in Fig. 18.

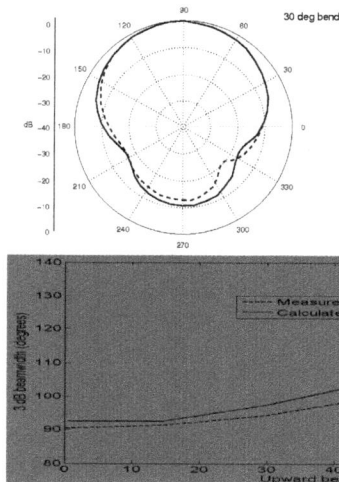

Figure 18. *E* plane elevation pattern and beamwidth behavior of the adjustable upward bend antenna.

With the completion of the initial set of simulations and measurements, we pursued the model with additional parameter variations. Bend locations at the 10 and 50 mm positions were tested for the $\varepsilon_r = 2.2$ substrate. Also, the $\varepsilon_r = 4.2$ substrate was examined when the bend was positioned at the 30 mm position. The results are summarized in Tables 4 and 5.

A changing upward bend produced little effect on the phase in the upper hemisphere. All phase curves above 90^0 elevation fell within 30^0 of each other over the full range of bend angles. Like the downward bend, the RMS phase error over the hemisphere did not vary substantially with bend angle, although a slightly greater differentiation is evident between different ground plane sizes. The elevation gain and cross polarization rejection for various upward bend angles are displayed in Figs. 19 and 20, respectively. When compared to Figure 15, it is apparent that the rejection is better by 2 dB near the horizon for upward bends, but the opposite is true below the horizon. In both cases, bending the ground plane reduces the cross-polarization discrimination near the horizon.

$\varepsilon_r=2.2$		Boresight gain (dBi)			3 dB Beamwidth (deg)			Near horizon gain Roll-off (dB)			RMS phase error (deg)	
Width-mm		60	80	100	60	80	100	60	80	100	60	80	100
Bend angle (10 mm top)	0	7.2	7.5	7.7	88	83	80	-11.	-12.	-13.	3.0	2.6	2.0
	15	4.0	6.2	6.4	158	134	118	-14.	-11.	-11.	6.2	6.8	9.1
	30	4.7	6.5	1.3	143	122	120	-9.4	-10.	-12.	3.6	3.5	5.5
	45	5.1	3.5	3.5	137	170	106	-7.2	-7.6	-9.4	4.5	3.3	3.3
	60	4.1	-.8	2.6	142	302	214	-3.8	-6.7	-7.1	6.6	5.2	4.6
	70	2.5	-4.4	-.42	127	327	300	0.44	5.8	5.7	35.8	6.5	11.2
	90	-1.5	-6.1	-6.4	188	360	360	-2.3	-6.6	0.93	11.0	69.9	80.1
Bend angle (30 mm top)	0	7.2	7.5	7.7	88	83	80	-11.	-12.	-13.	3.0	2.6	2.0
	15	7.7	8.4	8.7	75	69	63	-12.	-13.	-15.	1.8	1.6	3.9
	30	7.1	7.6	7.7	81	75	73	-11.	-12.	-13.	2.8	2.4	4.0
	45	6.6	7.0	7.3	89	83	79	-9.1	-9.8	-11.	2.3	1.5	2.0
	60	6.0	6.2	6.0	95	91	91	-8.3	-7.8	-7.9	2.2	1.3	1.5
	70	4.9	4.9	4.9	104	110	114	-6.7	-5.8	-5.3	4.4	2.8	1.9
	90	5.1	0.6	3.7	112	176	142	-6.1	-2.1	-3.9	6.8	4.7	1.5
Bend angle (50 mm top)	0	7.2	7.5	7.7	88	83	80	-11.	-12.	-13.	3.0	2.6	2.0
	15	7.3	7.6	7.9	82	75	73	-12.	-13.	-14.	1.2	1.5	2.5
	30	7.0	7.4	7.9	79	77	69	-12.	-12.	-13.	1.6	2.4	3.4
	45	7.3	7.5	7.8	83	79	73	-10.	-11.	-12.	1.1	1.5	1.9
	60	7.1	6.8	7.1	87	81	87	-9.7	-9.4	-10.	1.0	1.1	1.3
	70	6.9	6.8	6.9	83	87	81	-9.4	-5.4	-4.5	2.5	2.9	2.0
	90	6.5	5.8	5.1	89	99	110	-8.9	-7.2	-6.4	5.6	12.6	10.3

Table 4. Results for the drooped microstrip, substrate permittivity 2.2.

$\varepsilon_r=4.2$		Bore-sight gain (dBi)			3 dB Beamwidth (deg)			Near horizon gain Roll-off (dB)			RMS phase error (deg)	
Width-mm		60	80	100	60	80	100	60	80	100	60	80	100
Bend angle (30 mm top)	0	4.7	5.8	6.2	114	106	100	-11.	-14.	-14.	1.6	1.2	1.1
	15	6.6	7.0	7.5	89	81	75	-8.4	-9.4	-11.	0.8	1.0	1.7
	30	5.1	5.4	5.6	106	102	93	-7.8	-8.6	-9.4	4.2	3.8	4.7
	45	5.6	6.0	6.4	91	89	83	-7.1	-7.4	-8.1	2.1	2.0	2.6
	60	4.5	4.9	5.1	102	104	100	-7.7	-7.3	-7.3	5.2	3.9	3.8
	70	3.4	4.2	4.3	114	110	118	-7.2	-6.1	-5.0	5.0	2.7	1.6
	90	4.7	4.7	4.4	102	100	110	-6.7	-6.0	-5.1	2.5	2.1	1.4

Table 5. Results for the drooped microstrip, substrate permittivity 4.2.

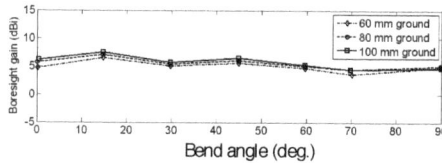

Figure 19. Gain variation with upward bend angle for three ground plane sizes, $\varepsilon_r = 4.2$. (Bend forming a 30 mm flat center section).

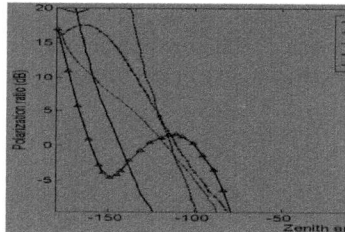

Figure 20. Cross polarization rejection (Defined as the ratio of the cross-polarized to the co-polarized components) for different bend angles (0°, 15°, 45°, 60°, and 90°), 100 mm ground plane, $\varepsilon_r = 2.2$.

4. Concluding remarks

In this chapter, we have presented numerical simulations and experimental measurements to analyze downward and upward drooped microstrip antennas with the intent of modifying the radiation pattern of the basic planar patch to accommodate the coverage requirements of GPS marine navigation and positioning. Magnitude and phase of the simulated and measured far-field radiation patterns are presented to reveal the tradeoffs in performance between patch geometry, ground plane size, and orientation. Results reported for the wide range of structural variations applied to the base antenna along with changes in the substrate material should be valuable to designers seeking to achieve a specific coverage performance.

It has been found that an accurate and stable phase center can be obtained over the entire hemisphere for moderate upward bends. Numerical simulations and measurements demonstrate that the 3-dB beamwidth of the flat microstrip patch can be increased by at least 15% and 60% for the downward and upward bends, respectively. The phase stability demonstrated by the slightly bent structures may be viewed as advantageous in cases where circumstances require distorting the element but where a significant alteration in the pattern is not desired. These were accomplished, however, at the expense of some loss of the low-profile character of the antenna.

Although not dramatically affected, the cross polarization discrimination of the bent antennas is reduced by 3 dB at the horizon compared to the equivalent flat case. Calculations of the RMS error in the spherical phase fit over the upper hemisphere showed little change with bend angle or position. In general, there is a tradeoff in achieving broad-beam pattern coverage, and maintaining high cross polarization discrimination.

The crossed dipole source, when used with the pedestal ground plane, demonstrated significant pattern improvements which inspired our interest in the drooped microstrip structure. It should be noted, however, that the crossed dipole is fundamentally different from the microstrip antenna, being itself a stand-alone radiator which operates in the presence of secondary sources created by the ground plane image. For the microstrip antenna, the ground plane is an integral part of the structure. The interior field distribution of the fundamental mode remains essentially unchanged with ground plane manipulation, and pattern modification can come about only by repositioning of the radiating edges in space.

Author details

Ken G. Clark[1], Hussain M. Al-Rizzo[2*], James M. Tranquilla[1], Haider Khaleel[3] and Ayman Abbosh[2]

*Address all correspondence to: hmalrizzo@ualr.edu

1 EMR Microwave Technology Corporation, 64 Alison Blvd., Fredericton, NB, Canada

2 Systems Engineering Department, Donaghey College of Engineering and Information Technology, University of Arkansas at Little Rock, USA

3 Department of Engineering Science, Sonoma State University, Rohnert Park, CA, USA

References

[1] Wells, D. Editor), Guide to GPS Positioning, Larry D Hothem, December (1986).

[2] Ho, C. H, Shumaker, P. K, Fan, L, Smith, K. B, & Liao, J. W. Printed Cylindrical Slot Antenna for GPS Commercial Applications. *Electron. Lett.*, Feb. (1996). , 32, 151-152.

[3] Ho, C. H, Shumaker, P. K, Fan, L, Smith, K. B, & Liao, J. W. A Novel GPS Avionics Antenna. *IEEE Antennas Propagat. Lett.*, (1996). , 1950-1953.

[4] Altshuler, E. E. Hemispherical Coverage Using a Double-Folded Monopole. *IEEE Trans. Antennas Propagat.*, Aug. (1996). , 44(8), 1112-1119.

[5]   Ying, Z, & Kildal, P. S. Improvement of Dipole, Helix, Spiral, Microstrip Patch and Aperture Antennas with Ground Planes by Using Corrugated Soft Surfaces. *IEE Proc. Microw. Antennas Propagat.*, Jun. (1996). , 143(3), 244-248.

[6]   Pozar, D. M, Duffy, S. M, & Dual-band, A. Circularly Polarized Aperture-Coupled Stacked Microstrip Antenna for Global Positioning Satellite. *IEEE Trans. Antennas Propagat.*, Nov. (1997). , 45(11), 1618-1625.

[7]   Padros, N, Ortigosa, J. I, Baker, J, Iskander, M. F, & Thornberg, B. Comparative Study of High-Performance GPS Receiving Antenna Designs. *IEEE Trans. Antennas Propagat.*, Apr. (1997). , 45(4), 698-706.

[8]   Kan, H. K, Waterhouse, R. B, & Small, A. CP-Printed Antenna Using 120 deg Sequential Rotation. *IEEE Trans. Antennas Propagat.*, Mar. (2002). , 50(3), 398-399.

[9]   Lee, S, Woo, J, Ryu, M, & Shin, H. Corrugated Circular Microstrip Patch Antennas for Miniaturization. *Electron. Lett.*, Mar. (2002). , 38, 262-263.

[10] Boccia, J. L, & Amendola, G. and G. Di Massa. A Shorted Elliptical Patch Antenna for GPS Applications. *IEEE Antennas Wireless Propagat. Lett.*, (2003). , 2, 6-8.

[11] Ozgun, O, Mutlu, S, Aksun, M. I, & Atalan, L. Design of Dual-Frequency Probe-Fed Microstrip Antennas with Genetic Optimization Algorithm. *IEEE Trans. Antennas Propagat.*, Aug. (2003). , 51(8), 1947-1954.

[12] Boccia, L, Amendola, G, & Massa, G. D. A Dual Frequency Microstrip Patch Antenna for High-Precision GPS Applications. *IEEE Antennas Propagat. Lett.*, (2004). , 3, 157-160.

[13] Oh, K, Kim, B, & Choi, J. Novel Integrated GPS/RKES/PCS Antenna for Vehicular Application. *IEEE Microwave Wireless Comp. Lett.*, Apr. (2005). , 15(4), 244-246.

[14] Basilio, L. I, Williams, J. T, Jackson, D. R, & Khayat, M. A. A Comparative Study of a New GPS Reduced-Surface-Wave Antenna. *IEEE Antennas Propagat. Letters*, (2005). , 4, 233-236.

[15] Zhang, Y, & Hui, H. T. A Printed Hemispherical Helical Antenna for GPS Receivers. *IEEE Microwave and Wireless Comp. Lett.*, January (2005). , 15(1), 10-12.

[16] Azaro, R, De Natale, F, Donelli, M, Zeni, E, & Massa, A. Synthesis of a Prefractal Dual-Band Monopolar Antenna for GPS Applications," *IEEE Antennas Wireless Propagat. Lett.*, (2006). , 5, 361-364.

[17] Zhou, Y, Koulouridis, S, Kiziltas, G, Volakis, J. L, & Novel, A. Quadruple Antenna for Tri-Band GPS Applications. *IEEE Antennas Wireless Propagat. Lett.*, (2006). , 5, 224-227.

[18] Basilio, L. I, Chen, R. L, Williams, J. T, & Jackson, D. R. A New Planar Dual-Band GPS Antenna Designed for Reduced Susceptibility to Low-Angle Multipath. *IEEE Trans. Antennas Propagat.*, Aug. (2007). , 55(8), 2358-2366.

[19] Boccia, L, Amendola, G, & Massa, G. D. Performance Evaluation of Shorted Annular Patch Antennas for High-Precision GPS Systems. *IET Microwave Antennas Propagat.*, (2007). , 1(2), 465-471.

[20] Zhou, Y, Chen, C. C, & Volakis, J. L. Dual Band Proximity-Fed Stacked Patch Antenna for Tri-Band GPS Applications. *IEEE Trans. Antennas Propagat.*, Jan. (2007). , 55(1), 220-223.

[21] Lachapelle, G, Casey, M, Eaton, R. M, Kleusberg, A, Tranquilla, J, & Wells, D. GPS Marine Kinematic Positioning Accuracy and Reliability. *The Canadian Surveyor*, Summer (1987). , 41(2), 143-172.

[22] Lachapelle, G, Liu, C, Lu, G, Cannon, M. E, Townsend, B, & Hare, R. Precise Marine DGPS Positioning Using P Code and High Performance C/A Code Technologies. *National Technical Metting, ION*, San Francisco, Jan. (1993).

[23] Yaesh, I, & Priel, B. Design of Leveling Loop for Marine Navigation System. *IEEE Trans. Aerospace Electr. Sys.*, Apr. (1993). , 29(2), 599-604.

[24] Deifes, D. GPS Based Attitude Determining System for Marine Navigation. *IEEE Position, Location, and Navigation Symposium*, Apr. (1994). , 806-812.

[25] Lachapelle, G, Cannon, M. E, Lu, G, & Loncarevic, B. Ship borne GPS Attitude Determination During MMST-93. *IEEE J. Oceanic Eng.*, Jan. (1996). , 21(1), 100-105.

[26] Shumaker, P. K, Ho, C. H, & Smith, K. B. Printed Half-Wavelength Quadrifilar Helix Antenna for GPS Marine Applications. *Electron. Lett.*, February (1996). , 32, 153-154.

[27] Beiter, S, Poquette, R, Filipo, B. S, & Goetz, W. Precision Hybrid Navigation System for Varied Marine Applications. *IEEE Position, Location, and Navigation Symposium*, April (1998). , 316-323.

[28] Tranquilla, J. M, Carr, J. P, & Al-rizzo, H. M. Analysis of a Choke Ring Ground Plane for Multipath Control in Global Positioning System (GPS) Applications. *IEEE Trans. Antennas Propagat.*, Oct. (1994). , 42, 905-911.

[29] Counselman, C. C. III. Multipath-Rejecting GPS Antennas. *Proc. of the IEEE*, Jan. (1999). , 87(1), 86-91.

[30] Townsend, B. R, & Fenton, P. C. A Practical Approach for the Reduction of Pseudorange Multipath Errors in a L1 GPS Receiver. ION-94, Salt Lake City, Sept. (1994). , 1-6.

[31] Tranquilla, J. M. The Experimental Study of Global Positioning Satellite Antenna Backplane Configurations. Technical Report, NASA Jet Propulsion Lab., Contract 957959, Radiating Systems Research Laboratory, University of New Brunswick, Fredericton, NB, Canada, (1988).

[32] Tranquilla, J. M, Colpitts, B. G, & Carr, J. P. Measurement of Low Multipath Antennas for Topex. Proceedings of the *5th International Geodetic Symposium on Satellite Positioning*, Las Cruces, NM, Mar. 13-1 7, (1989)., 356-361.

[33] Tanquilla, J. M, & Colpitts, B. G. Development of a Class of Antennas for Space-Based NAVSTAR GPS Applications. *International Conference on Antennas and Propagation (ICAP 89)*, 6th, Coventry, England, Apr. 4-7, 1989, Proceedings. Part 1 (A90-27776 11-32). London, England and Piscataway, NJ, Institution of Electrical Engineers, (1989)., 65-69.

[34] Tranquilla, J. M, & Colpitts, B. G. GPS Antenna Design Characteristics for High Precision Applications. Proceedings of the *ASCE Conference GPS-88 Engineering Applications of GPS Satellite Surveying Technology*, Nashville, TN, May (1988)., 11-14.

[35] Tranquilla, J. M, & Best, S. R. A Study of the Quadrifilar Helix Antenna for Global Positioning Systems (GPS) Applications. *IEEE Trans. Antennas Propagat.*, Oct.(1990)., 38, 1545-1550.

[36] Noghanian, S, & Shafai, L. Control of Microstrip Antenna Radiation Characteristics by Ground Plane Size and Shape. *IEE Proc. Microw. Antennas Propagat.*, Ju.(1998)., 145, 207-212.

[37] Clark, K. G. The Finite-Difference Time-Domain Technique Applied to the Drooped Microstrip Antenna. PhD thesis, Department of Electrical Engineering, University of New Brunswick, Fredericton, NB, Canada, Jul. (1996).

[38] Feller, W. Three Dimensional Microstrip Patch Antenna. US Patent Publication Apr. (1993). (5)

[39] Su, C. W, Huang, S. K, & Lee, C. H. CP Microstrip Antenna with Wide Beamwidth for GPS Band Application. *Electronics Lett.*, [th] Sept. (2007)., 43(20)

[40] Colpitts, B. G. The Uniform Theory of Diffraction Applied to Wedges and Curved Surfaces. PhD thesis, Department of Electrical Engineering, University of New Brunswick, Fredericton, NB, Canada, Apr.(1988).

[41] Taflove, A, & Umashankar, K. R. Finite-Difference Time-Domain Modeling of Electromagnetic Wave Scattering and Interaction Problems. *IEEE Trans. Antennas Propagat. Society Newslett.*, Apr. (1988)., 5-20.

[42] Jurgens, T. G, Taflove, A, Umashankar, K. R, & Moore, T. G. Finite-Difference Time-Domain Modeling of Curved Surfaces. *IEEE Trans. Antennas Propagat.*, Apr. (1992)., AP-40, 357-366.

[43] Tranquilla, J. M, & Best, S. R. Phase Center Considerations for the Monopole Antenna. *IEEE Trans. Antennas Propagat.*, May (1986)., AP-34, 741-744.

Chapter 12

Planar Microstrip-To-Waveguide Transition in Millimeter-Wave Band

Kazuyuki Seo

Additional information is available at the end of the chapter

1. Introduction

Many kind of millimeter-wave automotive radars have been developed [1], [2]. The microstrip antenna becomes a good candidate when radar sensors are widely used in vehicle due to its advantages of low cost and low profile. Generally microstrip antennas are placed on the surface of a radar sensor and are connected to millimeter-wave circuits inside of the sensor via waveguides. Therefore, transitions from waveguide to microstrip line are required, as shown in Figure 1.

Rectangular waveguides were one of the earliest types of transmission lines used to transport microwave signals and are still used today for many applications. Because of the recent trend toward miniaturization and integration, a lot of microwave circuitry is currently fabricated using planar transmission lines, such as microstrip or strip line, rather than waveguide. There is, however, still a need for waveguides in many applications such as millimeter wave systems, and in some precision test applications.

Various types of millimeter-wave transitions from waveguide to microstrip line have been proposed. The ridge waveguide type [3], quasi-Yagi type [4], and planar waveguide type [5] have been studied as longitudinal connection of waveguide with microstrip line. With regard to vertical transitions, a conventional type of probe feeding has a wideband characteristic [6], [7], but it needs a metal short block with a quarter-wavelength on the substrate. The replacement of the metal short block is a patch element in the waveguide to achieve sufficient coupling between waveguide and microstrip line. The slot coupling type [8] achieves coupling between the microstrip line and the patch element in the waveguide by means of a slot, it is composed of two dielectric substrates without a metal short block. The proximity coupling type [9] has been developed more recently. It can be composed of a single dielectric substrate attached to the waveguide. A rectangular patch element on the lower plane of the dielectric substrate couples with a microstrip line on the upper plane of the dielectric substrate. It is suitable for mass production. The proximity coupling type has been further developed for wideband [10].

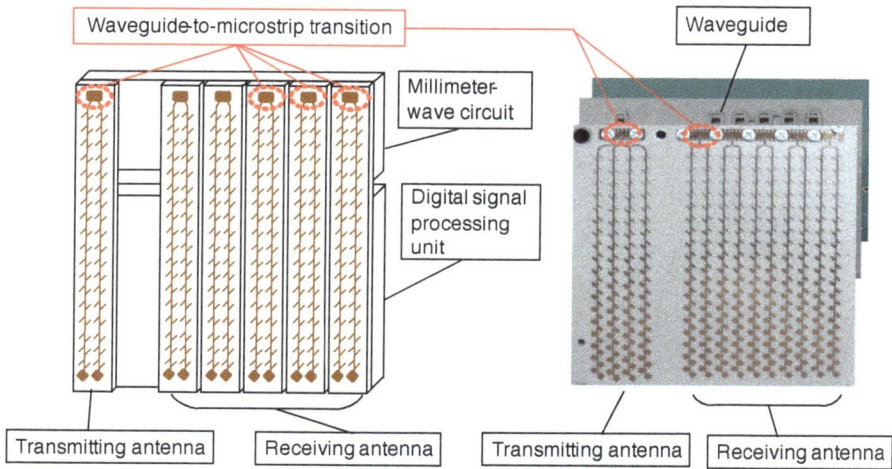

Figure 1. Construction of millimeter-wave automotive radar sensor and photograph for example

2. Probe transition with back-short

The transitions with short-circuited waveguide of 1/4 guided wavelength on the substrate are very popular [6], [7] because their principle of mode transformation is almost the same with that of ordinary transitions of a waveguide and a coaxial cable [11]. The probe transition connects a microstrip line and a waveguide as shown in Figure 2. A probe at one end of the microstrip line is inserted into the perpendicular waveguide whose one end is short-circuited by the back-short waveguide.

The configuration is shown in Figure 3. A dielectric substrate with conductor patterns on its both sides is placed on an open-ended waveguide (WR-12 standard waveguide). An aperture of the substrate is covered with an upper waveguide. A short circuit of the upper waveguide is essentially $\lambda_g/4$ (λ_g: guided wavelength of the waveguide) above the substrate. Consequently, the electric current on the probe couples to the magnetic field of TE$_{10}$ dominant mode of the waveguide as shown in Figure 4. Via holes are surrounding the waveguide in the structure in order to reduce the leakage of parallel plate mode transmitting into the substrate. Impedance matching could be achieved by controlling the length ρ of the probe and the length S_s of the upper waveguide. Each parameters in Figure 3 are shown in Table 1 for example.

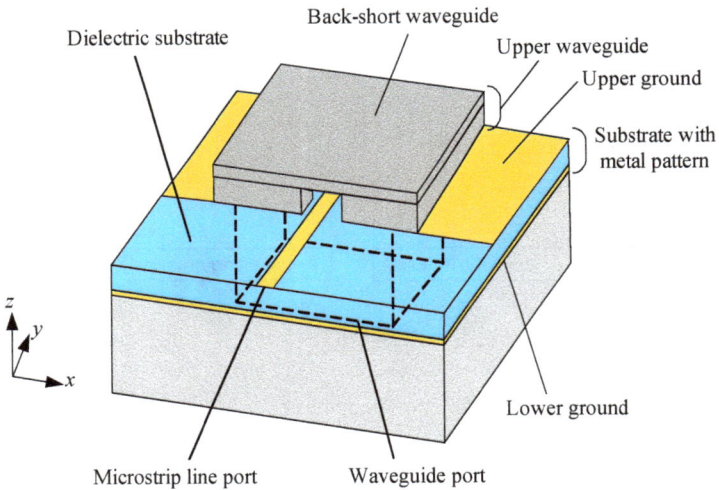

Figure 2. Probe transition with back-short

In order to reduce the leakage from the waveguide window at the insertion of the microstrip line, the width of the window should be narrow than the width of the cut off condition and is 0.9 mm in this case. S-parameters of the reflection S_{11} and the transmission S_{21} are calculated by using an electromagnetic simulator based on the finite element method (Ansys HFSS) as shown in Figure 5. From the simulated results, this transition has wide frequency bandwidth.

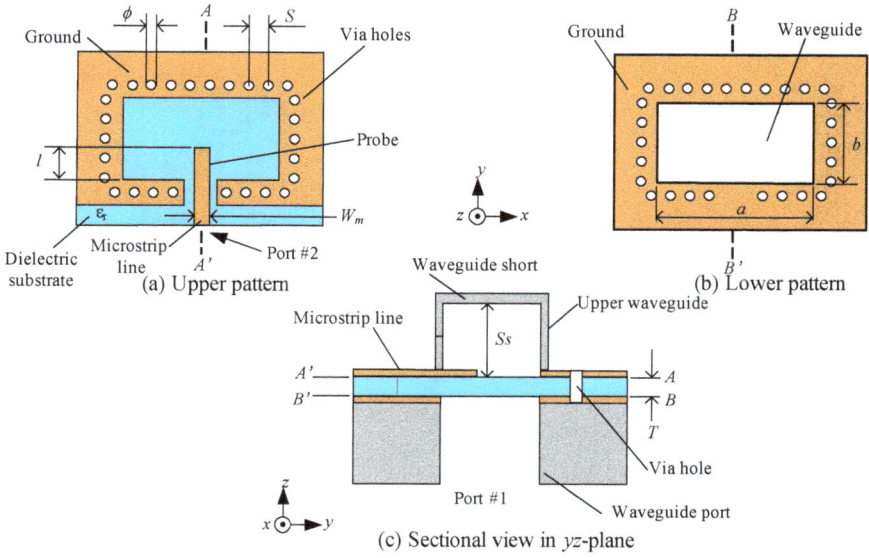

(a) Upper pattern

(b) Lower pattern

(c) Sectional view in yz-plane

Figure 3. Detailed configuration of the probe transition with back-short

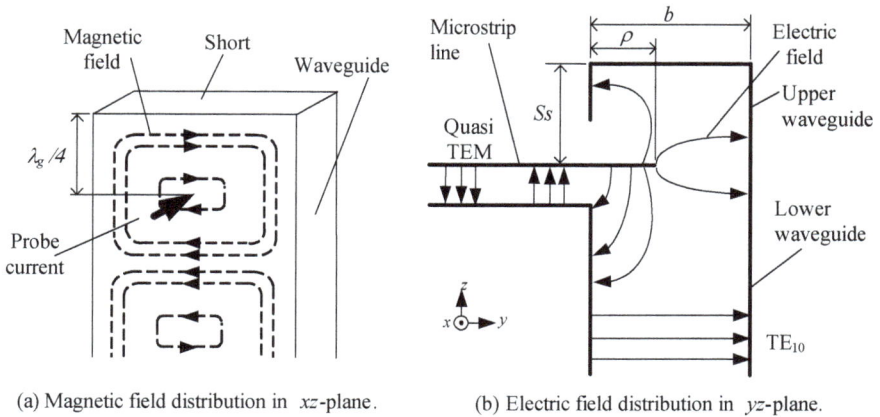

(a) Magnetic field distribution in xz-plane.

(b) Electric field distribution in yz-plane.

Figure 4. Magnetic and electric field lines

Description	Name	Value (mm)	Description	Name	Value (mm)
Broad wall length of waveguide	a	3.1	Narrow wall length of waveguide	b	1.55
Width of microstrip line	W_m	0.3	Length of inserted probe	p	0.675
Length of back short waveguide	S_s	0.61	Relative permittivity	ε_r	2.2
Thickness of substrate	T	0.127	Diameter of via hole	ϕ	0.2
Space between via holes	S	0.5			

Table 1. Parameters of probe transition with back-short

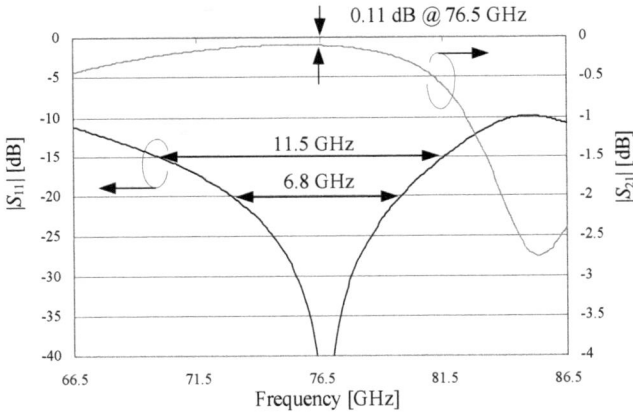

Figure 5. Reflection characteristic $|S_{11}|$ and insertion loss $|S_{21}|$ of probe transition with back-short

3. Planar proximity coupling transition

Planar proximity coupling transitions shown in Figure 6 and Figure 7 have been proposed [9]. This transition can be composed of only a single dielectric substrate attached to the waveguide end and suitable for mass production. The conductor pattern with a notch (it is named a waveguide short pattern because of its function) and the microstrip line are located on the upper plane of the dielectric substrate. A rectangular patch element and a surrounding ground are patterned on the lower plane of the dielectric substrate. Via holes are surrounding the aperture of the waveguide on the lower plane of the dielectric substrate to connect the surrounding ground and the waveguide short electrically.

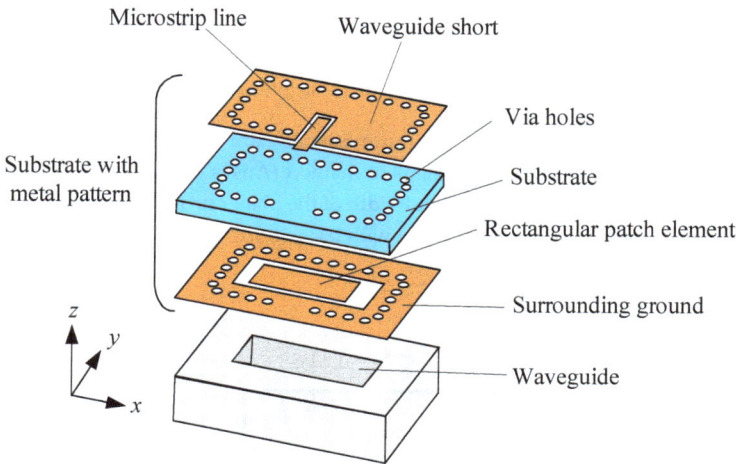

Figure 6. Planar proximity coupling transition

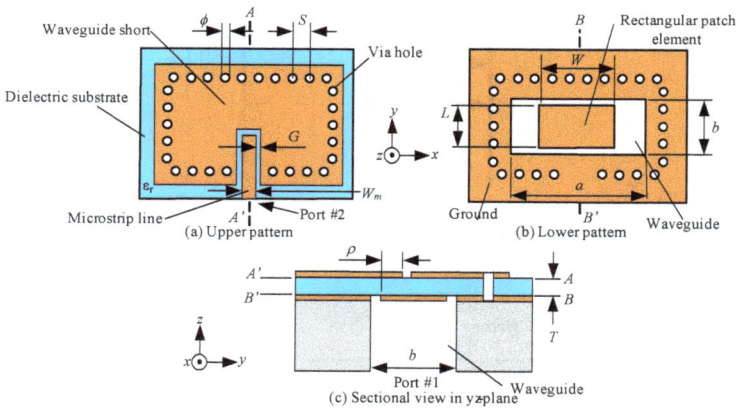

Figure 7. Detailed configuration of planar proximity coupling transition

The microstrip line is inserted into the waveguide and overlaps on the rectangular patch element with overlap length $\rho = 0.34$ mm. The parameters of the transition are presented in Table 2 for example.

Figure 8 shows the electric field distribution of each mode in yz-plane. The modes of the microstrip line, the rectangular patch element and the waveguide are quasi TEM transmission mode, TM_{01} fundamental resonant mode and TE_{10} fundamental transmission mode, respectively. Low transmission loss is realized by exchanging quasi TEM transmission mode and TE_{10} fundamental transmission mode with high efficiency utilizing TM_{01} fundamental resonant mode. Each parameters in Figure 8 are shown in Table 2 for example.

S-parameters of the reflection S_{11} and the transmission S_{21} are calculated by using an electromagnetic simulator based on the finite element method (Ansys HFSS) as shown in Figure 9. From the simulated results, frequency bandwidth of the planar proximity coupling transition is narrow than the ordinary probe transition with back-short.

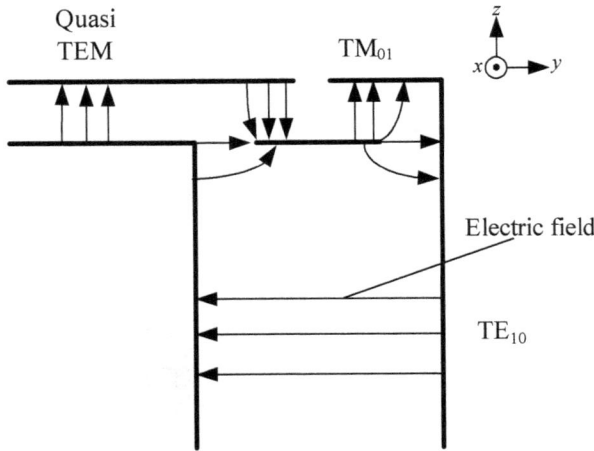

Figure 8. Electric field lines of each mode in yz-plane

Description	Name	Value (mm)	Description	Name	Value (mm)
Broad wall length of waveguide	a	3.1	Narrow wall length of waveguide	b	1.55
Width of patch element	W	2	Length of patch element	L	1.1
Width of microstrip line	W_m	0.3	Overlap length of inserted probe	p	0.34
Width of gap	G	0.1	Relative permittivity	ε_r	2.2
Thickness of substrate	T	0.127	Diameter of via hole	ϕ	0.2
Space between via holes	S	0.5			

Table 2. Parameters of planar proximity coupling transition

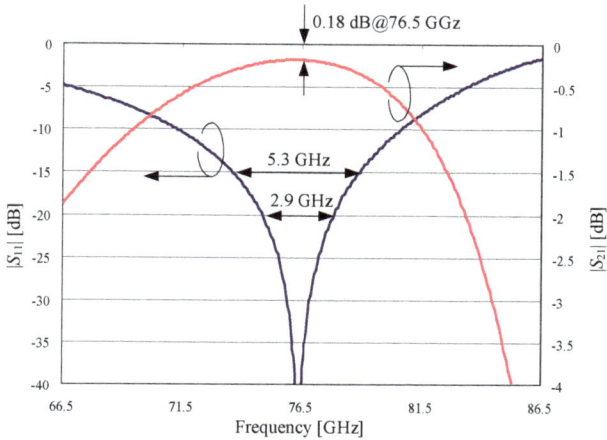

Figure 9. Reflection characteristic $|S_{11}|$ and insertion loss $|S_{21}|$ of planar proximity coupling transition

3.1. Bandwidth of planar proximity coupling transition

The relationships between the parameters and the bandwidth were investigated to specify the optimum parameters for wideband [10]. Figure 10 shows an analytical model that uses a cavity model, which is used for the design of microstrip patch antennas, and the dyadic Green's function of the waveguide. L_e and W_e are the effective length and width of the patch element, including the fringing effect. t and ε_e are the thickness and the effective relative permittivity of the dielectric substrate. The waveguide dimensions are a by b.

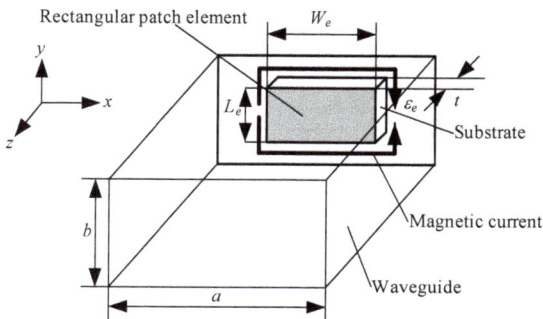

Figure 10. Analytical model using cavity model and dyadic Green's function of waveguide

The quality factor QE of the patch element is given by

$$\frac{1}{Q_E} = \frac{1}{Q_{WG}} + \frac{1}{Q_C} + \frac{1}{Q_D}$$ (1)

Where, Q_{wg}, QC, and QD are quality factors of the power transmitted into the waveguide, conductor loss, and dielectric loss.

The quality factor Q_{WG} is given with the cavity model and the dyadic Green's function of the waveguide [9] as follows:

$$Q_{WG} = \frac{15\omega\pi\varepsilon_0\varepsilon_e L_e ab}{2W_e t} \frac{1}{\sqrt{1 - \left(\frac{\lambda_g}{2a}\right)^2}} \frac{1}{\left(\frac{\sin\left(\frac{W_e\pi}{2a}\right)}{\left(\frac{W_e\pi}{2a}\right)}\right)^2}$$ (2)

where, ω, ε_0 and λ_g are angular frequency, permittivity in free space, and guided wavelength of waveguide. Relationship between the quality factor Q_E and the effective width W_e is solved and the maximum bandwidth is obtained when W_e is expressed in (3) as follows:

$$W_e = \frac{2ac}{\pi}$$ (3)

where, C is a constant value of 1.666. Equation (3) gives the minimum Q factor. Q_{WG} is then given by

$$Q_{WG}\Big|_{W_e = \frac{2aC}{\pi}} = \frac{15\omega\pi^2\varepsilon_0\varepsilon_e L_e b}{4t} \frac{1}{\sqrt{1 - \left(\frac{\lambda_g}{2a}\right)^2}} \frac{C}{(\sin C)^2}$$ (4)

The bandwidth increases with increasing a, while the effective width W_e is set to the optimum width for wideband.

The relationships between the parameters and the bandwidth are summarized in Table 3.

Parameters		Bandwidth
Effective width W_e of patch element	$W_e = \frac{2aC}{\pi}$	Max.
Broad wall length a of waveguide	↗	↗
Narrow wall length b of waveguide	↘	↗
Effective relative permittivity ε_e	↘	↗
Thickness of substrate t	↗	↗
Effective length L_e of patch element	$\lambda_e/2$	—

Table 3. Relations between parameters and bandwidth

4. Broadband microstrip-to-waveguide transition

This section presents broadband techniques of the proximity coupling type transition. Refer to the 79 GHz UWB applications, 4 GHz bandwidth is required [12]. The proximity coupling type transition has bandwidth of 6.9 % (5.29 GHz) for the reflection coefficient below -15 dB [10]. Considering the tolerance for the manufacturing accuracy, much wider bandwidth is required. The boradband transition was presented using waveguide with large broad-wall [13]. Maximum width of the waveguide where higher order mode dose not propagate is applied and the distance from the edge of broad-wall of the waveguide to via holes on the broad-wall side of the waveguide is examined to have optimum length for wideband.

4.1. Transition structure

Configuration of the transition is shown in Figure 11 and Figure 12. A microstrip line, a probe and a waveguide short are located on the upper plane of the dielectric substrate. A rectangular patch element and a surrounding ground are patterned on the lower plane of the dielectric substrate. Via holes surround the aperture of the waveguide on the lower plane of the dielectric substrate to connect the surrounding ground and the waveguide short electrically. The required operation bandwidth is from 77 GHz to 81 GHz.

In terms of the bandwidth, it becomes wider as broad-wall length a of the waveguide increases, and narrow-wall length b of the waveguide decreases [10]. First, standard waveguide WR-10 can be applied for dominant mode propagation at the design frequency (79 GHz). Therefore narrow-wall length b of the waveguide is determined to be 1.27 mm which is the same as the narrow-wall length of WR-10 standard waveguide.

Next, the broad-wall length a of the waveguide is increased as large as possible to 3.1 mm where higher order mode in the waveguide dose not propagate. A rectangular patch element with the width W = 2.26 mm and the length L = 0.98 mm is located on the lower plane of the dielectric substrate at the center of the waveguide. The width W_m of the microstrip line is 0.3 mm corresponding to approximately 56 ohm of characteristic impedance. The probe with the width W_p = 0.35 mm is inserted into the waveguide and overlaps on the rectangular patch element with length ρ = 0.32 mm. The distance from the edge of broad-wall of the waveguide to via holes on the broad-wall side V_y is 0.46 mm. The distance from the edge of narrow-wall of the waveguide to via holes on the narrow- wall side V_x is 0.4 mm. The thickness of dielectric substrate T is 0.127 mm with relative permittivity ε_r is 2.2. The parameters of the transition are presented in Table 4.

4.2. Design and numerical investigation

The transition is investigated numerically by using the electromagnetic simulator based on the finite-element method (Ansys HFSS). In this calculation, loss tangent tanδ = 0.001 and conductivity σ = 5.8 × 10^7 S/m of a copper clad are used as loss factors. The reflection characteristic $|S_{11}|$ and the insertion loss $|S_{21}|$ of the transition with parameters in Table 4 are presented in Figure 13.

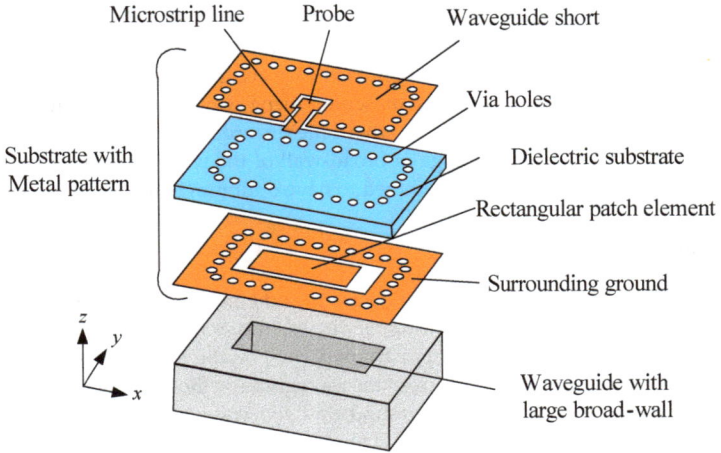

Figure 11. Configuration of broadband transition

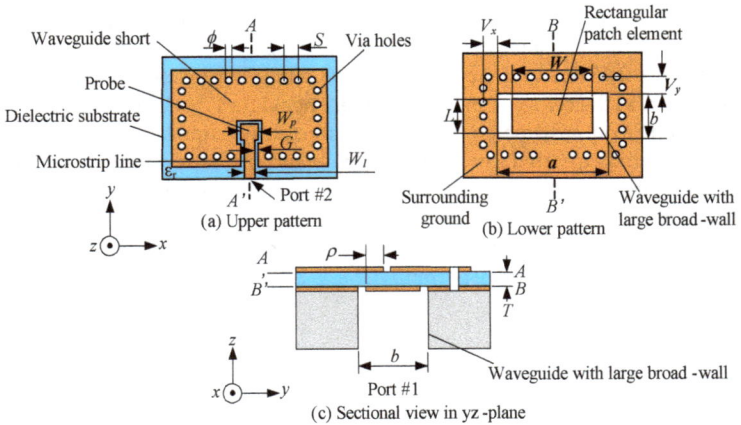

Figure 12. Detailed configuration of broadband transition

Description	Name	Value (mm)	Description	Name	Value (mm)
Broad wall length of waveguide	a	3.1	Narrow wall length of waveguide	b	1.27
Width of patch element	W	2.26	Length of patch element	L	0.98
Width of microstrip line	W_m	0.3	Width of probe	W_p	0.35
Overlap length of inserted probe	ρ	0.32	Width of gap	G	0.1
Thickness of substrate	T	0.127	Relative permittivity	ε_r	2.2
Space between via holes	S	0.4	Diameter of via hole	ϕ	0.2
Distance from broad wall to via hole	V_y	0.46	Distance from narrow wall to via hole	V_x	0.4

Table 4. Parameters of the broadband transition

It can be seen from the simulation results that the bandwidth for the reflection coefficient $|S_{11}|$ below -15 dB is 14.4 GHz, and the insertion loss $|S_{21}|$ is -0.28 dB from 77 GHz to 81 GHz. In this case, two different resonances are observed. Lower resonant frequency is 76.4 GHz and higher resonant frequency is 85.5 GHz.

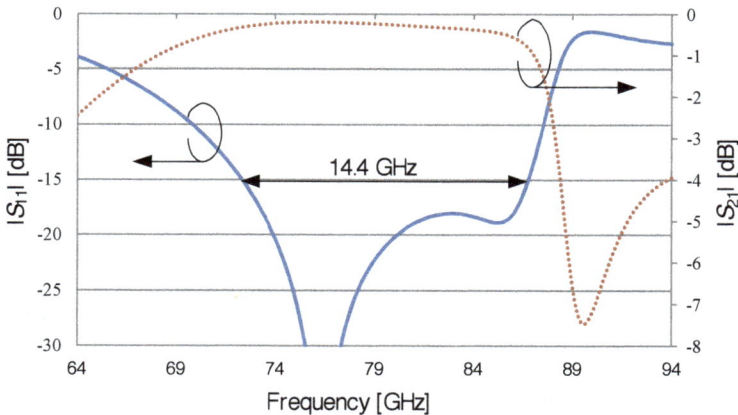

Figure 13. Reflection characteristic $|S_{11}|$ and insertion loss $|S_{21}|$ of broadband transition

Figure 14 shows the calculated electric field distributions in the xy-plane including BB'-line at 76.4 GHz, 79 GHz and 85.5 GHz. It is observed that fundamental mode of TM_{01} is excited at 76.4 GHz and 79 GHz in Figure 14 (a) and (b). On the other hand, a higher order mode is observed at 85.5 GHz as shown in Figure 14 (c).

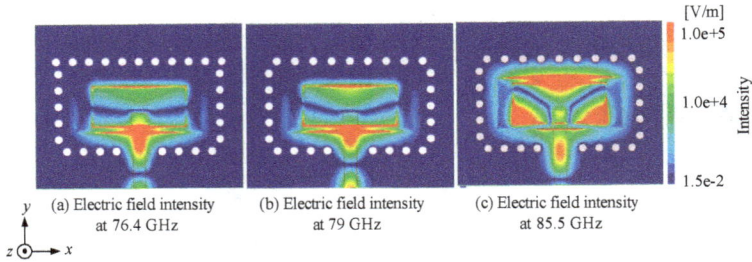

(a) Electric field intensity   (b) Electric field intensity   (c) Electric field intensity
at 76.4 GHz   at 79 GHz   at 85.5 GHz

Figure 14. Electric field intensity distributions in xy-plane.

4.2.1. Lower operation frequency by L

The length L of the rectangular patch element affects to the lower resonant frequency as shown in Figure 15. The lower resonant frequency can be controlled by the length L of the rectangular patch element.

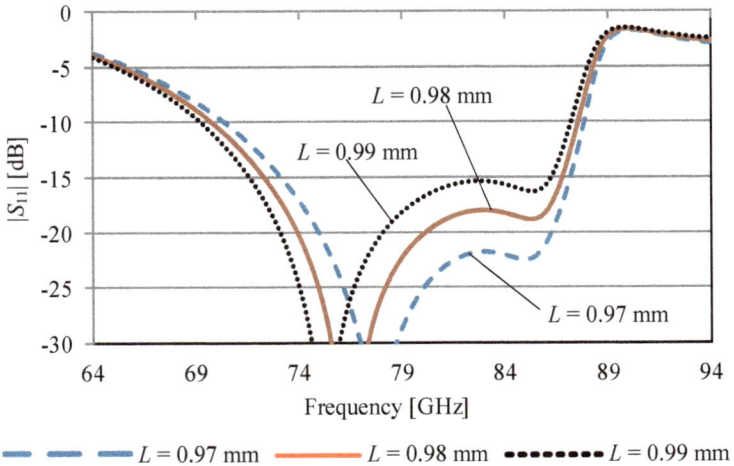

Figure 15. $|S_{11}|$ vs. length L of the patch element (Lower resonant frequency control)

4.2.2. Higher operation frequency by V_y

The distance V_y from the edge of the broad-wall of the waveguide to via holes affects to the higher resonant frequency as shown in Figure 16. The higher resonant frequency can be controlled by the distance V_y from the edge of the broad-wall of the waveguide to via holes.

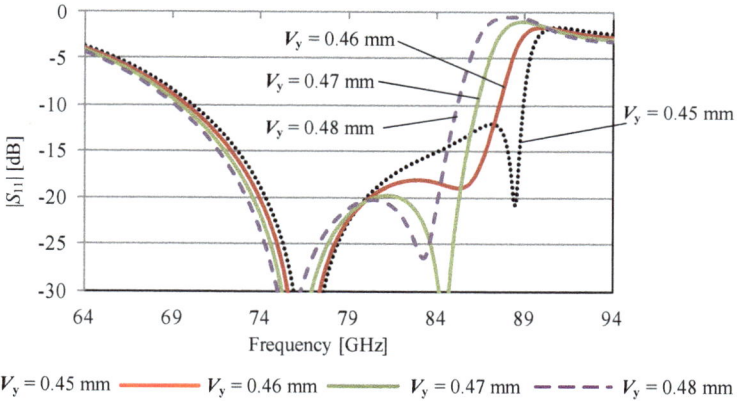

Figure 16. $|S_{11}|$ vs. distance V_y from the edge of the broad-wall of the waveguide to via holes (Higher resonant frequency control)|

4.2.3. Impedance matching by ρ and Wp

The overlap length ρ of the inserted probe is most effective for the impedance matching to the waveguide as shown in Figure 17. Increment of the overlap length ρ of the inserted probe with rectangular patch element causes increase of inductance.

The width W_p of the prove is also effective for the impedance matching to the waveguide as shown in Figure 18. Increment of width W_p of the prove causes decrease of resistance. These Smith charts are observed from the waveguide *port#1* in Figure 12 (c) at a distance of 2.0 mm under the surrounding ground on the lower plane of the substrate.

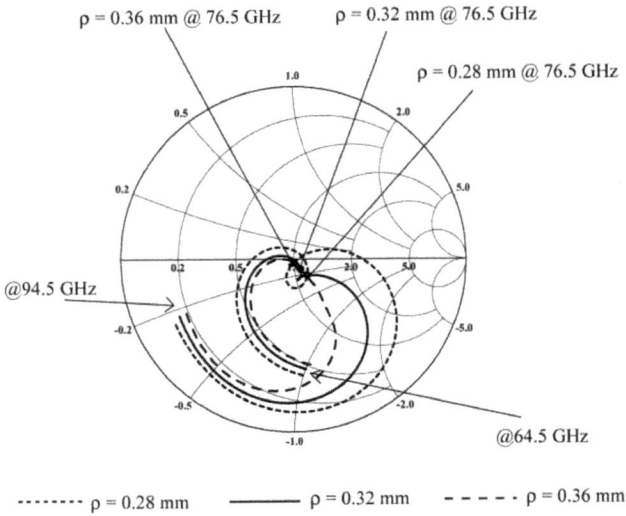

Figure 17. Impedance vs. overlap length ρ of the inserted probe

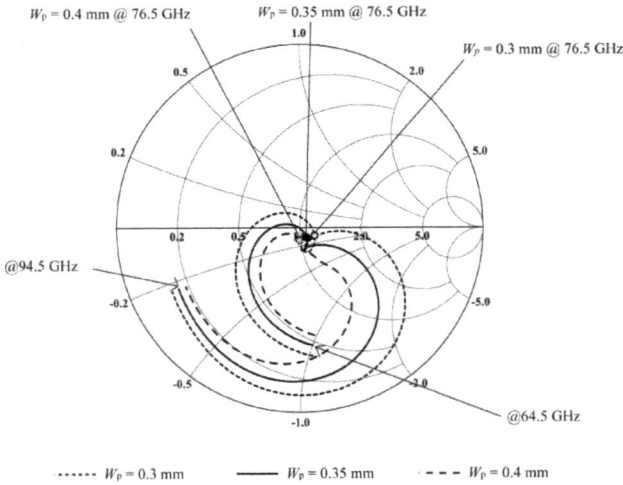

Figure 18. Impedance vs. width W_p of the probe

So, the impedance matching can be controlled by optimizing of the overlap length ρ of inserted probe and the width W_p of the prove.

4.2.4. Wideband impedance matching by ρ and W_p

For the wideband impedance matching, both of the length $ρ$ of the inserted probe and the width W_p of the probe are optimized as shown in Figure 19. It can be seen from the simulation results that both of $ρ$ and W_p affect the wideband impedance matching. In these design, other parameters except $ρ$ and W_p are same as in Table 4.

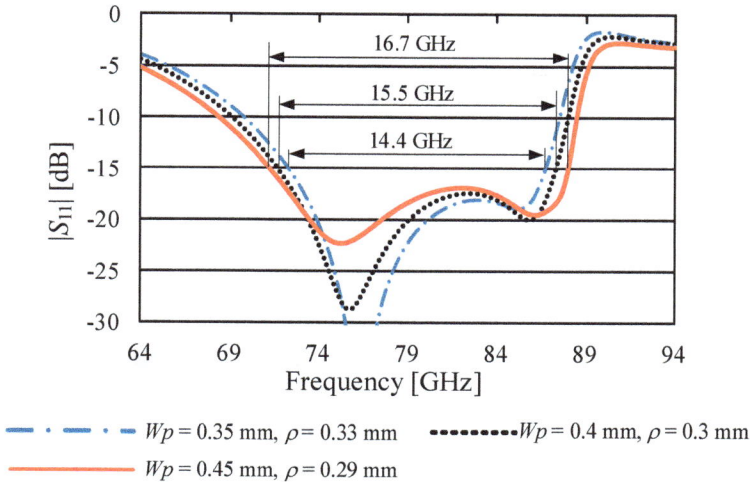

Figure 19. Comparison of three type transitions

4.3. Experiment

Three transitions for the results shown in Figure 19 are fabricated. The photograph of the fabricated transitions are in Figure 20. Figure 20 (a) shows the upper plane of the substrate and is common for each design except the width W_p of the probe and the overlap length $ρ$ of the inserted probe as each design. Figure 20 (b) shows the lower plane of the substrate and is common at each design.

4.3.1. Measured banwidth

Measured the reflection coefficient are shown in Figure 21. Maximum bandwidth for reflection coefficients below -15 dB is 15.1 GHz when W_p is 0.45 mm and $ρ$ is 0.29 mm. In this measurement, the device-under-test (DUT) was composed of a pair of transitions with one microstrip line between them as shown in Figure 21. The measured $|S_{11}|$ and $|S_{21}|$ in Figure 22 were given by taking the transmission coefficient of the DUT, subtracting the loss of the microstrip line, and dividing by two. The loss of the microstrip line was measured as 0.05 dB/mm from 77 GHz

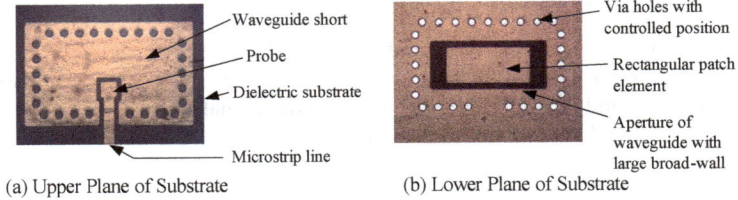

(a) Upper Plane of Substrate   (b) Lower Plane of Substrate

Figure 20. Fabricated transitions

to 81 GHz. A time gate function was used to exclude undesired waves, and high accuracy was achieved in this measurement. The distance between the center of the waveguides was set at 50 mm, which was long enough to distinguish between desired and undesired waves in the time domain.

Figure 21. DUT in measurement

4.3.2. Comparison of measured performance

Figure 23 shows the comparison of three designed transitions. Refer to the bandwidth, measured results are approximately 1.8 GHz decreased compared with the simulation results. For the insertion loss, the measured results are approximately 0.38 dB increased compared with the simulation results.

In these results, design of increased bandwidth causes increase of insertion loss. Therefore, the bandwidth and the insertion loss is in tradeoff relation. So, the transition required each application can be designed by optimizing of each parameters.

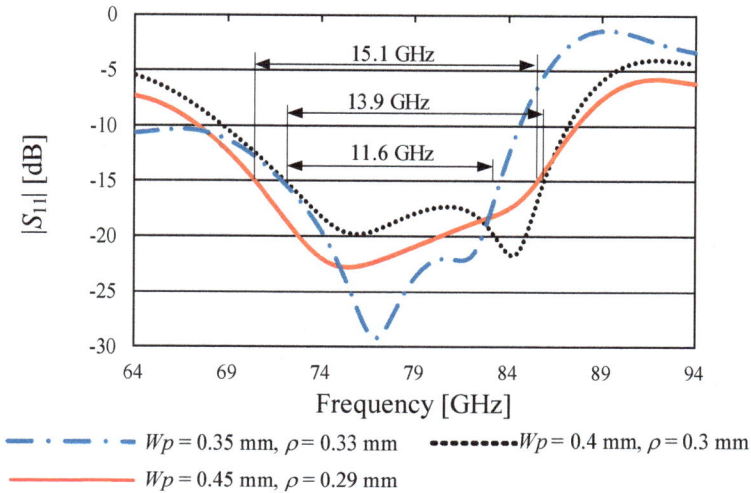

Figure 22. Measured bandwidth of three type transitions

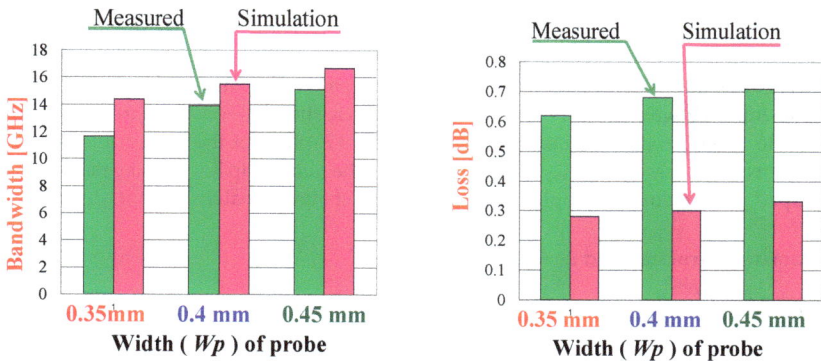

Figure 23. Comparison of measured performance

4.4. Conclusion

Broadband microstrip-to-waveguide transition using waveguide with large broad-wall were developed in millimeter-wave band. By applying large broad-wall, the bandwidth is extended. Moreover, the distance from the edge of the broad-wall of the waveguide to via holes are examined to create double resonances, consequently the bandwidth is extended.

Three types of design are presented. It is confirmed by experiments that the most wideband transition exhibits a bandwidth of 19.1 % (15.1 GHz) for the reflection coefficient below -15 dB and insertion loss of -0.71 dB from 77 GHz to 81 GHz.

5. Narrow-wall-connected microstrip-to-waveguide transition

Narrow-wall-connected microstrip-to-waveguide transition using V-shaped patch element in millimeter-wave band was proposed [14]. Since the microstrip line on the narrow-wall is perpendicular to the E-plane of the waveguide, the waveguide field does not couple directly to the microstrip line. The current on the V-shaped patch element flows along the inclined edges, then current on the V-shaped patch element couples to the microstrip line efficiently. Three types of the transitions are investigated. S-parameters of the reflection S_{11} and the transmission S_{21} are calculated by using an electromagnetic simulator based on the finite element method (Ansys HFSS). The numerical investigations of these transitions show some relations between the bandwidth and the insertion loss. It is confirmed that the improved transition exhibits an insertion loss of 0.6 dB from 76 to 77 GHz, and a bandwidth of 4.1 % (3.15 GHz) for the reflection coefficient below -15 dB.

5.1. Background

In some applications, narrow-wall-connected micro-strip-to-waveguide transition is required. Refer to the former developed proximity coupling type transition [9],[10], the microstrip line is located on the waveguide broad-wall and the microstrip line probe is parallel to E-plane of the waveguide, therefore, current on the rectangular patch element couples to the microstrip line efficiently. However, on the occasion of the microstrip line on the narrow-wall of the waveguide, the microstrip line probe is orthogonal to E-plane of the waveguide. Therefore, they do not couple essentially. To couple currents on the microstrip line and the patch element, a V-shaped patch element is applied instead of the rectangular patch element.

5.2. Transition structure and design

5.2.1. Transition structure

Configuration of the transition is shown in Figure 24 and Figure 25. The microstrip line and the waveguide short are located on the upper plane of the dielectric substrate. The V-shaped patch element and the surrounding ground are patterned on the lower plane of the dielectric substrate. Via holes are surrounding the aperture of the waveguide on the lower plane of the substrate to connect the surrounding ground and the waveguide short electrically. The design frequency range is from 76 GHz to 77 GHz. The V-shaped patch element is designed as follows. Refer to the conventional proximity coupling type [9],[10], the current on rectangular patch element has only y-component which is parallel to E-plane of the waveguide shown in Figure 26 (a). In this case, the current on the rectangular patch element is parallel to the microstrip line, therefore, the current on the rectangular patch element couples to the microstrip line. On the other hand, current on the patch element which is excited by electromagnetic field in the waveguide must have both of x-component and y-component in order to couple to the microstrip line on the waveguide narrow-wall. The current on the V-shaped patch element is divided to two directions along the side edge as shown in Figure 26 (b). Consequently, the current on the V-shaped patch element creates parallel component with the microstrip line, and effective coupling is achieved with the microstrip line.

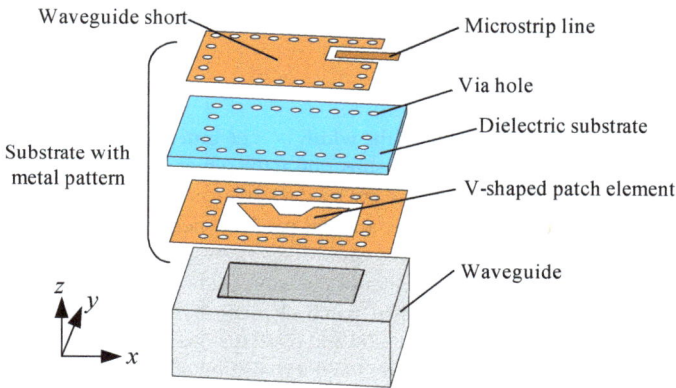

Figure 24. Configuration of transiton

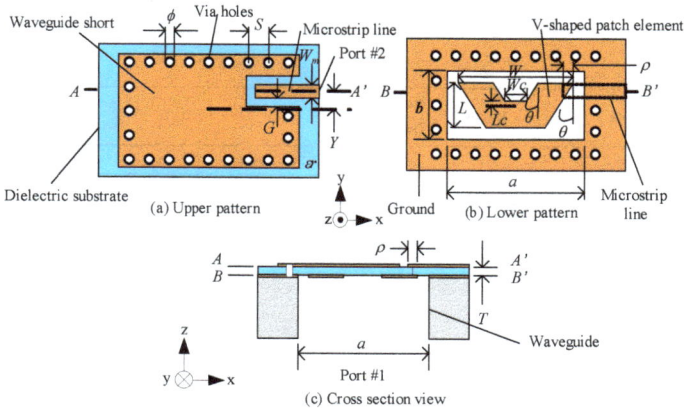

Figure 25. Detailed configurations of transition

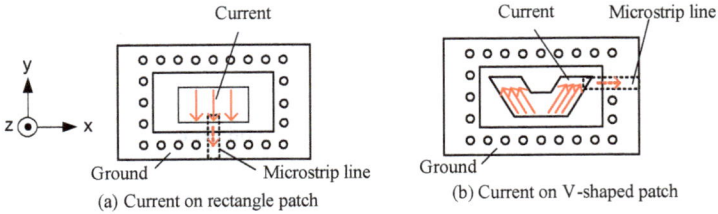

Figure 26. Current distributions on patch element

5.2.2. Transition design

First, the rectangular patch element with the width $W = 2.6$ mm and the length $L = 1.02$ mm is located on the lower plane of the dielectric substrate at the center of the waveguide. Then, both sides of the rectangular patch element are cut by the patch-cut-angle $\theta = 30$ degrees and also middle of upper horizontal edge of the rectangular patch element is cut as shown in Figure 25 (b). The microstrip line is located on the waveguide narrow-wall as shown in Figure 25 (a) with the shift length $Y = 0.34$ mm from the center of the waveguide. The microstrip line is located just above the side edge of the V-shaped patch element as shown in Figure 25 (b). The microstrip line is inserted into the waveguide and overlaps over the V-shaped patch element with the length $\rho = 0.23$ mm. The parameters of the transition are presented in Table 5.

Mode conversion from the waveguide to the microstrip line is achieved by the resonance of the V-shaped patch element. The dominant TE_{10} mode of the waveguide is converted to the quasi-TEM mode of the microstrip line. Figure 27 shows the calculated electric field intensity distribution in the xy-plane including BB'-line. The electric field intensity E includes x, y and z components of the electric field. The V-shaped patch element is resonated in two directions, by the resonance of current distribution along the both side edges of the V-shaped patch element.

Description	Name	Value (mm)	Description	Name	Value (mm)
Width of patch element	W	2.6	Width of gap	G	0.1
Length of patch element	L	1.02	Thickness of substrate	T	0.127
Patch cut angle	θ	30 deg.	Relative permittivity	ε_r	2.2
Overlap length of inserted microstrip line	ρ	0.23	Broad wall length of waveguide	a	3.1
Width of cut patch element	W_c	0.46	Narrow wall length of waveguide	b	1.55
Length of cut patch element	L_c	0.1	Diameter of via hole	ϕ	0.2
Width of microstrip line	W_m	0.3	Space between via holes	S	0.5
Shift length of microstrip line from center of waveguide	Y	0.34			

Table 5. Parameters of transition

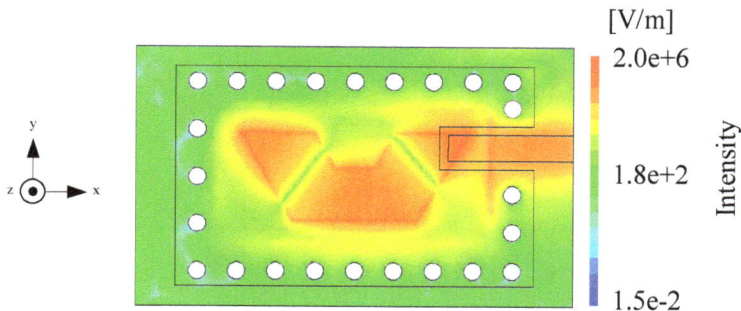

Figure 27. Electric field intensity distribution in *xy*-plane

5.3. Numerical investigation

5.3.1. Operating frequency by L

The reflection characteristic of the V-shaped patch element with the length L = 1.02 mm is presented in Figure 28. It can be seen from the simulation results that the bandwidth for $|S_{11}|$ below -15 dB is 2 GHz, and the insertion loss $|S_{21}|$ is 0.32 dB over the frequency range from 76 GHz to 77 GHz. The length L of the V-shaped patch element affects to the resonant frequency as shown in Figure 28. Increment of the length L of the V-shaped patch element causes the lower resonant frequency. So, operating frequency of this transition can be controlled by the length L of the V-shaped patch element.

5.3.2. Impedance matching by ρ

The overlap length ρ of the inserted microstrip line affects to the impedance as shown in Figure 29. Increment of the overlap length ρ causes increases of capacitive reactance at the desired frequency, and decrement of the overlap length ρ causes increases of inductive reactance. This Smith chart is observed from the waveguide *port#1* in Figure 25 (c) at a distance of 1.5 mm under the surrounding ground on the lower plane of the substrate. So, impedance matching can be controlled by adjusting the overlap length ρ to cancel reactive component.

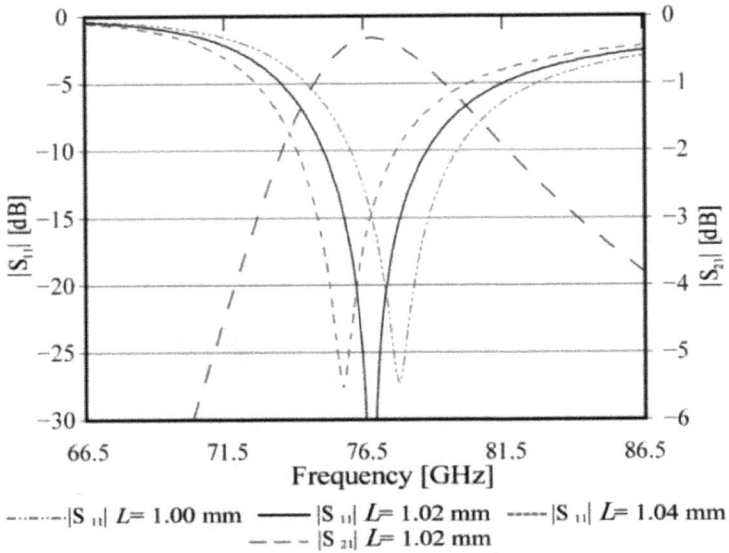

Figure 28. $|S_{11}|$ vs. length of V-shaped patch element L and transition characteristic $|S_{21}|$

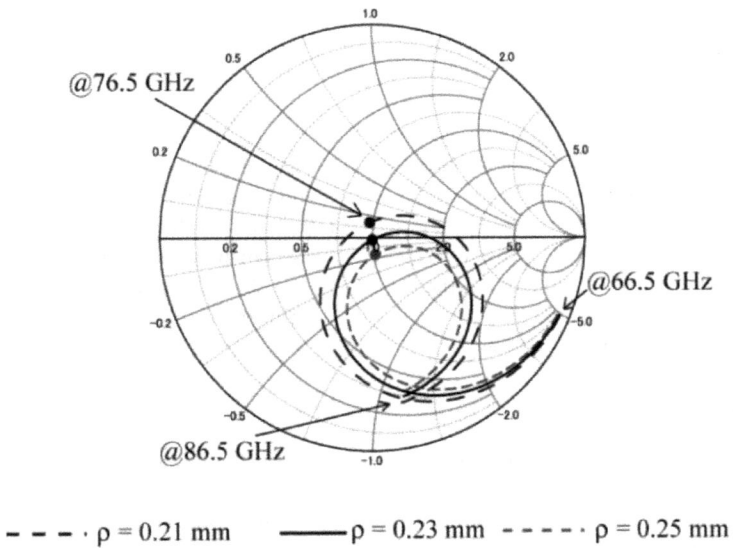

Figure 29. Relation between Impedance and Length of Inserted Microstrip Line ρ from 66.5 GHz to 76.5 GHz

5.3.3. Bandwidth by θ

The patch cut angle θ affects to the bandwidth for the reflection coefficient below -15 dB as shown in Figure 30. The transition characteristic change is investigated by change of the patch cut angle θ from 5 degrees to 50 degrees. Some parameters of W, L, ρ, W_c, L_c and Y are optimized at each patch cut angle θ. Least insertion loss $|S_{21}|$ is obtained at 30 degrees of the patch cut angle θ. In this case, the bandwidth for $|S_{11}|$ below -15 dB is 2 GHz and the insertion loss $|S_{21}|$ is -0.32 dB.

On the occasion of the patch cut angle θ = 10 degrees, the bandwidth is extended to 3.5 GHz, but the insertion loss $|S_{21}|$ is increased to -0.41 dB compared with the patch cut angle θ = 30 deg.

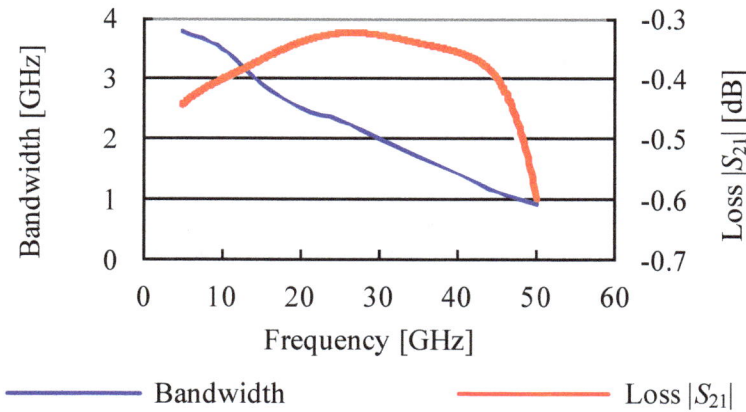

Figure 30. Bandwidth and transition characteristic $|S_{21}|$ related by patch cut angle θ

In the design with the small patch cut angle θ, current on the V-shaped patch element has small x-component, therefore the loss increases. The cause is that the magnetic field is small. This magnetic field is excited by the current of x-component on the V-shaped patch element and the magnetic field surrounds the microstrip line. Small magnetic field causes weak coupling between the V-shaped patch element and the microstrip line. Figure 4.31 shows calculated magnetic field distribution in the yz-plane at the x position of 2.48 mm. The magnetic field intensity H include x, y and z components of the magnetic field. On the other hand, in the design with large patch cut angle θ, the area of the V-shaped patch element decreases and the quality factor Q of the patch element increases, then bandwidth decreases.

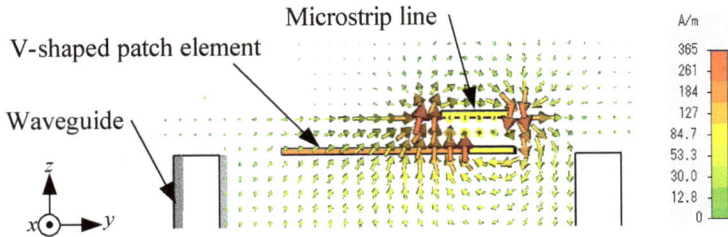

Figure 31. Magnetic field distribution in *yz*-plane at *x* = 2.48 mm

5.4. Design variety of transition

5.4.1. Low loss design

As shown in Figure 30, least insertion loss is obtained at the patch cut angle θ = 30 degrees. Configuration is shown in Figure 25 and design parameters are just the same as shown in Table 5. The bandwidth for $|S_{11}|$ below -15 dB is 2 GHz and the insertion loss $|S_{21}|$ is -0.32 dB.

5.4.2. Wideband design

In this design, the patch cut angle θ of 10 degrees is applied. Configuration is shown in Figure 30 and some parameters must be changed as Table 6, but other parameters are the same as Table 5. Characteristic of this transition is shown in Figure 4.13. The bandwidth for $|S_{11}|$ below -15 dB is 3.5 GHz and the insertion loss $|S_{21}|$ is -0.41 dB.

Description	Name	Value (mm)
Length of patch element	L	1.11
Patch cut angle	θ	10 deg.
Overlap length of inserted microstrip line	ρ	0.28
Width of cut patch element	W_c	0.5
Shift length of microstrip line from center of waveguide	Y	0.385

Table 6. Parameters of wideband design

5.4.3. Wideband and low loss design

The transition with the wideband design described before is modified. At the *y*-directional position C_y = 0.32 mm from the top of the V-shaped patch element, the V-shape patch element is cut to *x*-direction with the length C_x of 0.1mm as shown in Figure 32. The basic configuration is as shown in Figure 25 but the V-shaped patch element is modified as shown in Figure 32. Some parameters are optimized and changed as Table 7, although other parameters are the same as Table 5.

To get the wideband of the transition, the patch cut angle θ is kept to 10 degrees. To achieve strong coupling, this modification of V-shaped patch element is effective. Due to this structural modification, horizontal component of electric current on the patch element increases. Consequently strong coupling to the microstrip line is achieved. The bandwidth for $|S_{11}|$ below -15 dB is 3.1 GHz and the insertion loss $|S_{21}|$ is 0.34 dB. The bandwidth is 0.4 GHz narrow than the wideband design and the loss is approximately equal as the low loss design.

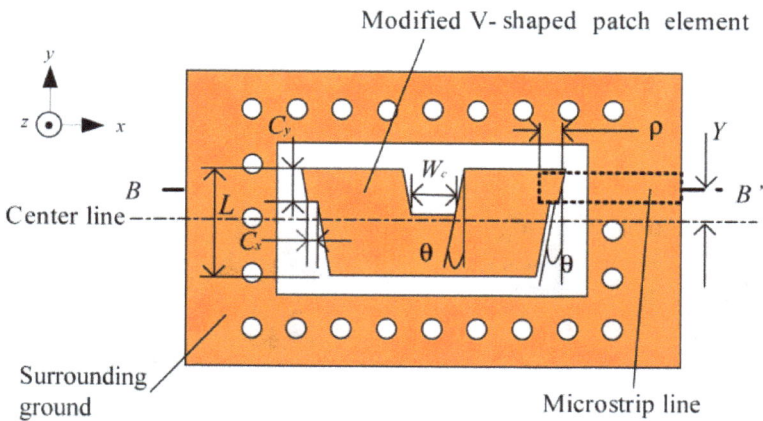

Figure 32. Lower pattern of the transition with modified V-shaped patch element

Description	Name	Value (mm)
Cut length in x-direction	C_x	0.1
Cut length in y-direction	C_y	0.32
Length of patch element	L	1.08
Patch cut angle	θ	10 deg.
Overlap length of inserted microstrip line	ρ	0.27
Width of cut patch element	W_c	0.45
Shift length of microstrip line from center of waveguide	Y	0.37

Table 7. Parameters of wideband and low loss design

5.5. Measured performance of three transitons

The photograph of the fabricated transitions are shown in Figure 33. Figure 33 (a) shows the upper plane of the substrate and is common for each design except y-position of the microstrip line(Y). As described before, the shift length Y of the microstrip line from the center of the waveguide is changed at each design. Figure 34 shows the comparison of three designed transitions. Refer to the bandwidth, measured results agree with the simulation results. For the insertion loss, the measured results are approximately 0.3 dB increased compared with the simulation results.

Three types of design are presented and as a compatible design of low loss and wideband, a new modified V-shape patch element is proposed. It is confirmed by experiments that the improved transition exhibits an insertion loss of 0.6 dB from 76 to 77 GHz, and a bandwidth of 4.1 % (3.15 GHz) for the reflection coefficient below -15 dB.

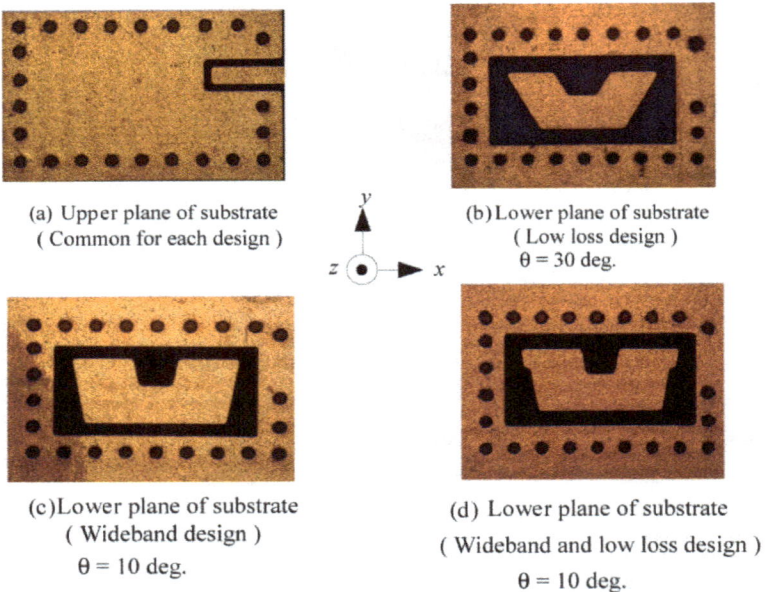

(a) Upper plane of substrate
(Common for each design)

(b) Lower plane of substrate
(Low loss design)
θ = 30 deg.

(c) Lower plane of substrate
(Wideband design)
θ = 10 deg.

(d) Lower plane of substrate
(Wideband and low loss design)
θ = 10 deg.

Figure 33. Fabricated transitons

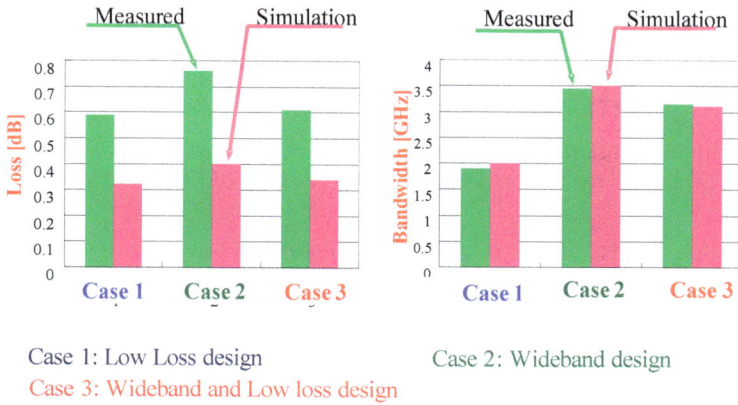

Figure 34. Comparison of each design

Author details

Kazuyuki Seo

Address all correspondence to: kazuyuki.seo@pillar.co.jp; k_seo1951@pure.ocn.ne.jp

Process Development Dept., Nippon Pillar Packing Co., Ltd. Sanda City, Japan

References

[1] Russel, M. E, Grain, A, Curran, A, Campbell, R. A, Drubin, C. A, & Miccioli, W. F. Millimeter-wave radar sensor for automotive intelligent cruise control(ICC)," IEEE Trans. Microw. Thory Tech., Dec. (1997). , 45(12), 2444-2453.

[2] Russel, M. E, Drubin, C. A, Marinilli, A. S, & Woodington, W. G. Integrated automotive sensors, "IEEE Trans. Microw. Theory Tech., Mar. (2002). , 50(3), 674-677.

[3] Yano, H. Y, Abdelmonem, A, Liang, J. F, & Zaki, K. A. Analysis and design of microstrip to waveguide transition, "IEEE Trans. Microw. Theory Tech., Dec. (1994). , 42(12), 2371-2379.

[4] Kaneda, N, Qian, Y, & Itoh, T. A broadband microstrip-to-waveguide transition using quasi-Yagi antenna, "IEEE Trans. Microw. Thory Tech., Dec. (1999). MWSYM. 1999.780218, 47(12), 2562-2567.

[5] Deslandes, D, & Wu, K. Integrated microstrip and rectangular waveguide in planar form, "IEEE Microw. Wireless Compon. Lett., Feb. (2001). , 11(2), 68-70.

[6] Ho, T. Q, & Shih, Y. C. Spectral-domain analysis of E-Plane waveguide to microstrip transitions, "IEEE Trans. Microw. Theory Tech., Feb. (1989). , 37(2), 388-392.

[7] Leong, Y, & Weinreb, S. Full band waveguide to microstrip probe transitions," IEEE MTT-S Int. Microw. Symp. Dig., Anaheim, CA, May (1999). MWSYM.1999.780219, 4, 1435-1438.

[8] Grabherr, W, Hudder, B, & Menzel, W. Microstrip to waveguide transition compatible with mm-wave integrated circuits, "IEEE Trans. Microw. Theory Tech., Sep. (1994). , 42(9), 1842-1843.

[9] Iizuka, H, Watanabe, T, Sato, K, & Nisikawa, K. Millimeter-wave microstrip line to waveguide transition fabricated on a single layer dielectric substrate," IEICE Trans. Commun., Jun. (2002). , E85-B(6), 1169-1177.

[10] Iizuka, H, Sakakibara, K, & Kikuma, N. Millimeter-Wave Transition From Waveguide to Two Microstrip Lines Using Rectangular Patch Element," IEEE Trans. Microw. Theory Tech., May. (2007). TMTT.2007.895139, 55(5), 899-905.

[11] Bahl, I. J, Trivedi, D. K, & , A. s Guide to Microstrip Line,"Microwaves, May (1977). , 174-182.

[12] Strohm, K. M, Bloecher, H. L, Schneider, R, & Wenger, J. Development of Future Short Range Radar Technology," Radar Conference, 2005, EURAD 2005, European, Oct. (2005). , 165-168.

[13] Seo, K, Sakakibara, K, & Kikuma, N. Microstrip-to-waveguide Transition using Waveguide with Large Broad-wall in Millimeter-wave Band," IEEE International Conference on Ultra-Wideband, ICUWB2010, Sep. (2010). ICUWB.2010.5614169, 1, 209-212.

[14] Seo, K, Sakakibara, K, & Kikuma, N. Narrow-Wall-Connected Microstrip-to-Waveguide Transition Using V-Shaped Patch Element in Millimeter-Wave Band," IEICE Trans. Commun., Oct. (2010). , E93-B(10), 2523-2530.

Chapter 13

Wearable Antennas for Medical Applications

Albert Sabban

Additional information is available at the end of the chapter

1. Introduction

Microstrip antennas are widely employed in communication system and seekers. Microstrip antennas posse's attractive features such as low profile, flexible, light weight, small volume and low production cost. In addition, the benefit of a compact low cost feed network is attained by integrating the RF frontend with the radiating elements on the same substrate. Microstrip antennas are widely presented in books and papers in the last decade [1-7]. However, the effect of human body on the electrical performance of wearable antennas at 434 MHz is not presented [8-13]. RF transmission properties of human tissues have been investigated in several articles [8-9]. Several wearable antennas have been presented in the last decade [10-14]. A review of wearable and body mounted antennas designed and developed for various applications at different frequency bands over the last decade can be found in [10]. In [11] meander wearable antennas in close proximity of a human body are presented in the frequency range between 800 MHz and 2700 MHz. In [12] a textile antenna performance in the vicinity of the human body is presented at 2.4 GHz. In [13] the effect of human body on wearable 100 MHz portable radio antennas is studied. In [13] the authors concluded that wearable antennas need to be shorter by 15% to 25% from the antenna length in free-space. Measurement of the antenna gain in [13] shows that a wide dipole (116 x 10 cm) has -13dBi gain. The antennas presented in [10-13] were developed mostly for cellular applications. Requirements and the frequency range for medical applications are different from those for cellular applications

In this chapter, a new class of wideband compact wearable microstrip antennas for medical applications is presented. Numerical results with and without the presence of the human body are discussed. The antennas VSWR is better than 2:1at 434 MHz + 5%. The antenna beam width is around 100º. The antennas gain is around 0 to 4 dBi. The antenna resonant frequency is shifted by 5% if the air spacing between the antenna and the human body is increased from 0 mm to 5 mm.

2. Dually polarized 434 MHz printed antenna

A new compact microstrip loaded dipole antennas has been designed to provide horizontal polarization. The antenna dimensions have been optimized to operate on the human body by employing Agilent Advanced Design System (ADS) software [16]. The antenna consists of two layers. The first layer consists of RO3035 0.8 mm dielectric substrate. The second layer consists of RT-Duroid 5880 0.8 mm dielectric substrate. The substrate thickness determines the antenna bandwidth. However, thinner antennas are flexible. Thicker antennas have been designed with wider bandwidth. The printed slot antenna provides a vertical polarization. In several medical systems the required polarization may be vertical or horizontal. The proposed antenna is dually polarized. The printed dipole and the slot antenna provide dual orthogonal polarizations. The dimensions of the dual polarized antenna presented in Figure 1are 26 x 6 x 0.16 cm. The antenna may be used as a wearable antenna on a human body. The antenna may be attached to the patient shirt, patient stomach, or in the back zone. The antenna has been analyzed by using Agilent ADS software. There is a good agreement between measured and computed results. The antenna bandwidth is around 10% for VSWR better than 2:1. The antenna beam width is around 100°. The antenna gain is around 2 dBi. The computed S_{11} and S_{22} parameters are presented in Figure 2. Figure 3 presents the antenna measured S_{11} parameters. The computed radiation patterns are shown in Figure 4. The co-polar radiation pattern belongs to the yz plane. The cross-polar radiation pattern belongs to the xz plane. The antenna cross polarized field strength may be adjusted by varying the slot feed location. The dimensions of the folded dually polarized antenna presented in Figure 5 are 7 x 5 x 0.16 cm. Figure 6 presents the antenna computed S_{11} and S_{22} parameters. The computed radiation patterns of the folded dipole are shown in Figure 7. The antennas radiation characteristics on human body have been measured by using a phantom. The phantom electrical characteristics represent the human body electrical characteristics.

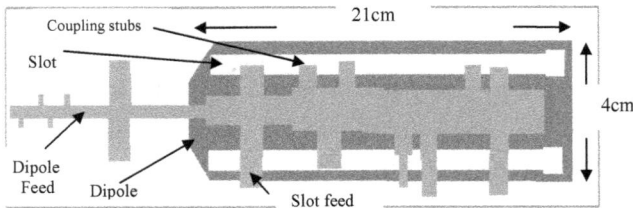

Figure 1. Printed dually polarized antenna, 26 x 6 x 0.16 cm.

S11, S22

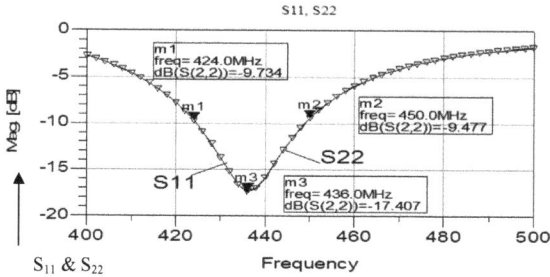

S_{11} & S_{22}

Figure 2. Computed S_{11} and S_{22} results

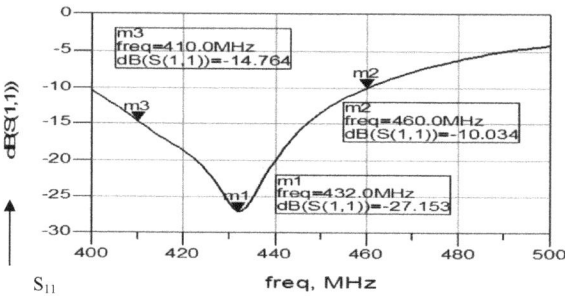

S_{11}

Figure 3. Measured S_{11} on human body

The phantom has a cylindrical shape with a 40cm diameter and a length of 1.5m. The phantom contains a mix of 55% water 44% sugar and 1% salt. The antenna under test was placed on the phantom during the measurements of the antennas radiation characteristics. S_{11} and S_{12} parameters were measured directly on human body by using a network analyzer. The measured results were compared to a known reference antenna.

3. New loop antenna with ground plane

A new loop antenna with ground plane has been designed on Kapton substrates with thickness of 0.25mm and 0.4mm. The antenna without ground plane is shown in Figure 8. The loop antenna VSWR without the tuning capacitor was 4:1. This loop antenna may be tuned by adding a capacitor or varactor as shown in Figure 8. Tuning the antenna allow us to work in a wider bandwidth. Figure 9 presents the loop antenna computed S_{11}on human body. There is good agreement between measured and computed S_{11}. The computed radiation pattern is shown in Fig 10.

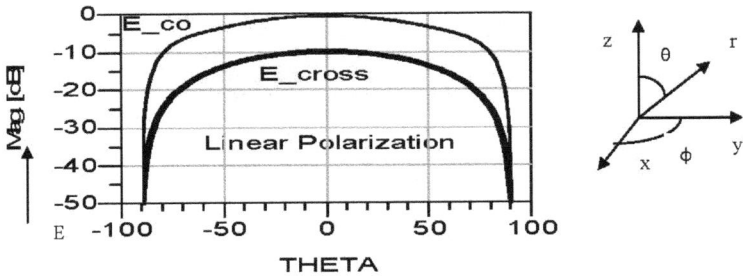

Figure 4. Antenna Radiation patterns

Table I compares the electrical performance of a loop antenna with ground plane with a loop antenna without ground plane. Tuning the antenna allow us to work in a wider bandwidth. Figure 9 presents the loop antenna computed S11on human body.

There is good agreement between measured and computed S_{11}. The computed radiation pattern is shown in Fig 10. Table I compares the electrical performance of a loop antenna with ground plane with a loop antenna without ground plane. There is a good agreement between measured and computed results of antenna parameters on human body. The results presented in Table I indicates that the loop antenna with ground plane is matched to the human body environment, without the tuning capacitor, better than the loop antenna without ground plane. The computed 3D radiation pattern is shown in Fig 11.

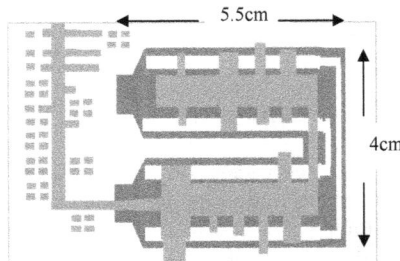

Figure 5. Folded dual polarized antenna, 7x5x0.16cm.

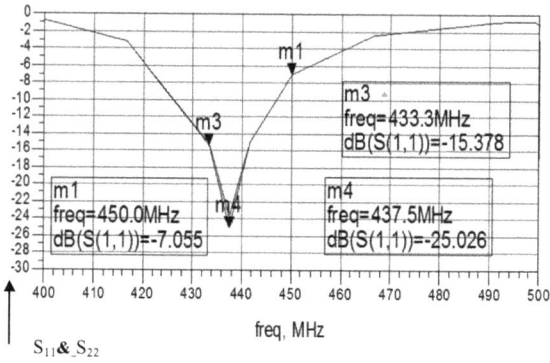

Figure 6. Folded antenna Computed S_{11} and S_{22} results

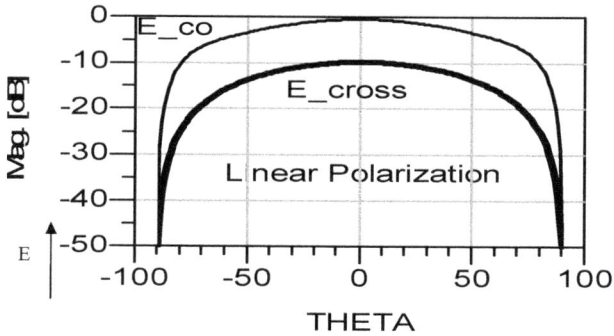

Figure 7. Folded antenna Radiation patterns

Antenna with no tuning capacitor	Beam width 3dB	Gain dBi	VSWR
Loop no GND	100°	0	4:1
Loop with GND	100°	0	2:1

Table 1. Comparison of Loop Antennas

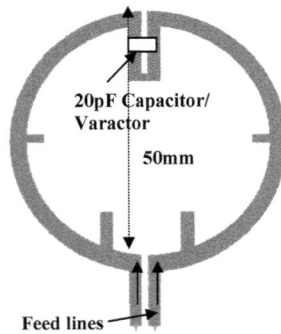

Figure 8. Tunable loop antenna without ground plane

The computed 3D radiation pattern and the coordinate used in this chapter are shown in Fig 11. Computed S_{11} of the Loop Antenna with a tuning capacitor is given in Figure 12.

4. Antenna S11 variation as function of distance from body

The Antennas input impedance variation as function of distance from the body had been computed by employing ADS software. The analyzed structure is presented in Figure 14. The patient body thickness was varied from 15mm to 300mm. The dielectric constant of the body was varied from 40 to 50. The antenna was placed inside a belt with thickness between 2 to 4mm with dielectric constant from 2 to 4. The air layer between the belt and the patient shirt may vary from 0mm to 8mm. The shirt thickness was varied from 0.5mm to 1mm. The dielectric constant of the shirt was varied from 2 to 4. Properties of human body tissues are listed in Table II see [8]. These properties were employed in the antenna design. Figure 15 presents S_{11} results (of the antenna shown in Figure 1) for different belt thickness, shirt thickness and air spacing between the antennas and human body.

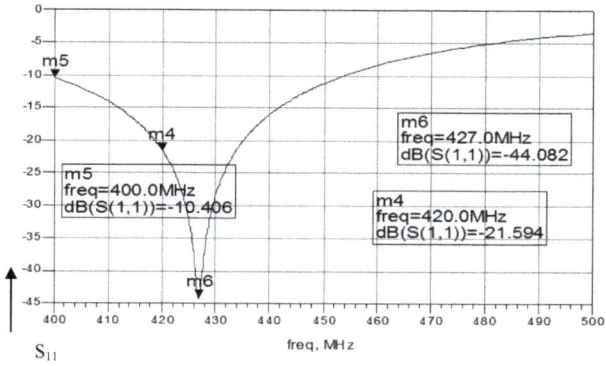

Figure 9. Computed S_{11} of new Loop Antenna

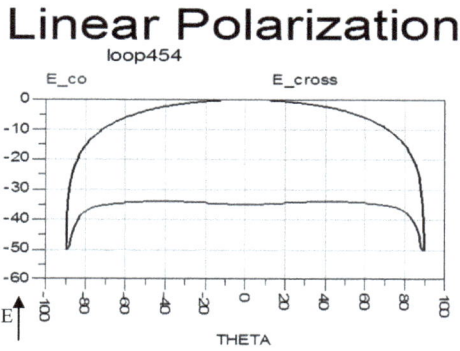

Figure 10. New Loop Antenna Radiation pattern on human body

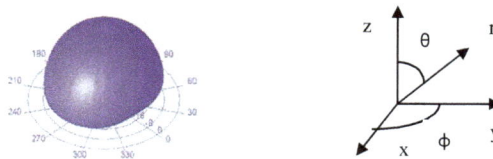

Figure 11. New Loop Antenna 3D Radiation pattern

Tissue	Property	434 MHz	600 MHz
Skin	σ	0.57	0.6
	ε	41.6	40.43
Stomach	σ	0.67	0.73
	ε	42.9	41.41
Colon, Muscle	σ	0.98	1.06
	ε	63.6	61.9
Lung	σ	0.27	0.27
	ε	38.4	38.4

Table 2. Properties of human body tissues

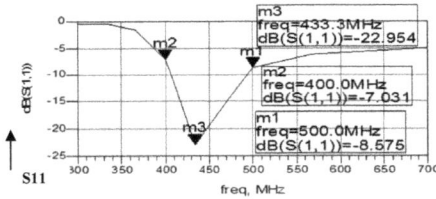

Figure 12. Computed S_{11} of Loop Antenna, without ground plane, with a tuning capacitor

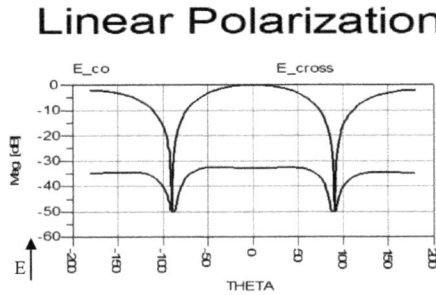

Figure 13. Radiation pattern of Loop Antenna without ground on human body

One may conclude from results shown in Figure 15 that the antenna has V.S.W.R better than 2.5:1 for air spacing up to 8mm between the antennas and patient body. For frequencies ranging from 415MHz to 445MHz the antenna has V.S.W.R better than 2:1 when there is no air spacing between the antenna and the patient body. Results shown in Figure 16 indicates that the folded antenna (the antenna shown in Figure 5) has V.S.W.R better than 2.0:1 for air spacing up to 5mm between the antennas and patient body. Figure 16 presents S_{11} results of the folded antenna results for different position relative to the human body. Explanation of Figure 16 is given in Table 3. If the air spacing between the sensors and the human body is increased from 0mm to 5mm the antenna resonant frequency is shifted by 5%. The loop antenna with ground plane has V.S.W.R better than 2.0:1 for air spacing up to 5mm between the antennas and patient body.

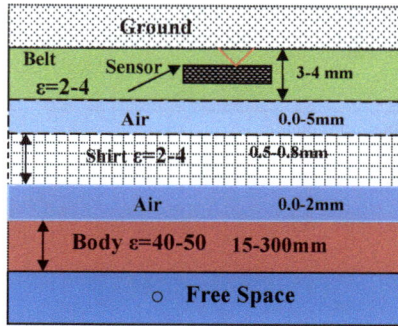

Figure 14. Analyzed structure for Impedance calculations

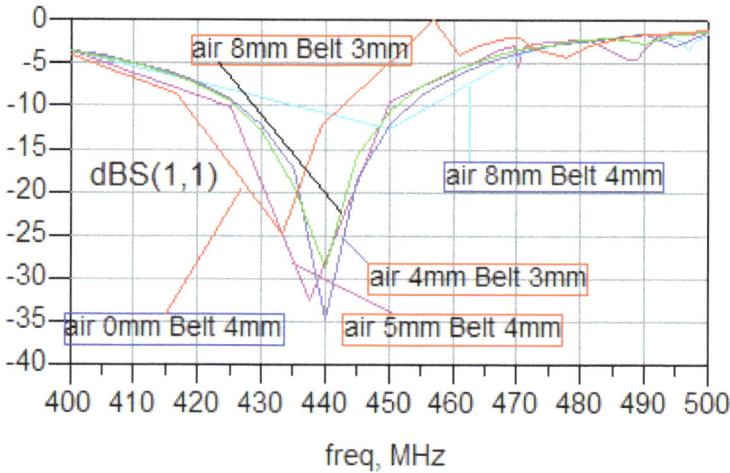

Figure 15. S_{11} results for different antenna positions relative to the human body

If the air spacing between the sensors and the human body is increased from 0mm to 5mm the computed antenna resonant frequency is shifted by 2%. However, if the air spacing between the sensors and the human body is increased up to 5mm the measured loop antenna resonant frequency is shifted by 5%. Explanation of Figure 17 is given in Table 4.

Plot colure	Sensor position
Red	Shirt thickness 0.5mm
Blue	Shirt thickness 1mm
Pink	Air spacing 2mm
Green	Air spacing 4mm
Sky	Air spacing 1mm
Purple	Air spacing 5mm

Table 3. Explanation of Figure 16

5. Wearable antenna

An application of the proposed antenna is shown in Figure 18. Three to four folded dipole or loop antennas may be assembled in a belt and attached to the patient stomach. The cable from each antenna is connected to a recorder. The received signal is routed to a switching matrix. The signal with the highest level is selected during the medical test. The antennas receive a signal that is transmitted from various positions in the human body. Folded antennas may be also attached on the patient back in order to improve the level of the received signal from different locations in the human body. Figure 19 and Figure 20 show various antenna locations on the back and front of the human body for different medical applications.

Plot colure	Sensor position
Red	Body 15mm air spacing 0mm
Blue	Air spacing 5mm Body 15mm
Pink	Body 40mm air spacing 0mm
Green	Body 30mm air spacing 0mm
Sky	Body 15mm Air spacing 2mm
Purple	Body 15mm Air spacing 4mm

Table 4. Explanation of Figure 17

transmitting and receiving antennas is less than $2D^2/\lambda$. D is the largest dimension of the radiator. In these applications the amplitude of the electromagnetic field close to the antenna may be quite powerful, but because of rapid fall-off with distance, the antenna do not radiate energy to infinite distances, but instead the radiated power remain trapped in the region near to the antenna. Thus, the near-fields only transfer energy to close distances from the receivers. The receiving and transmitting antennas are magnetically coupled. Change in current flow through one wire induces a voltage across the ends of the other wire through electromagnetic induction. The amount of inductive coupling between two conductors is measured by their mutual inductance. In these applications we have to refer to the near field and not to the far field radiation.

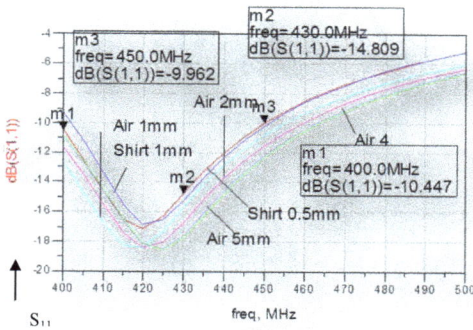

Figure 16. Folded antenna S_{11} results for different antenna position relative to the human body

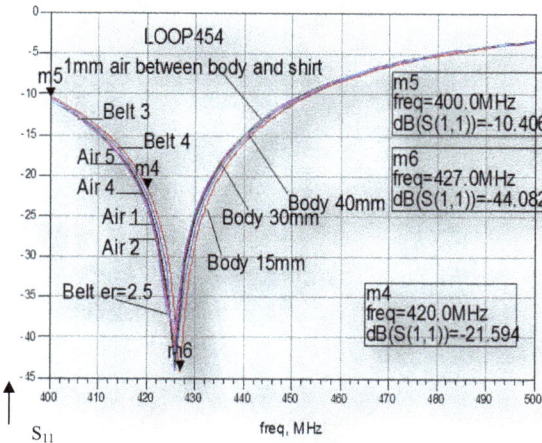

Figure 17. Loop antenna S_{11} results for different antenna position relative to the human body

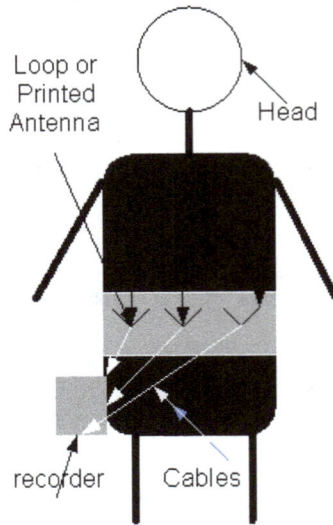

Figure 18. Printed Wearable antenna

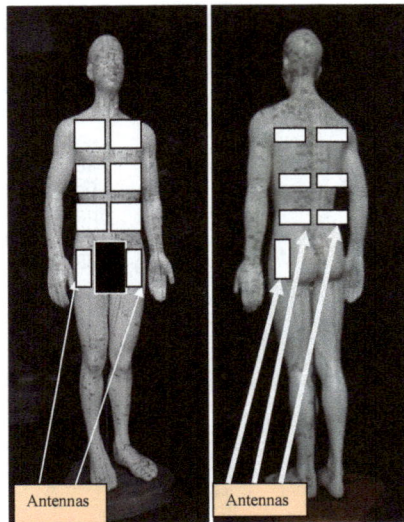

Figure 19. Printed Patch Antenna locations for various medical applications

In Figure 20 and 21 several microstrip antennas for medical applications at 434MHz are shown. The Backside of the antennas is presented in Figure 20.b. The diameter of the loop antenna presented in Figure 21 is 50 mm. The dimensions of the folded dipole antenna are 7x6x0.16cm. The dimensions of the compact folded dipole presented in Figure 21 are 5x5x0.5cm.

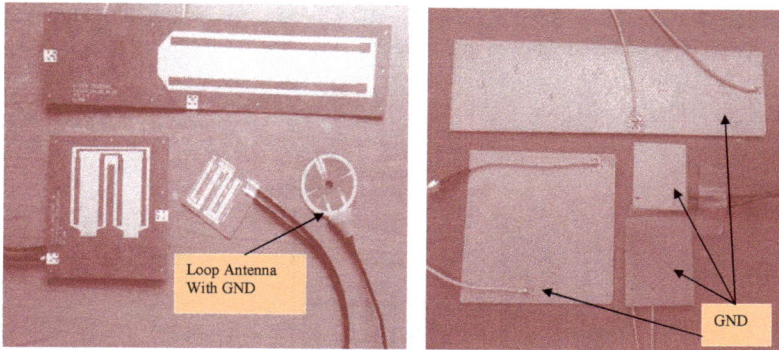

Figure 20. a. Microstrip Antennas for medical Applications b. Backside of the antennas6. Compact dual polarized printed antenna

A new compact microstrip loaded dipole antennas has been designed. The antenna consists of two layers. The first layer consistsof FR4 0.25mm dielectric substrate. The second layer con-sists of Kapton 0.25mm dielectric substrate. The substrate thickness determines the antenna bandwidth. However, with thinner substrate we may achieve better flexibility. The proposed antenna is dual polarized. The printed dipole and the slot antenna provide dual orthogonal polarizations. The dual polarized antenna is shown in Figure 22. The antenna dimensions are 5x5x0.05cm.

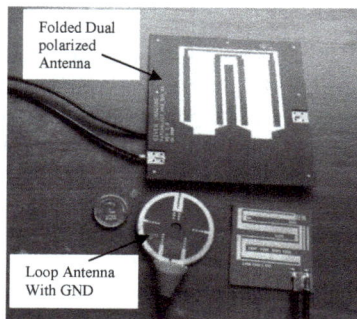

Figure 21. Microstrip Antennas for medical Applications

The antenna may be attached to the patient shirt in the patient stomach or back zone. The antenna has been analyzed by using Agilent ADS software. There is a good agreement between measured and computed results. The antenna bandwidth is around 10% for VSWR better than 2:1. The antenna beam width is around 100º. The antenna gain is around 0dBi. The computed S_{11} parameters are presented in Figure 23. Figure 24 presents the antenna measured S_{11} parameters. The antenna cross-polarized field strength may be adjusted by varying the slot feed location. The computed 3D radiation pattern of the antenna is shown in Figure 25. The computed radiation pattern is shown in Figure 26.

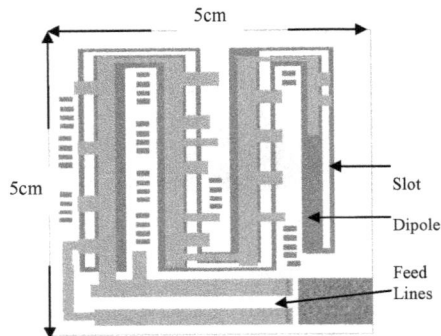

Figure 22. Printed Compact dual polarized antenna

6. Helix antenna performance on human body

In order to compare the variation of the new antenna input impedance as function of distance from the body to other antennas a helix antenna has been designed. A helix antenna with 9 turns is shown in figure 27. The backside of the circuit is copper under the microstrip matching stubs. However, in the helix antenna area there is no ground plane. The antenna has been designed to operate on human body. A matching microstrip line network has been designed on RO4003 substrate with 0.8mm thickness. The helix antenna has VSWR better than 3:1 at the frequency range from 440MHz to 460MHz. The antenna dimensions are 4x4x0.6cm. Figure 28 presents the measured S_{11} parameters on human body. The computed E and H radiation plane of the helix antenna is shown in Figure 29. The helix antenna input impedance variation as function of distance from the body is very sensitive. If the air spacing between the helix antenna and the human body is increased from 0mm to 2mm the antenna resonant frequency is shifted by 5%.

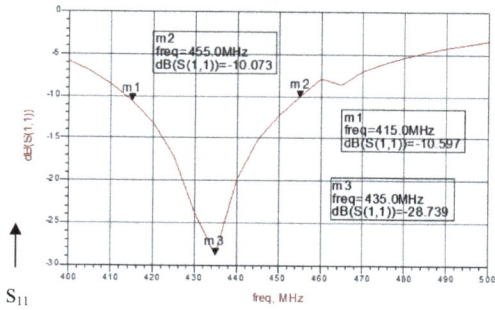

Figure 23. Computed S_{11} results of compact antenna

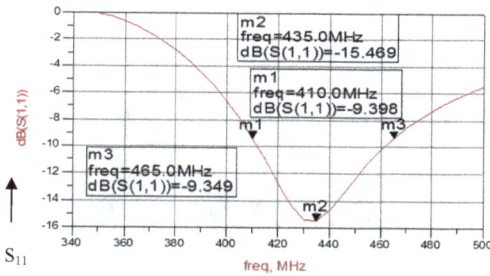

Figure 24. Measured S_{11} on human body

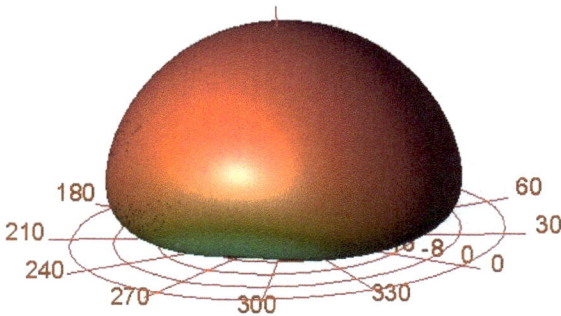

Figure 25. Compact Antenna 3D Radiation pattern

Linear Polarization

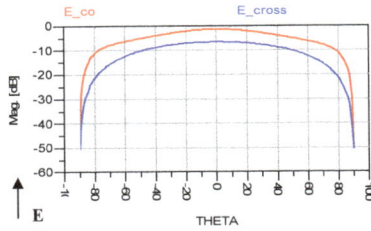

Figure 26. Antenna Radiation pattern

However, if the air spacing between the new dual polarized antenna and the human body is increased from 0mm to 5mm the antenna resonant frequency is shifted only by 5%.

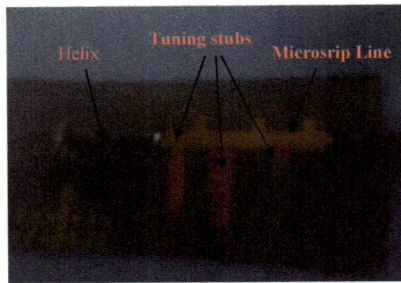

Figure 27. Helix Antenna for medical Applications

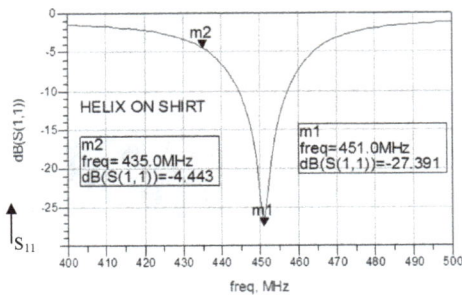

Figure 28. Measured S_{11} on human body

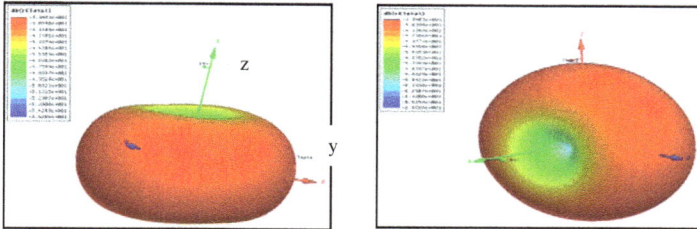

Figure 29. E and H plane radiation pattern of The Helix antenna

7. Wearable tunable printed antennas for medical applications

A new class of wideband tunable wearable microstrip antennas for medical applications is presented in this section. The antennas VSWR is better than 2:1at 434MHz+5%. The antenna beam width is around 100º. The antennas gain is around 0 to 2dBi. A voltage controlled varactor is used to control the antenna resonant frequency at different locations on the human body.

7.1. Dually polarized tunable printed antenna

A compact tunable microstrip dipole antenna has been designed to provide horizontal polarization. The antenna consists of two layers. The first layer consists of RO3035 0.8mm dielectric substrate. The second layer consists of RT-Duroid 5880 0.8mm dielectric substrate. The substrate thickness affects the antenna band width. The printed slot antenna provides a vertical polarization. The printed dipole and the slot antenna provide dual orthogonal polarizations. The dimensions of the dual polarized antenna are 26x6x0.16cm. Also tunable compact folded dual polarized antennas have been designed. The dimensions of the compact antennas are 5x5x0.05cm.Varactors are connected to the antenna feed lines as shown in Figure 30. The voltage controlled varactors are used to control the antenna resonant frequency. The varactor bias voltage may be varied automatically to set the antenna resonant frequency at different locations on the human body. The antenna may be used as a wearable antenna on a human body. The antenna may be attached to the patient shirt in the patient stomach or back zone. The antenna has been analyzed by using Agilent ADS software. There is a good agreement between measured and computed results. The antenna bandwidth is around 10% for VSWR better than 2:1. The antenna beam width is around 100º. The antenna gain is around 2dBi.

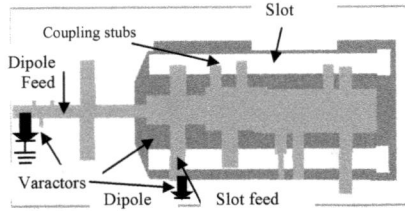

Figure 30. Dual polarized tunable antenna, 26x6x0.16 cm.

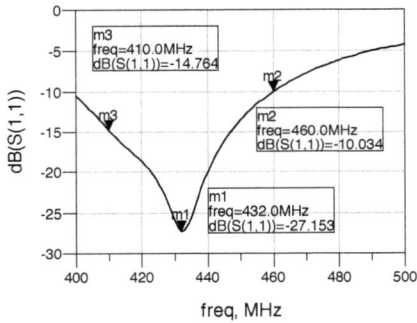

Figure 31. Measured S₁₁ on human body

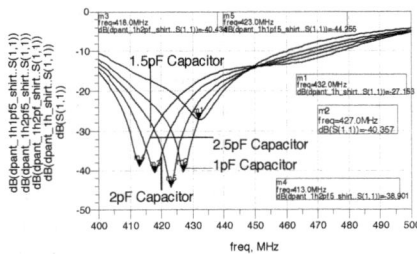

Figure 32. The Tunable S₁₁ parameter as function of varactor capacitance

Figure 31 presents the antenna measured S_{11} parameters without a varactor. Figure 32 presents the antenna S_{11} parameters as function of different varactor capacitances. Figure 33 presents the tunable antenna resonant frequency as function of the varactor capacitance. The antenna resonant frequency varies around 5% for capacitances up to 2.5pF. The antenna beam width is 100º. The antenna cross polarized field strength may be adjusted by varying the slot feed location.

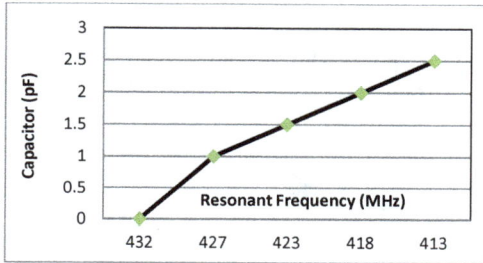

Figure 33. Resonant frequency as function of varactor capacitance

7.2. Antenna S11 varitaion as function of distance from body

The Antennas input impedance variation as function of distance from the body had been computed by employing ADS software. The analyzed structure is presented in Figure 14. Properties of human body tissues are listed in Table 2 see [8]. Figure 34 presents S_{11} results for different belt and shirt thickness, and air spacing between the antennas and human body.

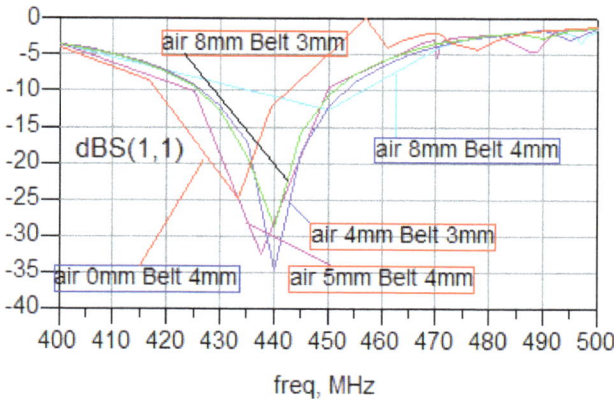

Figure 34. S_{11} results for different antenna positions

If the air spacing between the sensors and the human body is increased from 0mm to 5mm the antenna resonant frequency is shifted by 5%. There is good agreement between measured and calculated results. The voltage controlled varactor may be used to tune the antenna resonant frequency due to different antenna locations on a human body. Figure 35 presents several compact tunable Antennas for medical Applications. A voltage controlled varactor may be used also to tune the loop antenna resonant frequency at different antenna locations on the body.

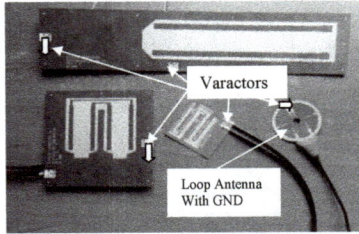

Figure 35. Tunable Antennas for medical Applications

7.3. Varactors

Tuning varactors are voltage variable capacitors designed to provide electronic tuning of microwave components. Varactors are manufactured on silicon and gallium arsenide sub-strates. Gallium arsenide varactors offer higher Q and may be used at higher frequencies than silicon varactors. Hyperabrupt varactors provide nearly linear variation of frequency with applied control voltage. However abrupt varactors provide inverse fourth root frequency dependence. MACOM offers several gallium arsenide hyperabrupt varactors such as MA46 series. Figure 36 presents the C-V curves of varactors MA46505 to MA46506.

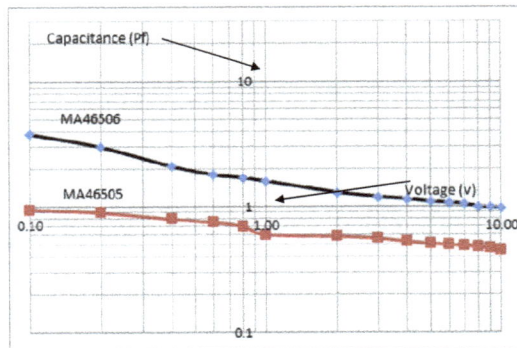

Figure 36. Varactor capacitance as function of bias voltage

Figure 37 presents the C-V curves of varactors MA46H070 to MA46H074

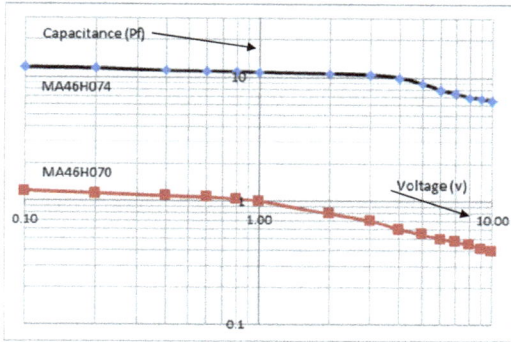

Figure 37. C-V curves of varactors MA46H070 to MA46H076

Figure 38 presents a compact tunable antenna with a varactor. Figure 39. presents measured S_{11} as function of varactor bias voltage. We may conclude that varactors may be used to compensate variations in the antenna resonant frequency at different locations on the human body.

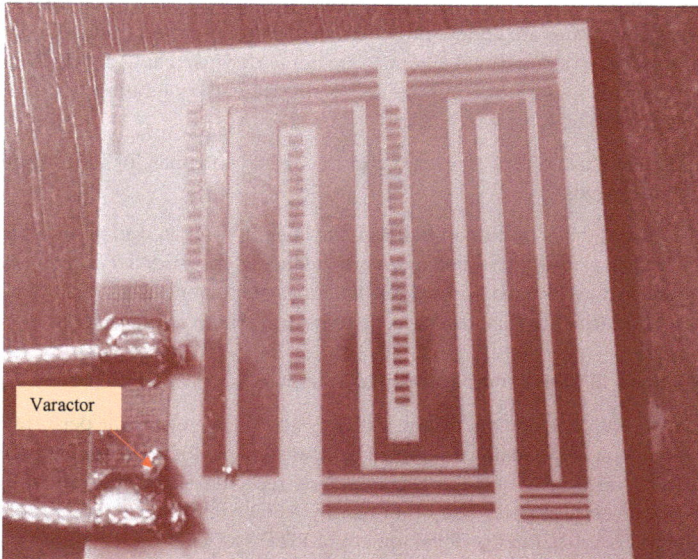

Figure 38. Tunable antenna with a varactor

Figure 38. Tunable antenna with a varactor

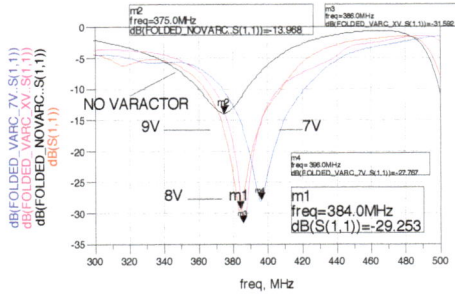

Figure 39. Measured S_{11} as function of varactor bias voltage

8. Compact wearable RFID antennas

RFID (**R**adio **F**requency **Id**entification) is an electronic method of exchanging data over radio frequency waves. There are three major components in RFID system: Transponder (Tag), Antenna and a Controller. The RFID tag, antenna and controller may be assembled on the same board. Microstrip antennas are widely presented in books and papers in the last decade [1-7]. However, compact wearable printed antennas are not widely used at 13.5MHz RIFD systems. HF tags work best at close range but are more effective at penetrating non-metal objects especially objects with high water content.

A new class of wideband compact printed and microstrip antennas for RFID applications is presented in this chapter.

RF transmission properties of human tissues have been investigated in several papers [8-9]. The effect of human body on the antenna performance is investigated in this chapter. The proposed antennas may be used as wearable antennas on persons or animals. The proposed antennas may be attached to cars, trucks, containers and other various objects.

8.1. Dual polarized 13.5MHz compact printed antenna

One of the most critical elements of any RFID system is the electrical performance of its antenna. The antenna is the main component for transferring energy from the transmitter to the passive RFID tags, receiving the transponder's replying signal and avoiding in-band interference from electrical noise and other nearby RFID components. Low profile compact printed antennas are crucial in the development of RIFD systems.

A new compact microstrip loaded dipole antennas has been designed at 13.5MHz to provide horizontal polarization. The antenna consists of two layers. The first layer consists of FR4

0.8mm dielectric substrate. The second layer consists of Kapton 0.8mm dielectric substrate. The substrate thickness determines the antenna bandwidth. A printed slot antenna provides a vertical polarization. The proposed antenna is dual polarized. The printed dipole and the slot antenna provide dual orthogonal polarizations. The dual polarized RFID antenna is shown in Figure 40. The antenna dimensions are 6.4x6.4x0.16cm. The antenna may be attached to the customer shirt in the customer stomach or back zone. The antenna has been analyzed by using Agilent ADS software.

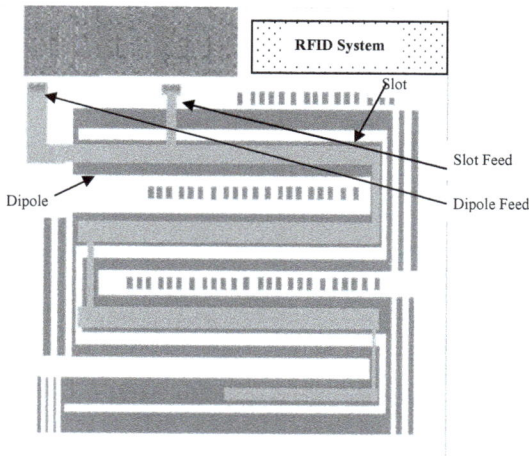

Figure 40. Printed Compact dual polarized antenna, 64x64x1.6mm

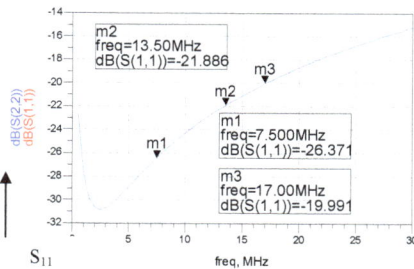

Figure 41. Computed S_{11} results

The antenna S11 parameter is better than -21dB at 13.5MHz. The antenna gain is around -10dBi. The antenna beam width is around 160°. The computed S_{11} parameters are presented in Figure

41. There is a good agreement between measured and computed results. Figure 42 presents the antenna measured S_{11} parameters. The antenna cross- polarized field strength may be adjusted by varying the slot feed location. The computed radiation pattern is shown in Figure 43. The computed 3D radiation pattern of the antenna is shown in Figure 44.

8.2. Varying the antenna feed network

Several designs with different feed network have been developed. A compact antenna with different feed network is shown in Figure 45. The antenna dimensions are 8.4x6.4x0.16cm. Figure 46 presents the antenna computed S11on human body. There is a good agreement between measured and computed results. The computed radiation pattern is shown in Fig 47. Table 5 compares the electrical performance of a loop antenna with the compact dual polarized antenna.

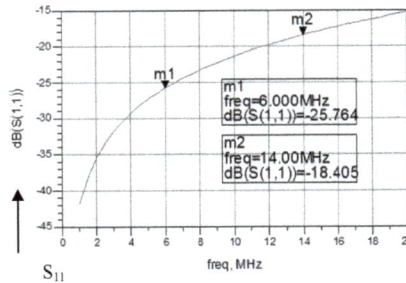

Figure 42. Measured S_{11} on human body

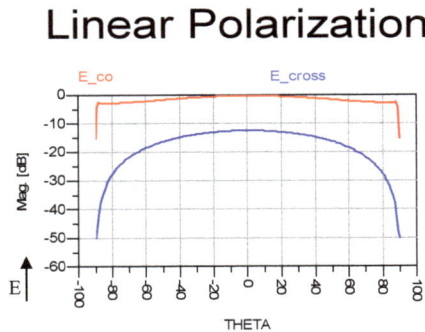

Linear Polarization

Figure 43. Antenna Radiation pattern

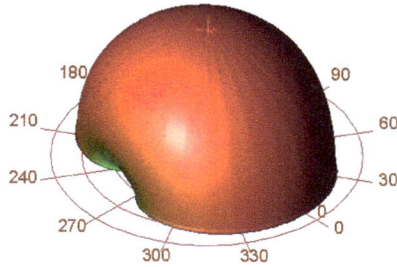

Figure 44. Antenna Radiation pattern

8.3. RFID wearable loop antennas

RFID loop antennas are widely used. Several RFID loop antennas are presented in [15]. RFID loop antennas have low efficiency and narrow bandwidth. As an example the measured impedance of a square four turn loop at 13.5MHz is

$0.47 + j107.5$ Ω. A matching network is used to match the antenna to 50 Ω. The matching network consists of a 56pF shunt capacitor, 1kΩ shunt resistor and another 56pF capacitor. This matching network has narrow bandwidth.

A transmitting antenna is placed 20cm from a square four turn loop antenna. 0dBm CW signal is applied to the transmitting antenna. The measured power level at the output of the loop antenna is -50dBm. A square four turn loop antenna has been designed at 13.5MHz by using Agilent ADS software.

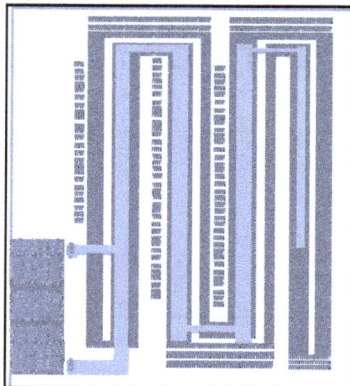

Figure 45. RFID printed antenna, 8.4x6.4x0.16cm.

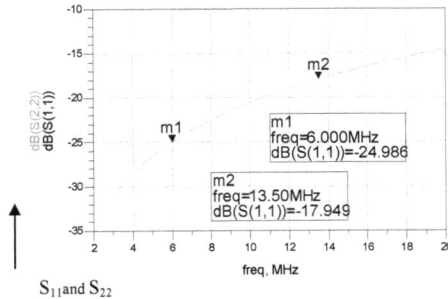

$S_{11}$and S_{22}

Figure 46. RFID Antenna Computed $S_{11}$and S_{22} results

The antenna is printed on a FR4 substrate. The antenna dimensions are 32x52.4x0.25mm. The antenna layout is shown in figure 48. S11 results of the printed loop antenna are shown in figure 49. The antenna S_{11} parameter is better than -9.5dB without an external matching network. The computed radiation pattern is shown in Figure 50. The computed radiation pattern takes into account an infinite ground plane.

Antenna	Beam width° 3dB	Gain dBi	VSWR
Loop Antenna	140	-25	2:1
Microstrip Antenna	160	-10	1.2:1

Table 5. Comparison of Loop Antenna and Microstrip antenna parameters

Linear Polarization

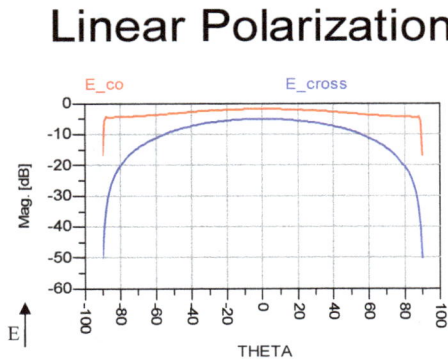

Figure 47. Compact Antenna, 8.4x6.4x0.16cm, Radiation pattern

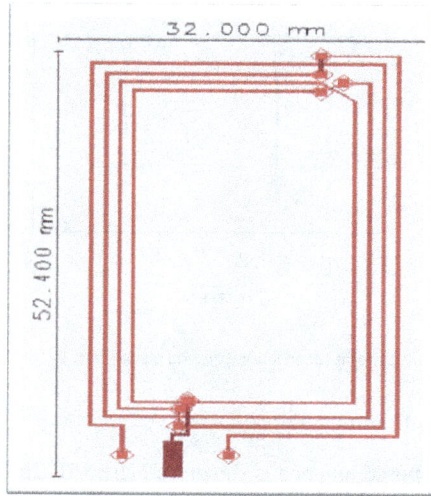

Figure 48. A square four turn loop antenna

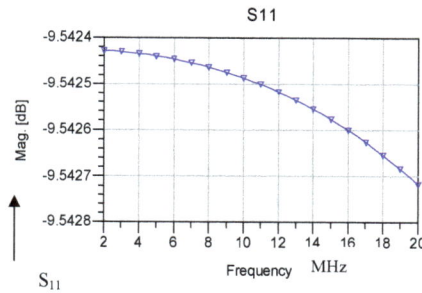

Figure 49. Loop Antenna Computed S_{11} results

The Microstrip Antenna input impedance variation as function of distance from the body has been computed by employing ADS software. The analyzed structure is presented in Figure 14 Properties of human body tissues are listed in Table 2 see [8]. These properties were used in the antenna design. S_{11} parameters for different human body thicknesses have been computed. We may note that the differences in the results for body thickness of 15mm to 100mm are negligible. S_{11} parameters for different position relative to the human body have been computed. If the air spacing between the antenna and the human body is increased from 0mm to 10mm the antenna S_{11} parameters may change by less than 1%. The VSWR is better than 1.5:1.

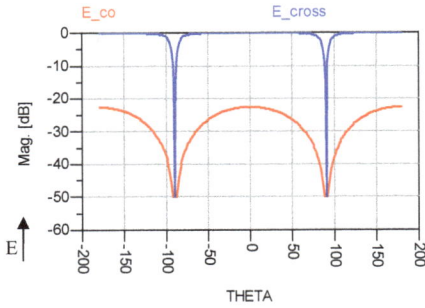

Figure 50. Loop Antenna Radiation patterns for an infinite ground plane

8.4. Proposed antenna applications

An application of the proposed antenna is shown in Figure 51. The RFID antennas may be assembled in a belt and attached to the customer stomach. The antennas may be employed as transmitting or as receiving antennas. The antennas may receive or transmit information to medical systems.

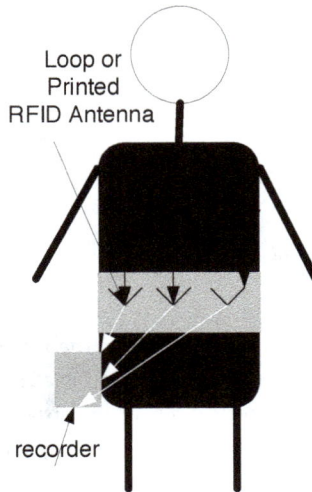

Figure 51. Wearable RFID antenna

In RFID systems the distance between the transmitting and receiving antennas is less than $2D^2/\lambda$, where D is the largest dimension of the antenna. The receiving and transmitting antennas are magnetically coupled. In these applications we refer to the near field and not to the far field radiation pattern.

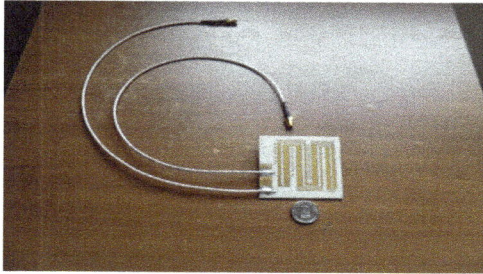

Figure 52. New Microstrip Antenna for RFID applications

Figure 52 and Figure 53 present compact printed antenna for RFID applications. The presented antennas may be assembled in a belt and attached to the patient stomach or back.

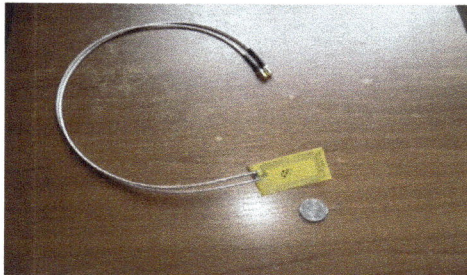

Figure 53. Loop Antenna for RFID applications

9. Conclusions

This chapter presents wideband microstrip antennas with high efficiency for medical applications. The antenna dimensions may vary from 26x6x0.16cm to 5x5x0.05cm according to the medical system specification. The antennas bandwidth is around 10% for VSWR better than 2:1. The antenna beam width is around 100°. The antennas gain varies from 0 to 4dBi. The antenna S11 results for different belt thickness, shirt thickness and air spacing between the antennas and human body are presented in this chapter. If the air spacing between the new dual polarized antenna and the human body is increased from 0mm to 5mm the antenna resonant frequency is shifted by 5%. However, if the air spacing between the helix antenna and the human body is increased only from 0mm to 2mm the antenna resonant frequency is shifted by 5%. The effect of the antenna location on the human body should be considered in the antenna design process. The proposed antenna may be used in Medicare RF systems.

A wideband tunable microstrip antennas with high efficiency for medical applications has been presented in this chapter. The antenna dimensions may vary from 26x6x0.16cm to 5x5x0.05cm according to the medical system specification. The antennas bandwidth is around 10% for VSWR better than 2:1. The antenna beam width is around 100º. The antennas gain varies from 0 to 2dBi. If the air spacing between the dual polarized antenna and the human body is increased from 0mm to 5mm the antenna resonant frequency is shifted by 5%. A varactor is employed to compensate variations in the antenna resonant frequency at different locations on the human body.

This chapter presents also wideband compact printed antennas, Microstrip and Loop antennas, for RFID applications. The antenna beam width is around 160º. The antenna gain is around -10dBdBi. The proposed antennas may be used as wearable antennas on persons or animals. The proposed antennas may be attached to cars, trucks and other various objects. If the air spacing between the antenna and the human body is increased from 0mm to 10mm the antenna S_{11} parameters may change by less than 1%. The antenna VSWR is better than 1.5:1 for all tested environments.

Author details

Albert Sabban[1,2,3*]

1 Ort Braude College, Karmiel, Israel

2 Tel Aviv University, Israel

3 Colorado University, Boulder, USA

References

[1] J.R. James, P.S Hall and C. Wood, "Microstrip Antenna Theory and Design",1981.

[2] Sabban and K.C. Gupta, "Characterization of Radiation Loss from Microstrip Discontinuities Using a Multiport Network Modeling Approach", I.E.E.E Trans. on M.T.T, Vol. 39,No. 4,April 1991, pp. 705-712.

[3] Sabban," A New Wideband Stacked Microstrip Antenna", I.E.E.E Antenna and Propagation Symp., Houston, Texas, U.S.A, June 1983.

[4] Sabban, E. Navon " A MM-Waves Microstrip Antenna Array", I.E.E.E Symposium, Tel-Aviv, March 1983.

[5] R. Kastner, E. Heyman, A. Sabban, "Spectral Domain Iterative Analysis of Single and Double-Layered Microstrip Antennas Using the Conjugate Gradient Algorithm", I.E.E.E Trans. on Antennas and Propagation, Vol. 36, No. 9, Sept. 1988, pp. 1204-1212.

[6] Sabban, "Wideband Microstrip Antenna Arrays", I.E.E.E Antenna and Propagation Symposium MELCOM, Tel-Aviv,1981.

[7] Sabban, "Microstrip Antenna Arrays", Microstrip Antennas, Nasimuddin Nasimuddin (Ed.), ISBN: 978-953-307-247-0, InTech, http://www.intechopen.com/articles/show/title/microstrip-antenna-arrays , pp..361-384, 2011.

[8] Lawrence C. Chirwa*, Paul A. Hammond, Scott Roy, and David R. S. Cumming, "Electromagnetic Radiation from Ingested Sources in the Human Intestine between 150 MHz and 1.2 GHz", IEEE Transaction on Biomedical eng., VOL. 50, NO. 4, April 2003, pp 484-492.

[9] D.Werber, A. Schwentner, E. M. Biebl, "Investigation of RF transmission properties of human tissues", Adv. Radio Sci., 4, 357–360, 2006.

[10] Gupta, B., Sankaralingam S., Dhar, S.,"Development of wearable and implantable antennas in the last decade", Microwave Symposium (MMS), 2010 Mediterranean 2010 , Page(s): 251 – 267.

[11] Thalmann T., Popovic Z., Notaros B.M, Mosig, J.R.," Investigation and design of a multi-band wearable antenna", 3rd European Conference on Antennas and Propagation, EuCAP 2009. Pp. 462 – 465.

[12] Salonen, P., Rahmat-Samii, Y., Kivikoski, M.," Wearable antennas in the vicinity of human body", IEEE Antennas and Propagation Society International Symposium, 2004. Vol.1 pp. 467 – 470.

[13] Kellomaki T., Heikkinen J., Kivikoski, M., " Wearable antennas for FM reception", First European Conference on Antennas and Propagation, EuCAP 2006 , pp. 1-6.

[14] Sabban, "Wideband printed antennas for medical applications" APMC 2009 Conference, Singapore, 12/2009.

[15] Youbok Lee, "Antenna Circuit Design for RFID Applications", Microchip Technology Inc., Microchip AN 710c.

[16] ADS software, Agilent http://www.home.agilent.com/agilent/product.jspx?cc=IL&lc=eng&ckey=1297113&nid=-34346.0.00&id=1297113

Chapter 14

Design, Fabrication, and Testing of Flexible Antennas

Haider R. Khaleel, Hussain M. Al-Rizzo and
Ayman I. Abbosh

Additional information is available at the end of the chapter

1. Introduction

Recent years have witnessed a great deal of interest from both academia and industry in the field of flexible electronics. In fact, this research topic tops the pyramid of research priorities requested by many national research agencies.

According to market analysis, the revenue of flexible electronics is estimated to be 30 billion USD in 2017 and over 300 billion USD in 2028 [1].

Their light weight, low-cost manufacturing, ease of fabrication, and the availability of inexpensive flexible substrates (i.e.: papers, textiles, and plastics) make flexible electronics an appealing candidate for the next generation of consumer electronics [2]. Moreover, recent developments in miniaturized and flexible energy storage and self-powered wireless components paved the road for the commercialization of such systems [3].

Consistently, flexible electronic systems require the integration of flexible antennas operating in specific frequency bands to provide wireless connectivity which is highly demanded by today's information oriented society.

Needless to say, the efficiency of these systems primarily depends on the characteristics of the integrated antenna. The nature of flexible wireless technologies requires the integration of flexible, light weight, compact, and low profile antennas. At the same time, these antennas should be mechanically robust, efficient with a reasonably wide bandwidth and desirable radiation characteristics.

This chapter deals with the design, numerical simulation, fabrication process and methods, flexibility tests, and measurements of flexible antennas. As a benchmark, a flexible, compact, and low profile (50.8 μm) printed monopole antenna intended for the ISM band applications at 2.45 GHz is presented and discussed in details. The antenna is based on a Kapton Polyimide substrate and fabricated using the ink-jet technology. Finally, the performance of the antenna is compared with different antenna types reported in the literature in terms of electromagnetic performance and physical properties.

Figure 1. Flexible printed monopole antenna based on Kapton Polyimide substrate.

2. Choice of Antenna Substrate

To comply with flexible technologies, integrated components need to be highly flexible and mechanically robust; they also have to exhibit high tolerance levels in terms of bending repeatability and thermal endurance. A plethora of design approaches of flexible and conformal antennas were reported in the literature including Electro-textile [4], paper-based [5], fluidic [6], and synthesized flexible substrates [7]. In [4], a 150 mm × 180 mm flexible Electro textile antenna based on a 4 mm felt fabric is proposed. The antenna operates in the ISM 2.45 GHz band. Although it is suitable for wearable and conformal applications, fabric substrates are prone to discontinuities, fluids absorption, and crumpling.

In [5], a flexible single band antenna printed on a 46mm × 30mm paper-based substrate was proposed for integration into flexible displays for WLAN applications. However, paper based substrates are found to be not robust enough and introduce discontinuities when used in applications that require high levels of bending and rolling. Moreover, they have a relatively high loss factor (loss tangent (tanδ) is around 0.07 at 2.45 GHz) which compromises the antenna's efficiency [8].

Kapton Polyimide film was chosen as the antenna substrate in [9] due to its good balance of physical, chemical, and electrical properties with a low loss factor over a wide frequency range (tan $\delta = 0.002$). Furthermore, Kapton Polyimide offers a very low profile (50.8 µm) yet very robust with a tensile strength of 165 MPa at 73°F, a dielectric strength of 3500-7000 volts/mil, and a temperature rating of -65 to 150°C [10]. Other Polymer based and synthesized flexible substrates have been also used in several designs [11-14].

It is worth mentioning that there are several techniques used to characterize the electromagnetic properties of thin and flexible films/substrates such as: the near field microscopy, coplanar waveguide approach, differential open resonator method, and goniometric time-domain spectroscopy method [15-18]. However, the most popular method based on measurements of deposited transmission lines incorporating the material to be characterized which determine the dielectric constants of thin films and the conductivities of the metallic lines over a broad frequency range [19].

3. Choice of Antenna Type

Needless to say, conventional microstrip antennas are not a practical solution for flexible electronics due to their inherently narrow bandwidth which is a function of the substrate's thickness. In [20], a flexible aperture coupled antenna is reported. This technique is known to enhance the impedance bandwidth significantly, however, it leads to an increase in the overall profile; moreover, it involves multi layers, which complicates the fabrication process.

Planar Inverted-F antennas (PIFA) are widely used in mobile phones due the fact that wider impedance bandwidth is obtained despite the presence of a ground plane. Also, antennas incorporating a ground plane promote reduced Specific Absorption Rate (SAR); furthermore, their matching is less affected by the proximity of the human body.

In [21], a 50mm × 19mm textile based broadband PIFA fabricated using conductive textiles is proposed for Wireless Body Area Network (WBAN) applications. Although the antenna exhibits a good impedance bandwidth and radiation characteristics, its overall thickness is 6mm which is considered high for the technology under consideration; moreover, it involves a multi-layer complex, and inaccurate fabrication process.

On the other hand, planar monopole and dipole antennas have received much interest over other antenna types due to their relatively large impedance bandwidth, low profile, ease of fabrication, and omni directional radiation pattern which is highly preferred in many wireless schemes.

Given the technology envisioned in this chapter, Co-Planar Waveguide (CPW) is preferred over other feeding techniques since no via holes or shorting pins are involved, in addition to several useful characteristics such as: low radiation losses, larger bandwidth, improved impedance matching, and more importantly, both radiating element and ground plane are printed on the same side of the substrate, which promotes low fabrication cost and complexity in addition to the capability of roll to roll production.

4. Choice of Fabrication Method

This section reviews the currently available methods for fabrication of flexible and wearable antennas. Method overview, advantages, and drawbacks of each technique are discussed.

4.1. Screen Printing

Screen printing is one of the simplest and most cost effective techniques used by electronics manufacturers. This technique is based on a woven screen that has different thicknesses and thread densities. To produce a printed pattern, a squeegee blade is driven down forcing the screen into contact with the affixed substrate. This in turn forces the ink to be ejected through the exposed areas of the screen on the substrate, and thus, the desired pattern is formed [22]. Polyester and stainless steel are the most common materials used in this technology.

Three different screen printing methods are currently used: flat bed, cylinder, and rotary. Flat bed is the simplest and most common screen printing method. Cylinder screen printing is quite similar to the flat bed except the pattern is deposited as the substrate rotates while attached to the screen roll. In rotary screen, ink and squeegee assembly are rotated inside a rolled screen where impression cylinder produces pressure to substrate [23]. Rotary screen enables much higher throughput capacity than flat bed screen; hence, it is often integrated into a roll to roll production line.

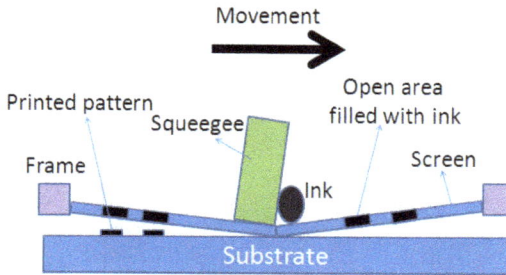

Figure 2. Illustration of the screen printing process.

Screen printing is an additive process as opposed to the subtractive process of chemical etching which makes it a more cost-effective and environmentally friendly. Rather than masking a screen, the patterned mask is applied onto the substrate directly where the conductive ink is administered and thermally cured. Several RFIDs and flexible transparent antennas have been prototyped successfully using this technique [24-26]. However, there are some problems associated with this technique including the limited control over the thickness, number of passes, and resolution of the printed patterns. Layer consistency is also a challenge, as thermal curing of solvent based inks could leave behind artifacts that change with ink viscosity and surface energy of the substrate [22]. Figure 2 depicts a scheme of the screen printing process.

4.2. Chemical Etching

Chemical etching often accompanied by photolithography is the process of fabricating metallic patterns using a photoresist and etchants to mill out a selected area corrosively. This technique has emerged in the 1960s as a branch-out of the Printed Circuit Board (PCB) industry. Chemical etching gained a wide popularity since it can produce highly complex patterns with high resolution accurately [27].

Photoresist materials are organic polymers whose chemical characteristics change when exposed to ultraviolet light. When the exposed area becomes more soluble in the developer, the photoresist is positive. While if it becomes less soluble, the compound is considered a negative resist.

A major drawback of negative resists is that the exposed regions swell as the counterpart is dissolved by the developer, which compromises the resolution of the process. Swelling occurs due to the penetration of the developer solution into the photoresist material which in turn leads to a distortion in the patterned region [27]. Hence, current practice in the photolithography based antenna and RF circuits Industry relies mainly on positive resists since they present higher resolution than negative resists.

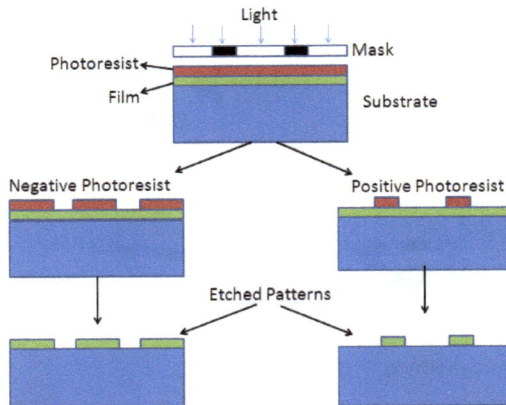

Figure 3. Process flow of the chemical etching (photolithography) process.

Although patterns with high complexity and fine details can be produced using this technique, its lengthy process, low throughput, involvement of dangerous chemicals (neutralization is required), clean room requirement, in addition to byproduct and waste leftovers are major drawbacks of this technology. The reader is referred to [28] for further information on this method. The chemical etching process is illustrated in Figure 3.

4.3. Flexography

Flexography is a type of relief printing. An image is produced by a printmaking process where a protuberating surface of the printing plate matrix is inked while the recessed areas are free of ink. Image printing is a simple process since it only involves inking the protruding surface of the matrix and bringing it in contact with the substrate [29]. Due to its relatively fine resolution, low cost, and high throughput, flexography gained a great interest by RFID antenna manufacturers. Moreover, this technique requires a lower viscosity ink than screen printing inks, and yields imaged (printed) dry films of a thickness of less than 2.5 μm. Hence, flexography inks need to posses higher bulk conductivity than those used in screen printing to compensate for the increase in sheet resistance since the efficiency of printed antennas depends mainly on the electrical conductivity of the traced pattern. Substrate parameters like surface porosity, hydrophobicity, and surface energy have a direct influence on the ink film thickness of the printed trace [23]. The consistency in ink film thickness and line width has also a profound impact on the sheet resistance. The process scheme is demonstrated in Figure 4.

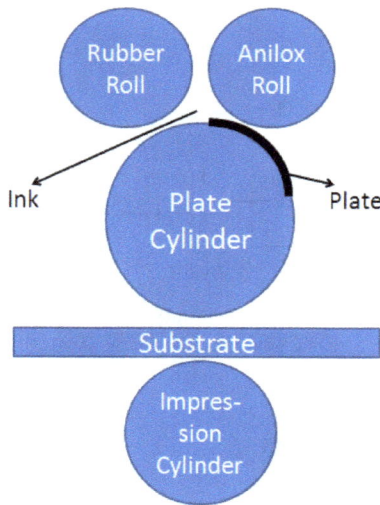

Figure 4. Illustration of the flexography printing process based on flexible relief plate.

4.4. Ink Jet Printing

Inkjet printing of RF circuits and antennas using highly conductive inks have become extremely popular in recent years. New inkjet material printers operate by depositing ink droplets of a size down to few pico liters, and hence, these high resolution printers can produce compact designs with tiny details accurately [30].

This new technology utilizes conductive inks based on different nano-structural materials such as silver nano-particle based ink, which is widely used due to its high conductivity. This type of printing technique can be categorized into two types: drop-on-demand and continuous inkjet. Drop on demand print heads apply pressurized pulses to ink with either a piezo or thermo element in which drives a drop from a nozzle when needed. Most printed electronics manufacturers utilize the piezo pulse type [31]. Printing quality depends mainly on the ink characteristics such as viscosity, surface tension, and particle size. The surface topology of the substrate, the platen temperature and the print head parameters are also important factors.

Printing processes and setups are completely controlled from the user's computer, and do not require a clean-room environment which reduces the levels of environmental contamination [31].

Unlike photolithography, which is a subtractive method since it involves removing unwanted pattern from the substrate's metallic/conductive side; inkjet printing deposits a controlled amount of ink droplets from the nozzle to the specified position. Hence, no waste or byproduct is produced resulting in an economical, clean, and fast solution. Figure 5 depicts the ink jet printing process.

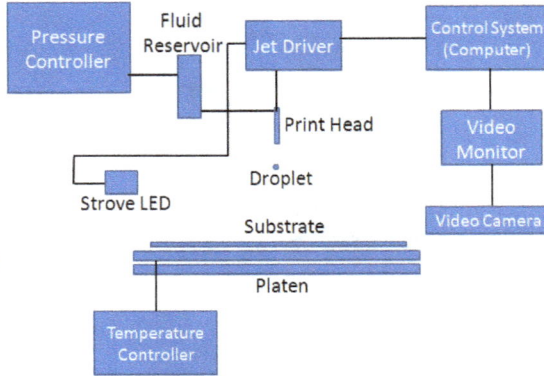

Figure 5. Illustration of the ink jet (droplet on-demand) printing process.

5. Benchmarking Prototype

To benchmark the performance of flexible antennas, a single band printed monopole is presented in this section which has the merits of light weight, ultra low profile (50.8 μm), large bandwidth, robustness, compactness, and high efficiency. The antenna design which is fabricated using the inkjet printing technology covers the ISM 2.45 GHz and fed by a CPW feed. Moreover, the performance of the antenna is evaluated under bending effects in terms of impedance matching and shift in resonant frequency. Finally, the characteristics of the antenna under study are compared to several flexible antenna types reported in the literature.

5.1. ISM Band Printed Monopole Antenna

The ISM 2.45 GHz band is internationally recognized as one the most commonly used standards in wireless communication systems [32]. For example, all of Wireless local-area networks (WLAN), IEEE 802.11/WiFi, Bluetooth and Personal Area Network (PAN) IEEE 802.15.4, ZigBee utilize the ISM 2.45 GHz band. Additionally, several potential applications based on these technologies are possibly applied in the future.

Obviously, the integration of a wireless connectivity based on the abovementioned technologies within flexible devices triggers the need for ultra light/thin/flexible antennas. At the same time, these antennas should be robust, cost effective, and highly efficient with desirable radiation characteristics.

In response to such needs, several design approaches of flexible and conformal antennas based on flexible substrates were reported in the literature [33-39]. In [32], a flexible antenna printed on a 46mm x 30mm paper-based substrate was proposed for integration into flexible displays for WLAN applications. However, paper based substrates are found to be not robust enough and introduce discontinuities when used in applications that require high levels of bending and rolling as mentioned earlier. Moreover, they have a high loss factor (loss tangent (tanδ) is around 0.07 at 2.45 GHz) which compromises the antenna's efficiency. In [34], a stretchable antenna based on an elastic substrate is presented. The design offers a good solution in terms of flexibility and stretchability; however, it involves a complex manufacturing process where the conductors are realized by injecting a room temperature liquid metal alloy into molded micro-structured channels on an elastic dielectric material followed by channels encapsulation. In [35], a conformal exponentially tapered slot antenna based on a 200 μm Liquid Crystal Polymer (LCP) substrate is reported. The design exhibits excellent radiation characteristics; however, the dimensions (130mm × 43mm) are too large for integration within modern compact and flexible electronics. In this section, a flexible compact split ring printed monopole antenna intended for flexible/wearable/conformal applications is presented. The antenna is printed on a 50.8 μm Kapton substrate and fed by a CPW. Both radiating element and ground plane are printed on the same side of the substrate which promotes low fabrication cost and complexity in addition to roll to roll production.

5.1.1. Antenna Design

As shown in Figure 2, the antenna consists of a square split ring shaped radiating element fed by a CPW. The winding lengthens the current path which in turn reduces the structure size without significant efficiency degradation or disturbance to the radiation pattern. The separation distance between the arms is optimized as 5 mm to achieve the least return loss. It is worth mentioning that a smaller separation leads to an increased capacitive coupling between the arms which in turn degrades the impedance matching. The split ring monopole is fed by CPW feed, which adds the merit of fabrication simplicity since both the radiating element and ground plane are printed on the same side of the substrate.

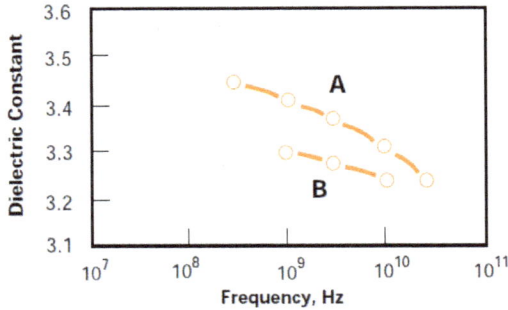

Figure 6. Dielectric constant Vs. frequency for the HN type Kapton polyimide (125 µm). The characteristics are similar to the 50 µm used in the reported prototype. Curve A is for measurement at 25°C (77°F) and 45% RH with the electric field in the plane of the sheet, while Curve B is the same measurement after conditioning the film at 100°C [10].

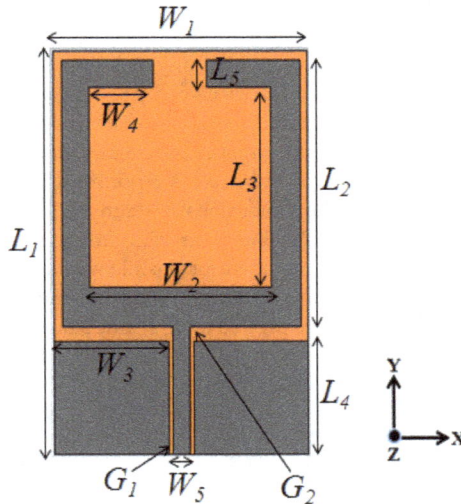

Figure 7. Geometry and dimensions of the reported Split Ring printed monopole antenna (the grey colored area represents the ground plane and the radiating element).

The antenna structure is printed on a 38 mm × 25 mm Kapton Polyimide substrate with a dielectric constant of 3.4 and a loss tangent of 0.002 (dielectric constant versus frequency is provided in Figure 6). The geometry and dimensions of the antenna are depicted in Figure 7 and Table 1.

L_1	38	W_1	25
L_2	26	W_2	18
L_3	19	W_3	10.5
L_4	9.5	W_4	6.5
L_5	3	W_5	2
G1	2	G2	0.5

Table 1. ISM Band Printed Monopole Antenna Dimensions in Millimeter.

5.1.2. Simulations, Fabrication and Measurements

Design and analysis of the reported printed monopole antennas have been carried out using the full wave simulation software CST Microwave Studio which is based on the Finite Integration Technique (FIT) [40].

To ensure a high simulation accuracy, the number of mesh cells was mainly determined through sufficient meshing of the antenna element where the smallest geometric detail (i.e. CPW gap, microstrip line, etc,..) is covered by at least three mesh cells both horizontally and vertically. The total number of the mesh cells generated for the antenna under study is 498,750 cells.

Figure 8. Installing the polyimide substrate on the platen of the material printer.

Before starting the printing process which is performed using the Dimatix DMP 2831 Fujifilm material printer [41], the final simulated design is exported to the printer using Dimatix Drop Manager Software in a Gerber file format which contains all the geometrical dimensions of the antenna design. Moreover, all printing processes and setup conditions can be controlled using the Dimatix Drop Manager software such as the number of layers to be printed, heating temperature of the platen desk, number of nozzles used in operation and height of the cartridge head with respect to the substrate.

A conductive ink based on sliver nano particles is deposited over the substrate utilizing 16 nozzles with 25 μm drop spacing. After the printing process is completed, thermal annealing is required to evaporate excess solvent and to remove ink impurities. Furthermore, the thermal annealing process provides an increased bond of the deposited material. The reported antenna is cured at 100° for 4 hours by a LPKF Protoflow industrial oven. It is worth mentioning that 2 layers of ink were deposited on the substrate to achieve a robust and continuous radiating element and more importantly to increase the electrical conductivity. It should be noted that due to the excellent thermal rating of kapton polyimide (-65 to 150°C), no shrinking was experienced during the annealing process. For measurement purposes, the antenna is fed by a 50 Ω SubMiniature Version A (SMA) coaxial RF connector.

Figure 9. Final printed Polyimide based antenna prototype after thermal annealing.

5.2.1. Reflection Coefficient S_{11}

The S-parameters were measured using an Agilent PNA-X series N5242A Vector Network Analyzer (VNA) with (10 MHz-26.5 GHz) frequency range. As can be seen in Figure 10, a good agreement is achieved between the simulated and measured reflection coefficient S_{11} for the split ring antenna. The simulated return loss for the antenna is 27 dB at 2.45 GHz, with a -10 dB bandwidth of 430 MHz. The measured return loss is -28.5 dB at 2.39 GHz with a -10 dB bandwidth of 540 MHz. The increase in the measured bandwidth is attributed to the decreased electrical conductivity caused by the solvent and impurities found in the silver nanoparticle ink, which in turn increases the quality factor and leads to bandwidth enlargement.

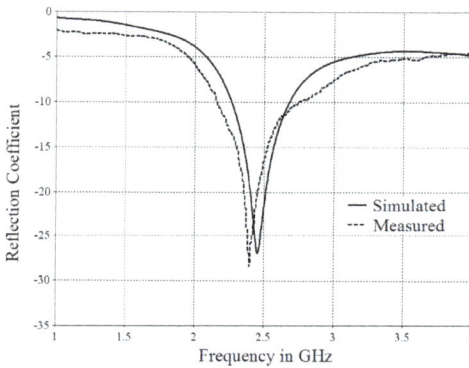

Figure 10. Measured and simulated reflection coefficient S_{11} for the split ring antenna.

5.2.2 Far-field Radiation Patterns

The far-field radiation patterns of the principal planes (E and H) were measured in a fully equipped anechoic chamber. The Antenna Under Test (AUT) was placed on an ETS Lindgren 2090 positioner and aligned to a horn antenna with adjustable polarization.

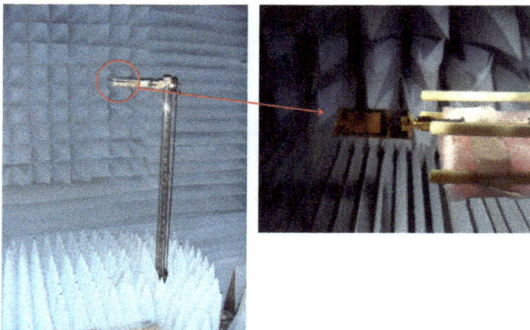

Figure 11. Radiation pattern measurement setup inside an anechoic chamber.

E-plane (*YZ* cut) and H-plane (*XZ* cut) far-field radiation patterns are shown in Figure 12. It can be seen that the radiation power is omni-directional at the resonant frequency. The antenna achieved a measured gain of 1.65 dBi which fairly agrees with the simulated value.

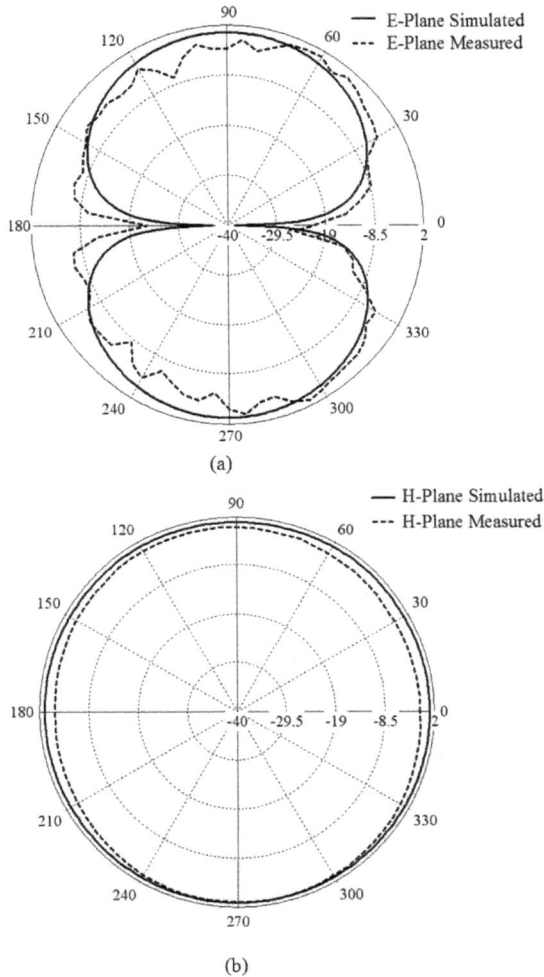

Figure 12. Measured and simulated radiation patterns for the split ring printed monopole at 2.45 GHz (a) E-plane (*YZ*) and (b) H-plane (*XZ*).

5.2.3. Flexibility Tests

Since the antenna is expected to be bent and rolled when worn or integrated within flexible devices, three tests need to be conducted for operative validation:

• Durability and robustness tests are required, which is performed by repeated testing of the fabricated antennas under bending, rolling and twisting to monitor the deposited conductive ink for any deformations, discontinuities, and to ensure there are no cracks wrinkles or permanent folds are introduced, which might compromise the antennas performance.

• Resonant frequency and return loss need to be evaluated under bending conditions since they are prone to shift/decrease due to impedance mismatch and a change in the effective electrical length of the radiating elements.

• Radiation patterns and gain of the antenna are required to be tested for distortion and/or degradation when conformed on a curved surface.

As stated before, Polyimide Kapton substrate was chosen for this technology mainly due to its physical robustness and high flexibility. Furthermore, the fabricated prototype demonstrated an excellent performance as it was tested repeatedly against bending, twisting, and rolling effects.

AUT is conformed on foam cylinders with different radii (the first is r=10mm and the second is r=8mm) to emulate different extents of bending while it is connected to the network analyzer.

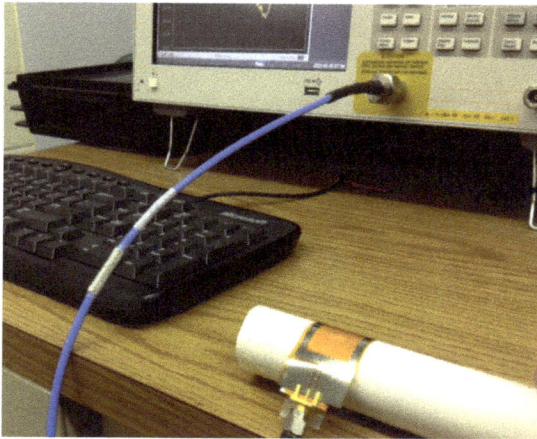

Figure 13. Flexibility test setup (AUT is conformed over a cylindrical foam with different radii to reflect different extents of bending).

As can be seen in Figure 14, around 35 MHz shift to a higher resonant frequency is experienced when the antenna is horizontally conformed on a 10 mm radius cylinder which mimics a moderate extent of bending; while a shift of 80 MHz is observed in the extreme case where the antenna is curved on a 8mm radius foam cylinder, while the antenna is less affected when bent in the vertical plane. However, the impedance bandwidth of the AUT is relatively large, which could overcome the shift caused by the bending effect. Figure 13 shows the flexibility test setup for the dual band antenna rolled on a foam cylinder with an 8 mm radius. Figure 14 depicts the reflection coefficient of the bent cases in both horizontal and vertical planes compared to the flat case.

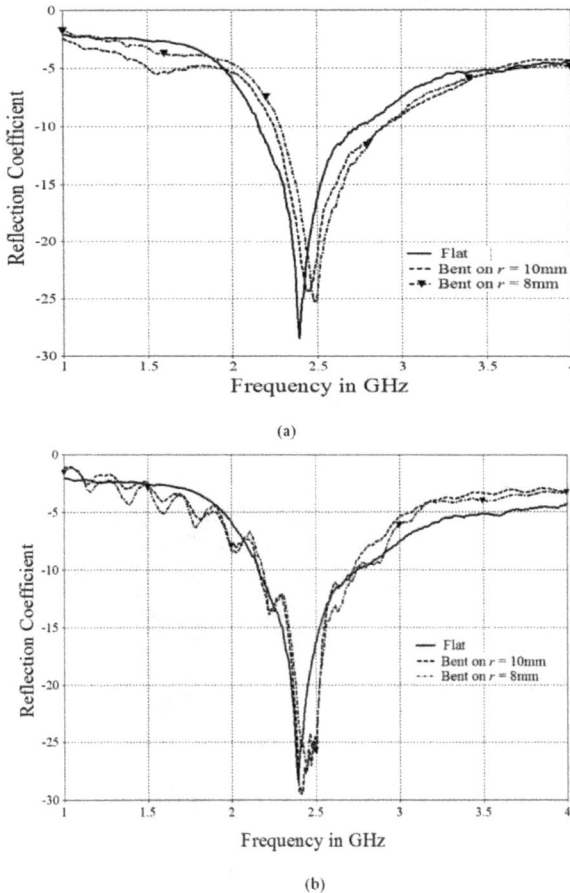

(a)

(b)

Figure 14. Measured S_{11} for the reported antenna when bent on a foam cylinder with different radii (r=10 mm and r=8 mm) to mimic different bending extents. (a) horizontal plane, (b) vertical plane.

5.2.4 Comparative Study

The split ring antenna design is compared to different types of flexible antennas reported in [4]-[7]. Given the applications envisioned in this study, the comparative study is focused on compactness (size and thickness), electrical properties and robustness. Robustness encompasses the major mechanical properties related to flexible/conformal electronic devices such as tensile strength, flexural strength, deformability, and thermal stability. Fabrication complexity criterion is also considered in this comparative study. Table 2 depicts these characteristics of the antenna under study.

Character-istics	Polyimide based antenna	Textile antenna [4]	Paper based antenna [5]	Fluidic antenna [6]	Flexible Bow-tie antenna [7]
Size in mm	38 x 27	180 x 150	46 x 35	65 x 10	39 x 25
Thickness mm	0.05	4	0.25	1	0.13
Band/f	Single/ 2.45 GHz	Dual/2.2, 3 GHz	Single/2.4 GHz	Variable	Single/7.6 GHz
Substrate	Polyimide 2=3.4	Felt fabric 2=1.5	Paper 2= 3.4	PDMS 2= 2.67	PEN film 2=3.2
Dielectric loss	Low loss tan δ=0.002	Low loss tan δ=0.02	Medium loss tan δ=0.065	High loss tan δ=0.37	Low loss tan δ=0.015
Tensile strength	High (165 MPA)	Low (2.7 MPA)	Low (30 MPA)	Low (3.9 MPA)	High (74 MPA)
Flexural strength	High (50000 p.s.i)	Low (8900 p.s.i)	Low (7200 p.s.i)	Low (650 p.s.i)	High (13640 p.s.i)
Deform-ability	Low	High	High	High	Low
Thermal stability	High	Low	Low	Low	High
Fabrication complexity	Simple/ printable	Complex/Non-printable	Simple/Printable	Complex/Non-printable	Simple/Printable

Table 2. Comparative Study of Different Types of Flexible Antennas.

As shown in Table 2, the antenna reported in this chapter offers a relatively smaller size, highly robust and flexible design. Furthermore, the antenna is printable and provides low cost and roll to roll production.

6. Conclusion

In this chapter, the design, fabrication, and measurement of flexible antennas are discussed in details. Types of substrates and available fabrication methodologies for flexible antennas are reviewed. As a benchmark, a single band printed monopole antenna operating in the

2.45 GHz ISM band is presented which has the merits of light weight, ultra low profile, wide bandwidth, robustness, compactness, and high efficiency. The reported design is based on a Kapton Polyimide substrate which is known for its flexibility, robustness and low dielectric losses. The prototype was fabricated using the inkjet printing technology. Furthermore, the antenna is tested under bending effects since it is expected to be flexed or conformed on curved surfaces. Flexibility, robustness, compactness, fabrication simplicity along with good radiation characteristics suggest that the reported methodology, antenna type and substrate is a reasonable candidate for integration within flexible electronics.

Author details

Haider R. Khaleel*, Hussain M. Al-Rizzo and Ayman I. Abbosh

*Address all correspondence to: hrkhaleel@ualr.edu

Department of Systems Engineering, University of Arkansas at Little Rock,, USA

References

[1] Hu, J. (2010). Overview of flexible electronics from ITRI's viewpoint. VLSI Test Symposium (VTS), 2010 28th, 19-22 April , 84.

[2] Nathan, A., & Chalamala, B. R. (2005, July). Special Issue on Flexible Electronics Technology, Part 1: Systems and Applications. *Proceedings of the IEEE*, 93(7), 1235-1238.

[3] Yongan, Huang, Chen, Jiankui, Yin, Zhouping, & Xiong, Youlun. (2011). Roll-to-Roll Processing of Flexible Heterogeneous Electronics With Low Interfacial Residual Stress. Components, Packaging and Manufacturing Technology IEEE Transactions on Sept., 1(9), 1368-1377.

[4] Salonen, P., Kim , Jaehoon, & Rahmat-Samii, Y. (2004). Dual-band E-shaped patch wearable textile antenna. *IEEE Antennas and Propagation Society Symposium*, 1, 466-469.

[5] Anagnostou, D. E., Gheethan, A. A., Amert, A. K., & Whites, K. W. (2010, Nov). A Direct-Write Printed Antenna on Paper-Based Organic Substrate for Flexible Displays and WLAN Applications. *Display Technology, Journal of*, 6(11), 558-564.

[6] Masahiro, Kubo., Xiaofeng, Li., Choongik, Kim., Michinao, Hashimoto., Wiley, Benjamin J., Ham, Donhee., & George, M. (2010). White sides Stretchable Microfl uidic Radio frequency Antennas,. *Wiley Inter science, Adv. Mater.*, 22, 2749-2752.

[7] Durgun, A. C., Reese, M. S., Balanis, C. A., Birtcher, C. R., Allee, D. R., & Venugopal, S. (2010, Jul). Flexible bow-tie antennas. *Antennas and Propagation Society International Symposium (APSURSI), 2010 IEEE*, 1.

[8] Yang, L., Rida, A., Vyas, R., & Tentzeris, M. M. (2007). RFID tag and RF structures on a paper substrate using inkjet-printing technology. IEEE Trans. Microw. Theory Techn. pt. 2, Dec., 55(12), 2894-2901.

[9] Khaleel, H. R., Al-Rizzo, H. M., & Rucker, D. G. (2012, Feb). Compact Polyimide-Based Antennas for Flexible Displays. *Display Technology, IEEE Journal of*, 8(2), 91-97.

[10] Du Pont Dupont Kapton Polyimide specification sheet, www2.dupont.com/kapton.

[11] Gheethan, A., & Anagnostou, D. (2012). Dual Band-Reject UWB Antenna with Sharp Rejection of Narrow and Closely-Spaced Bands. Antennas and Propagation, IEEE Transactions on 0.(99), 1.

[12] Scarpello, M. L., Kurup, D., Rogier, H., Vande, Ginste. D., Axisa, F., Vanfleteren, J., Joseph, W., Martens, L., & Vermeeren, G. (2011, Oct). Design of an Implantable Slot Dipole Conformal Flexible Antenna for Biomedical Applications. *Antennas and Propagation, IEEE Transactions on*, 59(10), 3556-3564.

[13] Choi, S. H., Jung, T. J., & Lim, S. (2010, August 19). Flexible antenna based on composite right/left-handed transmission line. *Electronics Letters*, 46(17), 1181-1182.

[14] De Jean, G., Bairavasubramanian, R., Thompson, D., Ponchak, G. E., Tentzeris, M. M., & Papapolymerou, J.. (2005). Liquid Crystal polymer (LCP): a new organic material for the development of multilayer dual-frequency/dual-polarization flexible antenna arrays. *Antennas and Wireless Propagation Letters, IEEE*, 4, 22-26.

[15] Karbassi, A., Ruf, D., Bettermann, A. D., Paulson, C. A., van der Weide, Daniel. W., Tanbakuchi, H., & Stancliff, R. (2008, Sep). Quantitative scanning near-field microwave microscopy for thin film dielectric constant measurement. *Review of Scientific Instruments*, 79(9), 094706-094706-5.

[16] Fratticcioli, E., Dionigi, M., & Sorrentino, R. (2004, Aug). A simple and low-cost measurement system for the complex permittivity characterization of materials. *Instrumentation and Measurement, IEEE Transactions on*, 53(4), 1071-1077.

[17] Dudorov, S. N., Lioubtchenko, D. V., Mallat, J. A., & Raisanen, A. V. (2005, Oct). Differential open resonator method for permittivity measurements of thin dielectric film on substrate. *Instrumentation and Measurement, IEEE Transactions on*, 54(5), 1916-1920.

[18] Ming, Li., Fortin, J., Kim, J. Y., Fox, G., Chu, F., Davenport, T., Toh-Ming, Lu., & Xi-Cheng, Zhang. (2001, Jul/Aug). Dielectric constant measurement of thin films using goniometric terahertz time-domain spectroscopy. *Selected Topics in Quantum Electronics, IEEE Journal of*, 7(4), 624-629.

[19] Janezic, M. D., Williams, D. F., Blaschke, V., Karamcheti, A., & Chi, Shih. Chang. (2003, Jan). Permittivity characterization of low-k thin films from transmission-line measurements. *Microwave Theory and Techniques, IEEE Transactions on*, 51(1), 132-136.

[20] Hertleer, C., Tronquo, A., Rogier, H., Vallozzi, L., & Van Langenhove, L. (2007). Aperture-Coupled Patch Antenna for Integration Into Wearable Textile Systems. *Antennas and Wireless Propagation Letters, IEEE*, 6, 392-395.

[21] Soh, P. J., Vandenbosch, G. A. E., Soo, Liam Ooi, & Rais, N. H. M. (2012, Jan). Design of a Broadband All-Textile Slotted PIFA. *Antennas and Propagation, IEEE Transactions on*, 60(1), 379-384.

[22] Gamota, D. R., Brazis, P., Kalyanasundaram, K., & Zhang, J. (2004). Printed organic and molecular electronics. USA 2004, Kluwer Academic Publishers. 10.1007/978-1-4419-9074-7 , 695.

[23] Halonen, E., Kaija, K., Mantysalo, M., Kemppainen, A., Osterbacka, R., & Bjorklund, N. (2009, 15-18 June). Evaluation of printed electronics manufacturing line with sensor platform application. Paper presented at Microelectronics and Packaging Conference, 2009. EMPC 2009. European,. 1-8.

[24] Kirsch, N. J., Vacirca, N. A., Plowman, E. E., Kurzweg, T. P., Fontecchio, A. K., & Dandekar, K. R. (2009, 27-28 April). Optically transparent conductive polymer RFID meandering dipole antenna. Paper presented at RFID, 2009 IEEE International Conference on. 278-282.

[25] Kirsch, N. J., Vacirca, N. A., Kurzweg, T. P., Fontecchio, A. K., & Dandekar, K. R. (2010, 11-13 Oct.) Performance of transparent conductive polymer antennas in a MIMO ad-hoc network. Paper presented at Wireless and Mobile Computing, Networking and Communications (WiMob), 2010 IEEE 6th International Conference on. 9-14.

[26] Leung, S. Y. Y., & Lam, D. C. C. (2007, July). Performance of Printed Polymer-Based RFID Antenna on Curvilinear Surface. *Electronics Packaging Manufacturing, IEEE Transactions on*, 30(3), 200-205.

[27] Kirschman, R. (1999). *Fabrication of Passive Components for High Temperature Instrumentation*, Wiley-IEEE, 679-682.

[28] Mitzner, Kraig. (2009). Complete pcb design using orcad capture and pcb editor. Publisher: Newnes; Pap/Cdr edition

[29] Niir Board. (2011). *Handbook On Printing Technology (offset, Gravure, Flexo, Screen) 2nd.*

[30] Lakafosis, V., Rida, A., Vyas, R., Li, Yang., Nikolaou, S., & Tentzeris, M. M. (2010, Sept). Progress Towards the First Wireless Sensor Networks Consisting of Inkjet-Printed, Paper-Based RFID-Enabled Sensor Tags. *Proceedings of the IEEE*, 98(9), 1601-1609.

[31] Orecchini, G., Alimenti, F., Palazzari, V., Rida, A., Tentzeris, M. M., & Roselli, L. (2011, June 6). Design and fabrication of ultra-low cost radio frequency identification

antennas and tags exploiting paper substrates and inkjet printing technology. *Microwaves, Antennas & Propagation, IET*, 5(8), 993-1001.

[32] Mitilineos, S. A., & Capsalis, C. N. (2007, Sept). A New, Low-Cost, Switched Beam and Fully Adaptive Antenna Array for 2.4 GHz ISM Applications. *Antennas and Propagation, IEEE Transactions on*, 55(9), 2502-2508.

[33] Anagnostou, D. E., Gheethan, A. A., Amert, A. K., & Whites, K. W. (2010, Nov). A Direct-Write Printed Antenna on Paper-Based Organic Substrate for Flexible Displays and WLAN Applications. *Display Technology, Journal of*, 6(11), 558-564.

[34] Shi, Cheng., Zhigang, Wu., Hallbjorner, P., Hjort, K., & Rydberg, A. (2009, Dec). Foldable and Stretchable Liquid Metal Planar Inverted Cone Antenna. *Antennas and Propagation, IEEE Transactions on*, 57(12), 3765-3771.

[35] Symeon, Nikolaou., Ponchak, G. E., Papapolymerou, J., & Tentzeris, M. M. (2006, June). Conformal double exponentially tapered slot antenna (DETSA) on LCP for UWB applications. *Antennas and Propagation, IEEE Transactions on*, 54(6), 1663-1669.

[36] Srifi, M. N., Podilchak, S. K., Essaaidi, M., & Antar, Y. M. M. (2011, Dec). Compact Disc Monopole Antennas for Current and Future Ultrawideband (UWB) Applications. *Antennas and Propagation. IEEE Transactions on*, 59(12), 4470-4480.

[37] Bae, Su Won, Yoon, Hyung Kuk, Kang, Woo Suk, Yoon, Young Joong, & Lee, Cheon-Hee. (2007). A Flexible Monopole Antenna with Band-notch Function for UWB Systems. Microwave Conference, 2007. APMC 2007. Asia-Pacific 11-14 Dec., 1-4.

[38] Jung, J., Lee, H., & Lim, Y. (2009, Mar). Broadband flexible comb-shaped monopole antenna. *Microwaves, Antennas & Propagation, IET*, 3(2), 325-332.

[39] Shaker, G., Safavi-Naeini, S., Sangary, N., & Tentzeris, M. M. (2011). Inkjet Printing of UWB Antennas on Paper-Based Substrates. *Antennas and Wireless Propagation Letters, IEEE*, 10, 111-114.

[40] CST. http/www.cst.com.

[41] Fuji Film. http://www.fujifilmusa.com/products/industrial_inkjet_printheads/deposition-Products/dmp-2800/index.html.

[42] Haga, N, Takahashi, M., Ito, K., & Saito, K. (2009). Characteristics of cavity slot antenna for Body-Area Networks. *IEEE Trans. on Antennas and Propag.*, 57(4), 837-843.

[43] Lin, C. C., Kuo, L. C., & Chuang, H. R. (2006, Nov). A Horizontally Polarized Omnidirectional Printed Antenna for WLAN Applications. *Antennas and Propagation, IEEE Transactions on*, 54(11), 3551-3556.

[44] Qi, Luo., Pereira, J. R., & Salgado, H. M. (2011). Compact Printed Monopole Antenna With Chip Inductor for WLAN. *Antennas and Wireless Propagation Letters, IEEE*, 10, 880-883.

[45] Wu, J. W., Hsiao, H. M., Lu, J. H., & Chang, S. H. (2004, Nov). Dual broadband design of rectangular slot antenna for 2.4 and 5 GHz wireless. *IEEE Electron. Lett.*, 40(23).

[46] Raj, R. K., Joseph, M., Aanandan, C. K., Vasudevan, K., & Mohanan, P. (2006, Dec). A new compact microstrip-fed dual-band coplanar antenna for WLAN applications. *IEEE Trans. Antennas Propag.*, 54(12), 3755-3762.

[47] Chen, H. D., Chen, J. S., & Cheng, Y. T. (2003, Oct). Modified inverted-L monopole antenna for 2.4/5 GHz dual-band operations. *IEEE Electron. Lett.*, 39(22).

Chapter 15

Reconfigurable Microstrip Antennas for Cognitive Radio

Mohammed Al-Husseini, Karim Y. Kabalan,
Ali El-Hajj and Christos G. Christodoulou

Additional information is available at the end of the chapter

1. Introduction

An increasing demand for radio spectrum has resulted from the emergence of feature-rich and high-data-rate wireless applications. The spectrum is scarce, and the current radio spectrum regulations make its use inefficient. This necessitates the development of new dynamic spectrum allocation policies to better exploit the existing spectrum.

According to the current spectrum allocation regulations, specific bands are assigned to particular services, and only licensed users are granted access to licensed bands. Cognitive radio (CR) is expected to revolutionize the way spectrum is allocated. In a CR network, the intelligent radio part allows unlicensed users (secondary users) to access spectrum bands licensed to primary users, while avoiding interference with them.

Two approaches to sharing spectrum between primary and secondary users have been considered: spectrum underlay and spectrum overlay. In the underlay approach, secondary users should operate below the noise floor of primary users, and thus severe constraints are imposed on their transmission power. Ultra-wideband (UWB) technology is very suitable as the enabling technology for this approach. In spectrum overlay CR, secondary users search for unused frequency bands, called white spaces, and use them for communication.

In this chapter, we report and discuss antenna designs for overlay and underlay CR. We start by studying techniques employed in the design of UWB antennas. This is done in Section 3. Such antennas are used for underlay CR, but also for channel sensing in overlay CR. We then move in Section 4 to antennas that allow the use of UWB in overlay CR. These are basically UWB antennas, but have the ability to selectively induce frequency notches in the bands of primary services, thus preventing any interference to them and giving the UWB transmitters used by the secondary users the chance to increase their power, and hence to achieve long-distance communication. In Section 5, we investigate the design of antennas for overlay CR. In this scheme, an antenna should be able to monitor the spectrum (sensing), and communicate over a chosen white space (communication). For the latter operation, the antenna must be frequency-reconfigurable. Single- and dual-port antennas for overlay CR can be designed. In the dual-port case, one port has UWB frequency response and is used for channel sensing, and the second port, which is frequency-reconfigurable, is used for communicating. In the more challenging single-port design, the same port can have UWB response for sensing, and can be reconfigured for tunable narrowband operation when required to communicate over a white space.

2. Dynamic spectrum access and cognitive radio

The increasing demand for wireless connectivity and current crowding of licensed and unlicensed spectra necessitate a new communication paradigm to exploit the existing spectrum in better ways. The current approach for spectrum allocation is based on assigning a specific band to a particular service. The FCC Spectrum Policy Task Force [1] reported vast temporal and geographic variations in the usage of allocated spectrum with utilization ranging from 15 to 85% in the bands below 3 GHz. In the frequency range above 3 GHz the bands are even more poorly utilized. In other words, a large portion of the assigned spectrum is used sporadically, leading to an under utilization of a significant amount of spectrum. This inefficiency arises from the inflexibility of the regulatory and licensing process, which typically assigns the complete rights to a frequency band to a primary user. This approach makes it extremely difficult to recycle these bands once they are allocated, even if these users poorly utilize this valuable resource. A solution to this inefficiency, which has been highly successful in the ISM (2.4 GHz), the U-NII (5–6 GHz), and microwave (57–64 GHz) bands, is to make spectra available on an unlicensed basis. However, in order to obtain spectra for unlicensed operation, new sharing concepts have been introduced to allow use by secondary users under the requirement that they limit their interference to pre-existing primary users.

2.1. Cognitive radio

Cognitive radio (CR) technology is key enabling technology which provides the capability to share the wireless channel with the licensed users in an opportunistic way. CRs are foreseen to be able to provide the high bandwidth to mobile users via heterogeneous wireless architectures and dynamic spectrum access techniques.

In order to share the spectrum with licensed users without interfering with them, and meet the diverse quality of service requirements of applications, each CR user in a CR network must [2]:

• Determine the portion of spectrum that is available, which is known as Spectrum sensing.

• Select the best available channel, which is called Spectrum decision.

• Coordinate access to this channel with other users, which is known as Spectrum sharing.

• Vacate the channel when a licensed user is detected, which is referred as Spectrum mobility.

To fulfill these functions of spectrum sensing, spectrum decision, spectrum sharing and spectrum mobility, a CR has to be cognitive, reconfigurable and self-organized. An example of the cognitive capability is the CR's ability to sense the spectrum and detect spectrum holes (also called white spaces), which are those frequency bands not used by the licensed users. The reconfigurable capability can be summarized by the ability to dynamically choose the suitable operating frequency (frequency agility), and the ability to adapt the modulation/coding schemes and transmit power as needed. The self-organized capability has to do with the possession of a good spectrum management scheme, a good mobility and connection management, and the ability to to support security functions in dynamic environments.

2.2. Spectrum sharing approaches

Dynamic spectrum access (DSA) represents the opposite direction of the current static spectrum management policy. It is broadly categorized under three models: the dynamic exclusive use model, the open sharing model, and the hierarchical access model. The taxonomy of DSA is illustrated in Fig. 1 [3].

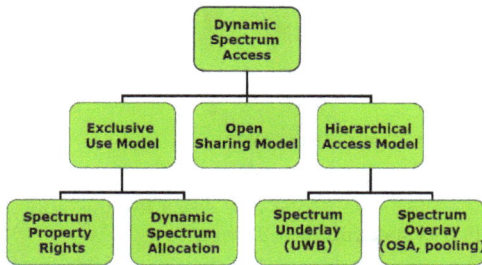

Figure 1. Dynamic spectrum access models [3]

In the dynamic exclusive use model, the spectrum bands are still licensed to services for exclusive use, as in the current spectrum regulation policy, but flexibility is introduced to improve spectrum efficiency. Two approaches have been proposed under this model: spectrum property rights and dynamic spectrum allocation. The first approach, the spectrum property rights, allows licensees to sell and trade spectrum and to freely choose technology. In the second approach, the dynamic spectrum allocation, the aim is to improve spectrum efficiency through dynamic spectrum assignment by exploiting the spatial and temporal traffic statistics of different services.

The open sharing model employs open sharing among peer users as the basis for managing a spectral region. Supporters of this model rely on the huge success of wireless services operating in the ISM band.

A hierarchical access structure with primary and secondary users is adopted by the third model. Here, the spectrum licensed to primary users is open to secondary users while limiting interference to the primary users. Two approaches to spectrum sharing between primary and secondary users have been considered: spectrum underlay and spectrum overlay.

In the underlay approach, secondary users should operate below the noise floor of primary users, and thus severe constraints are imposed on their transmission power. One way to achieve this is to spread the transmitted signals of secondary users over an ultra-wide frequency band (UWB), leading to a short-range high data rate with extremely low

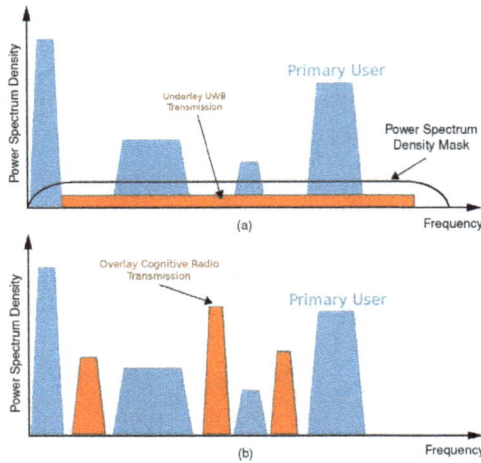

Figure 2. Underlay (a) and overlay (b) spectrum sharing approaches

transmission power (less than -42 dBm/MHz in the 3.1–10.6 GHz band). Assuming that primary users transmit all the time (worst case scenario), this approach does not rely on detection and exploitation of spectrum white space.

The spectrum overlay approach, also termed opportunistic spectrum access or OSA, imposes restrictions on when and where secondary users may transmit rather on their transmission power. In this approach, secondary users avoid higher priority users through the use of spectrum sensing and adaptive allocation. They identify and exploit the spectrum holes defined in space, time, and frequency.

The underlay and overlay approaches in the hierarchical model are illustrated in Fig. 2. They can be employed simultaneously for further spectrum efficiency improvement. Furthermore, the hierarchical model is more compatible with current spectrum management policies and legacy wireless systems as compared to the other two models.

3. UWB antennas

UWB antennas are required for underlay CR, and for sensing in overlay CR. UWB antennas were originally meant to radiate very short pulses over short distances. They have been used in medical applications, GPRs, and other short-range communications requiring high throughputs. The literature is rich with articles pertaining to the design of UWB antennas [4–9]. For example, the authors in [4] present a UWB knight's helm shape antenna fabricated on an FR4 board with a double slotted rectangular patch tapered from a 50-Ω feed line, and a partial ground plane flushed with the feed line. Three techniques are applied for good impedance matching over the UWB range: 1) the dual slots on the rectangular patch, 2) the tapered connection between the rectangular patch and the feed line, and 3) a partial ground plane flushed with the feed line. Consistent omnidirectional radiation patterns and a small group delay characterize this UWB antenna.

In general, the guidelines to design UWB antennas include:

- The proper selection of the patch shape. Round shapes and round edges lead to smoother current flow, and as a result to better wideband characteristics,

- The good design of the ground plane. Partial ground planes, and ground planes with specially designed slots, play a major role in obtaining UWB response. This property is discussed in 3.2,

- The matching between the feed line and the patch. This is achieved using either tapered connections, inset feed, or slits under the feed in the ground plane,

- The use of fractal shapes, which are known for their self-repetitive characteristic, used to obtain multi- and wide-band operation, and their space-filling property, which leads to increasing the electrical length of the antenna without tampering with its overall physical size.

To investigate the above guidelines, several UWB antennas have been designed [10–14]. Only two of them will be described in this Section.

3.1. Combination of UWB techniques

The UWB design presented in [12] features a microstrip feed line with two 45° bends and a tapered section for size reduction and matching, respectively. The ground plane is partial and comprises a rectangular part and a trapezoidal part. The patch is a half ellipse with the cut made along the minor axis. Four slots whose location and size relate to a modified Sierpinski carpet, with the ellipse as the basic shape, are incorporated into the patch. The configuration of this antenna is shown in Fig. 3.

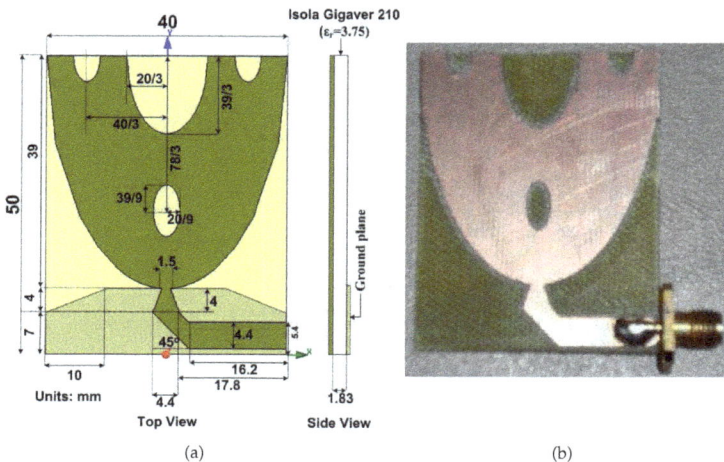

Figure 3. (a) Configuration and (b) photo of the UWB antenna in [12]. The antenna combines several bandwidth enhancement techniques.

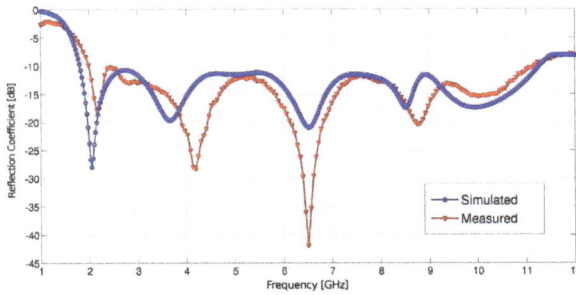

Figure 4. Reflection coefficient of the UWB antenna in Fig. 3

Four techniques are applied for good impedance matching over the UWB range: 1) the specially selected patch shape, 2) the tapered connection between the patch and the feed line, 3) the optimized partial ground plane, and 4) the slots whose design is based on the knowledge of fractal shapes. As a result, this antenna has an impedance bandwidth over the 2–11 GHz range, as shown in Fig. 4, and thus can operate in the bands used for UMTS, WLAN, WiMAX, and UWB applications. Consistent omnidirectional radiation patterns, and good gain and efficiency values characterize this UWB antenna. The radiation patterns are shown in Fig. 5.

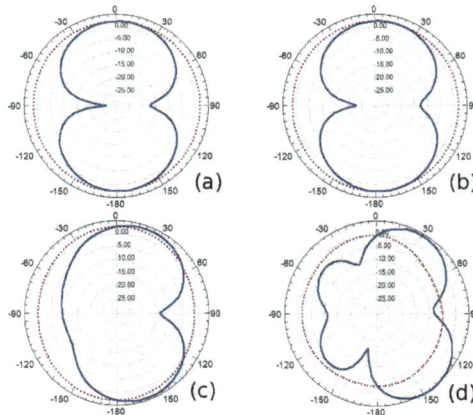

Figure 5. Patterns of the antenna in Fig. 3 in the X–Z plane (dotted line) and Y–Z plane (solid line) for (a) 2.1 GHz, (b) 2.4 GHz, (c) 3.5 GHz, and (d) 5.1 GHz

3.2. Effect of ground plane

The effect of the ground plane on the performance of UWB antennas is studied in [14–16]. The design in [14] is a coplanar-waveguide-fed antenna based on an egg-shaped conductor, and is taken as an example. The shape of the patch is suitable for UWB response. A large egg-shaped slot, with parametrized dimensions, was made in the ground, as shown in Fig. 6.

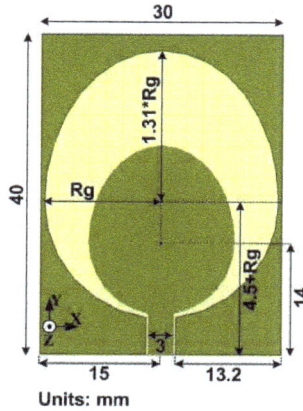

Figure 6. Design with parametrized ground plane slot [14]

The effect of changing the parameter R_g on the reflection coefficient is shown in Fig. 7. The results show that a slot of a specific size ($R_g = 14.5$ mm) results in a UWB response, so does a partial rectangular ground plane (corresponds to $R_g = \infty$). The configurations of these two optimal designs are shown in Fig. 8.

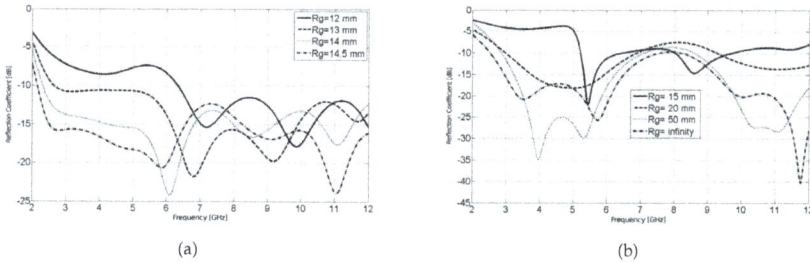

(a)   (b)

Figure 7. Reflection coefficient of the antenna in Fig. 6 for different R_g values

The measured and computed reflection coefficients of the two optimal designs are given in Fig. 9. Since the studied antenna is the type of a printed monopole, both optimal cases (with ground slot or partial rectangular ground) have omnidirectional radiation patterns, as shown in the measured patterns of Fig. 10, taken at 4 GHz.

4. Antennas with reconfigurable band rejection

UWB technology is usually associated with the CR underlay mode [17]. It can, however, be implemented in the overlay mode. The difference between the two modes is the amount of transmitted power. In the underlay mode, UWB has a considerably restricted power, which is spread over a wide frequency band. In the overlay mode, however, the transmitted power can be much higher. It actually can be increased to a level that is comparable to the power

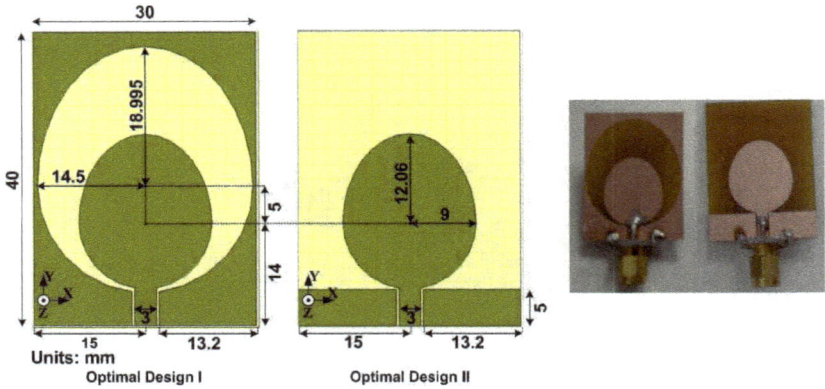

Figure 8. Two UWB antennas with optimized ground planes [14]

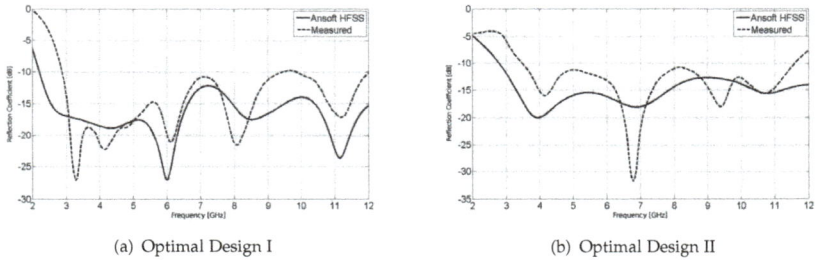

(a) Optimal Design I   (b) Optimal Design II

Figure 9. Measured and computed reflection coefficients of the designs in Fig. 8

of licensed systems, which allows for communication over medium to long distances. But this mode is only applicable if two conditions are met: 1) if the UWB transmitter ensures that the targeted spectrum is completely free of signals of other systems, or shapes its pulse to have nulls in the bands used by these systems, and 2) if the regulations are revised to allow this mode of operation [18]. Pulse adaptation for overlay UWB CR has been discussed in [19]. UWB can also operate in both underlay and overlay modes simultaneously. This can happen by shaping the transmitted signal so as to make part of the spectrum occupied in an underlay mode and some other parts occupied in an overlay mode. In the overlay UWB scenario, the antenna at the front-end of the CR device should be capable of operating over the whole UWB range, for sensing and determining the bands that are being used by primary users, but should also be able to induce band notches in its frequency response to prevent interference to these users. Even if the UWB power is not increased, having these band notches prevent raising the noise floor of primary users.

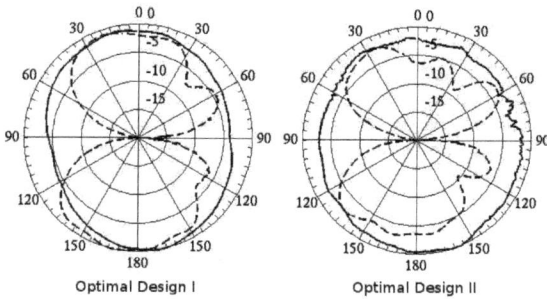

Figure 10. Measured radiation patterns of Optimal Design I (left column) and Optimal Design II (right column) in the XZ-plane (solid line) and the YZ-plane (dotted line) [14]

Antennas that allow the use of UWB in overlay CR are discussed in this Section. Several band-notching techniques are used in such antennas, the most famous of which being the use of split-ring resonators (SRRs) and the complementary split-ring resnators (CSRRs), which are discussed in 4.1. UWB antennas with fixed band notches are reported in [20–23]. Some UWB antennas with reconfigurable band notches, which are suitable for CR, are discussed below.

4.1. SRRs and CSRRs

Split-ring resonators (SRRs), originally proposed by Pendry *et al.* [24], have attracted great interest among electromagneticians and microwave engineers due to their applications to the synthesis of artificial materials (metamaterials) with negative effective permeability. From duality arguments, it has been shown that negative permittivity media can also be generated by means of resonant elements, namely, complementary split-ring resonators (CSRRs) [25]. These particles are simply the negative image of SRRs, and roughly behave as their dual counterparts.

The basic topologies of the SRR and the CSRR, and their equivalent-circuit models, are shown in Fig. 11. The equivalent-circuit models are reported in [26]. The SRR consists of two concentric metallic split rings printed on a microwave dielectric circuit board. The complementary of a planar metallic structure is obtained by replacing the metal parts of the original structure with apertures, and the apertures with metal plates. According to their lumped element models, SRRs and CSRRs do resonate. At the resonance frequency, SRRs have negative permeability, and CSRRs give negative permittivity, properties that lead to band rejection.

Single-ring SRRs and CSRRs are discussed in [27], where the relationship between the SRR and CSRR dimensions and their resonance frequencies are studied by simulations. The band rejection they cause about their resonance frequency can be controlled by mounting electronic switches across them and activating/deactivating these switches. This is illustrated in the example of Fig. 12, where two configurations of a reconfigurable bandstop filter are shown. In both configurations, a rectangular single-ring CSRR is incorporated in a 50-Ω microstrip line, which is printed on a 1.52mm-thick Rogers RO3203 substrate with $\varepsilon_r = 3.02$.

In the first configuration, an electronic switch (PIN diode or RF MEMS) is mounted on one side of the ring slot, as shown. Setting the switch ON leads to a resonating CSRR, which creates a stop band around the resonance frequency. When the switch is OFF, we end up with a complete unsplit ring, which does not have the characteristics of a CSRR, and as a result the stop band is removed. In the second configuration, a hard connection (or a capacitor with high capacitance) is present on one side of the ring slot, and a switch is mounted on the opposite side. When the switch is OFF, we have a CSRR with one gap, resonating at a certain frequency, and when the switch is ON, the CSRR will have two gaps, thus resonating at the higher frequency. The dimensions of the ring slot are chosen such that the stop band (of the single-gap case) is centered at 3.5 GHz. These dimensions are the same for both configurations.

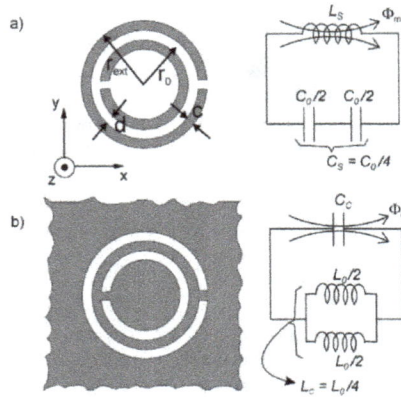

Figure 11. Topologies of the: (a) SRR and (b) CSRR, and their equivalent-circuit models. Grey zones represent the metallization. [26]

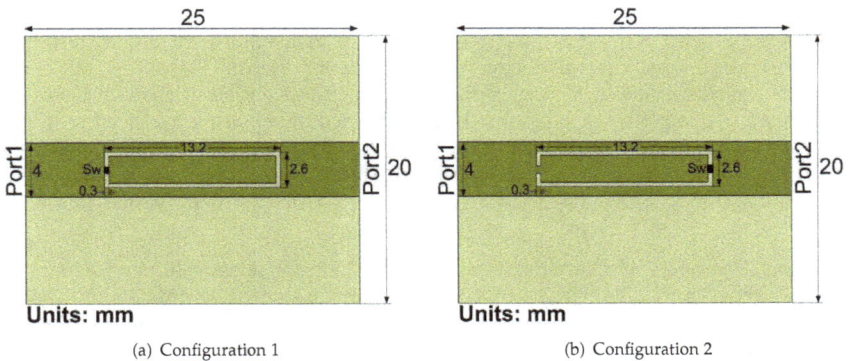

(a) Configuration 1   (b) Configuration 2

Figure 12. Reconfigurable bandstop filter based on a CSRR. (a) Configuration 1: electronic switch mounted over ring slot. (b) Configuration 2: hard connection on one side and an electronic switch over the CSRR slot on the other side.

The computed reflection and transmission coefficients for the first configuration are shown in Fig. 13. For the switch-ON case, a stop band, centered at 3.5 GHz, is created (Fig. 13(a)). When the switch is OFF, the stop band disappears, and the all-pass behavior is retrieved (Fig. 13(b)). Fig. 14 plots the computed reflection and transmission coefficients for the second configuration. A stop band, centered at 3.5 GHz, results when the switch is OFF, as shown in Fig. 14(b). A narrower stop band is created at a higher frequency, 6.6 GHz, when the switch is ON (Fig. 14(a)). It is to note that theses result hold if the CSRR is instead incorporated in the ground plane below the microstrip line. Similar properties hold for SRRs, as reported in [27]. The switching components can be replaced with varactors, to obtain notch tunability.

4.2. Antennas with a single reconfigurable rejection band

The design of a UWB antenna with a single switchable band rejection is reported in [28]. Two inverted T-shaped slits are embedded on the ground plane to allow band rejection characteristic from 5 to 6 GHz, and a PIN diode is connected to each slit to enable the switching capability for this band rejection function.

In [29], a wideband antenna with reconfigurable rejection within the operation band is presented. The antenna is a CPW-fed bow-tie, where a slot etched along the bow-tie upper edge provides the rejection of a certain band. Six PIN diodes mounted across the slots are used as switching elements, but only four switching cases are of use: one results in a notch-free wideband response, and the remaining three result each in a rejection in separate bands.

A UWB design with a single reconfigurable band notch is proposed in [30]. The configuration of this design and a photo of its fabricated prototype are shown in Fig. 15. Originally, the antenna is a UWB monopole, printed on a $30 \times 30 \times 1.6$ mm^3 Rogers RO3006 substrate with a dielectric constant $\varepsilon_r = 6.15$. It has a microstrip line feed and a partial ground plane. The patch is rectangular and is 14 mm $\times$ 15.5 mm in size, the ground is 30 mm $\times$ 10 mm, and the feed line is 2.4 mm $\times$ 10.5 mm. For better matching, the corners of the patch are rounded, by intersecting it with a circle of radius 8.75 mm, and a slit is etched in the ground below the

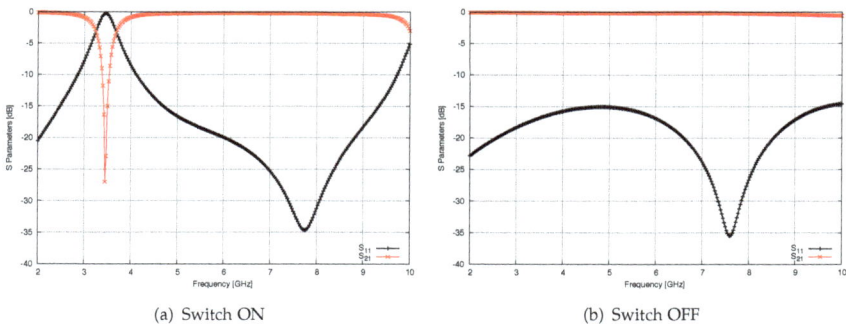

(a) Switch ON   (b) Switch OFF

Figure 13. Reflection coefficient and transmission of CSRR-based filter in Fig. 12(a). (a) Switch is ON. (b) Switch is OFF.

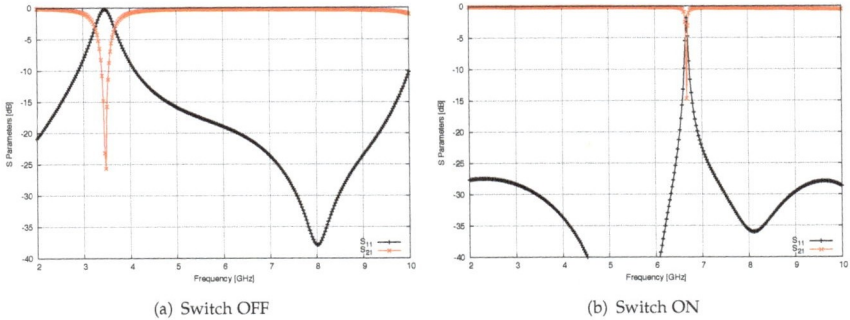

(a) Switch OFF   (b) Switch ON

Figure 14. Reflection coefficient and transmission of CSRR-based filter in Fig. 12(b). (a) Switch is OFF. (b) Switch is ON.

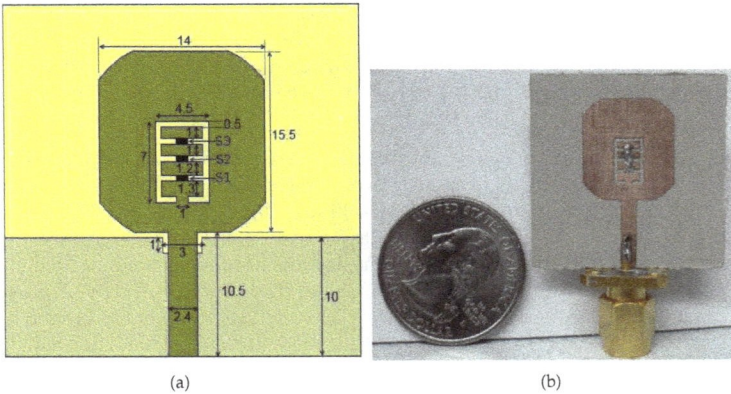

(a)   (b)

Figure 15. Configuration and photo of an antenna with one reconfigurable rejection band [30]

feed. The slit is 3 mm × 1 mm. As a result, this antenna has an impedance bandwidth that covers the whole UWB frequency range. Four nested CSRRs are incorporated in the patch. Three electronic switches, 1mm×0.5mm in size, are mounted across the slots. The sequential activation (deactivation) of the switches leads to the functioning of a larger (smaller) CSRR, and thus results in a notch at a lower (higher) frequency. The following switching cases are considered: Case 1 when all three switches are ON, Case 2 when only S3 is deactivated, Case 3 when only S1 is ON, and finally Case 4 when all switches are OFF. The resulting reflection coefficient plots, corresponding to the different switching states, are shown in Fig. 16. The plots show one notch, which can occur in one of 3 bands, or can completely disappear. In the latter case, the antenna retrieves its UWB response, which enables it to sense the whole UWB range.

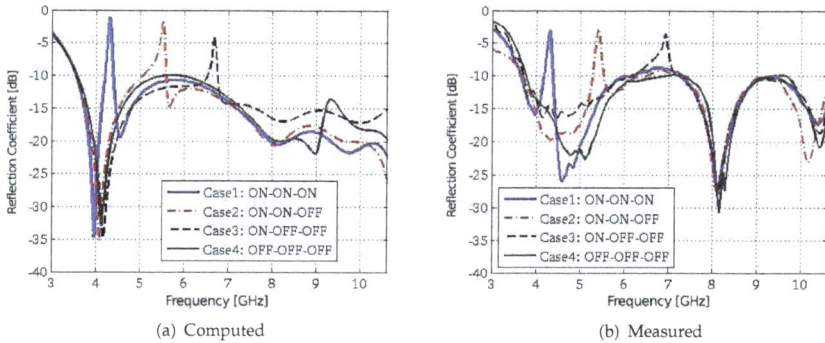

(a) Computed   (b) Measured

Figure 16. Computed and measured reflection coefficient for the different switching cases of the antenna in Fig. 15

4.3. Antennas with multiple reconfigurable rejection bands

The antenna reported in [31] is capable of inducing three band notches, which are independently controllable, using only three RF switches. Illustrated in Fig. 17, the antenna is a monopole printed on a 1.6-mm-thick Taconic TLY substrate with $\varepsilon_r = 2.2$, and features a partial rectangular ground plane. The patch is rectangular in shape, but the corners of the rectangle around the feed line are rounded to create a matching section. The dimensions of the different parts are optimized for an impedance bandwidth covering the 2–11 GHz range.

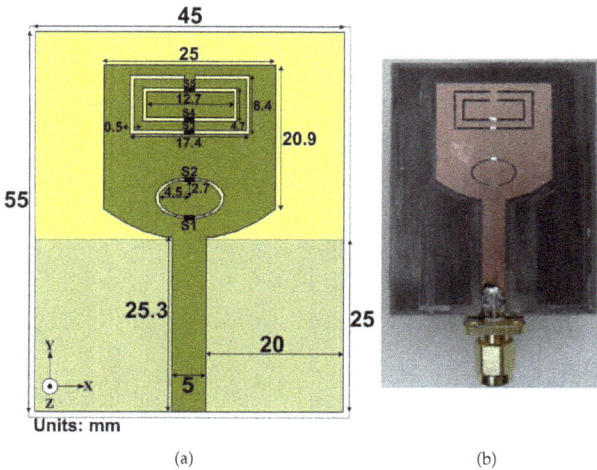

(a)   (b)

Figure 17. (a) Configuration of a UWB antenna with three independently reconfigurable band notches, and (b) photo of its prototype [31]

To create the band notches, two rectangular and one elliptical CSRRs are etched on the patch. Their shapes are selected to suit the part of the patch they are fitted in. Their sizes

Case	Notch bands (GHz)	S1	S2	S3
1	None (UWB operation)	ON	ON	ON
2	2.4	ON	OFF	ON
3	3.5	ON	ON	OFF
4	5.2	OFF	ON	ON
5	2.4, 3.5	ON	OFF	OFF
6	2.4, 5.2	OFF	OFF	ON
7	3.5, 5.2	OFF	ON	OFF
8	2.4, 3.5, 5.2	OFF	OFF	OFF

Table 1. The 8 switching cases for the design in Fig. 17 and the corresponding notched bands.

are optimized so that the larger rectangular CSRR causes a notch in the 2.4 GHz band, the smaller one in the 3.5 GHz band, and the elliptical one in the 5.2 GHz band. To enable band notch reconfigurability, three electronic switches (S1, S2, and S3) are mounted across the CSRRs.

The state of a switch controls the notch causing by the corresponding CSRR. When S1 is OFF, the elliptical CSRR induces a notch in the 5.2 GHz band. When S2 is OFF, the large rectangular CSRR causes a notch in the 2.4 GHz band. For the smaller rectangular CSRR, a notch appears at 3.5 GHz when S3 is OFF. When a switch is ON, the corresponding CSRR behaves as one with two gaps, and its resonance moves up in frequency and becomes too weak to affect the UWB response of the antenna. The different switching cases lead to different band notch combinations. These include the scenarios of one, two, three concurrent notches, or no notch at all. In the latter case, the antenna has a UWB response, which is required for channel sensing.

There are eight possible switching scenarios for this antenna, which are listed in Table 1. Fig. 18 shows the computed and measured reflection coefficient plots for some of the switching cases. For Case 1, a UWB response is obtained. The results for Case 2 reveal a single notch in the 2.4 GHz band, those for Case 4 show a single notch in the 5.2 GHz band, the results for Case 5 show two notches in the 2.4 and 3.5 GHz bands, and those for Case 8 show three notches in the 2.4, 3.5, and 5.2 GHz bands. A notch in a certain band helps to prevent interference to a primary user or the service operated in that band.

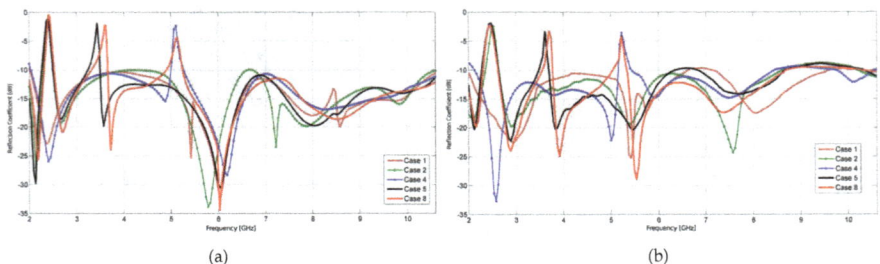

(a)   (b)

Figure 18. (a) Simulated and (b) measured reflection coefficient plots for the antenna in Fig. 17 for some of the adopted switching cases

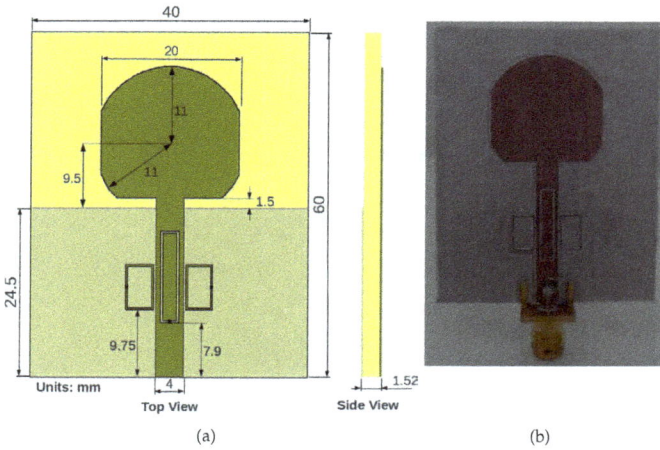

Figure 19. (a) Configuration of a filter antenna with two reconfigurable band notches, and (b) photo of its prototype [32]

The antenna has omnidirectional radiation patterns. It also has good gain values in its band(s) of operation. In a notched band, the gain drops to negative values due to strong reflections at the antenna's port.

4.4. Filter antennas with reconfigurable band notches

A UWB antenna with reconfigurable band notches can be designed by incorporating a bandstop filter in the feed line of a UWB antenna. With this structure, the switching elements will be mounted on the feed line, away from the radiating patch, which makes the bias circuit simpler to design.

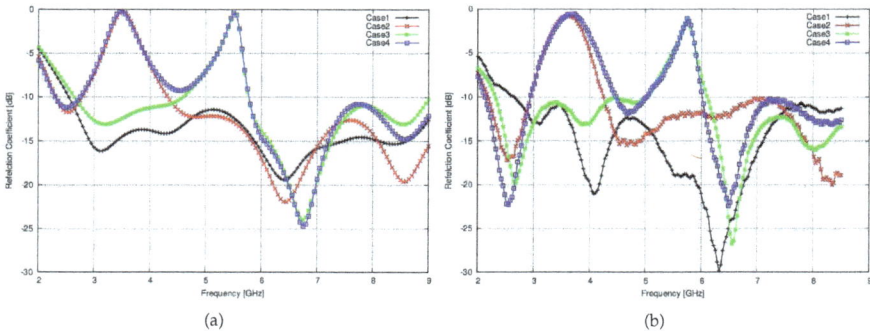

Figure 20. (a) Simulated and (b) measured reflection coefficient plots for the antenna in Fig. 19 for the four adopted switching cases.

A filter antenna with two reconfigurable rejection bands is presented in [32]. Its structure is shown in Fig. 19. The UWB antenna is based on a rounded patch and a partial rectangular

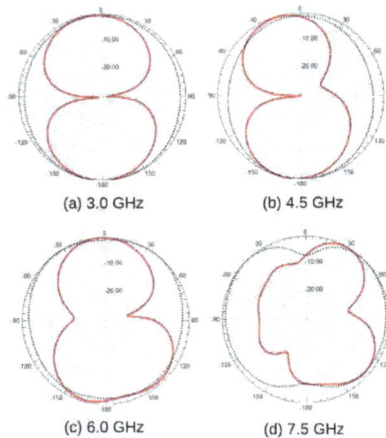

Figure 21. Radiation patterns of the antenna in Fig. 19 in the H-plane (dotted line) and E-plane (solid line)

ground plane. The Rogers RO3203 material is used for the 1.52mm-thick substrate. A reconfigurable filter with two stop bands is incorporated along its microstrip feed line. The filter is based on one rectangular single-ring CSRR etched on the line, and two identical rectangular single-ring SRRs placed in close proximity to it. The resonance of the CSRR is controlled via a switch, and that of the two SRRs via two switches that are operated in parallel. As a result, there are four switching scenarios. The simulated and measured S_{11} plots are shown in Fig. 20. Case 1, where no band notches exist, allows the antenna to sense the UWB range to determine the narrowband primary services that are transmitting inside the range. In the other three cases, the notches block the UWB pulse components in the 3.5 GHz band, the 5.5 GHz band, or both. It should be noted that notches due to the SRRs and the CSRR around the feed are stronger than those due to CSRRs or any notching structures implemented in the patch. This is because energy is concentrated in a smaller area in the feed, and coupling with the SRRs/CSRR is higher.

The normalized gain patterns of the filter antenna, for Case1, are shown in Fig. 21. The antenna has good omnidirectional patterns, and this is expected since its is a printed monopole with a small ground plane not covering the radiating patch. Since the patch is untouched, the patterns are independent of the switching cases. The realized peak gain of the antenna is plotted in Fig. 22 for Case 1 and Case 4. In Case 4, the gain drops sharply, to below -10 dB, at 3.5 GHz and at 5.5 GHz. In these two bands, very high reflections occur at the antenna's input. The gain drop in the notch band is large, as the coupling the CSRR and the SRRs cause is high. Due to the location of the switches, connecting the DC bias lines, especially to the SRRs, which are DC-separated from anything else, is an easy task. A wire can be used to drive the switch on the CSRR. A note is that extra band notches can be obtained by placing more SRRs around the feed line.

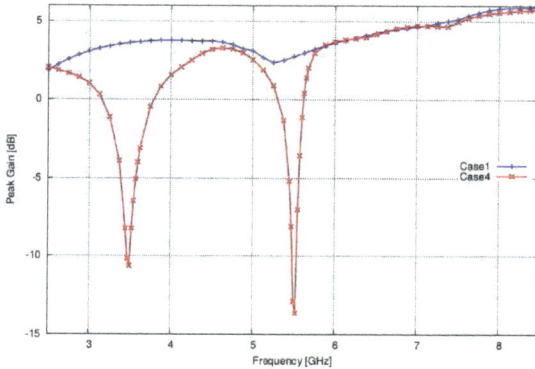

Figure 22. Realized peak gain of the antenna in Fig. 19. Case 1: gain is positive. Case 4: gain is negative in the 3.5 and 5.5 GHz bands.

5. Antennas for overlay CR

Antennas designed for overlay CR should have the capability to sense the channel and communicate over a small portion of it. These antennas can be implemented as dual-port, where one port is UWB, and the other is narrowband and frequency reconfigurable. The design of UWB antennas was discussed in Section 3. They can also be designed as single-port, where the same port is used for both sensing and communicating, and thus should switch between wideband and narrowband operations. The advantages of each design will be revealed in the rest of this Section.

5.1. Dual-port antennas for overlay CR

A dual-port antenna for overlay was proposed by Ebrahimi *et al.* [33, 34]. The structure consists of two printed antennas namely a wide- and a narrow-band antenna. Because the two antennas are in close proximity, high coupling exists between them, and the patterns of the NB antenna are affected by the presence of the UWB one. The authors successfully designed modified versions of the antenna system to solve the coupling issue.

A simpler design that offers good isolation between the two antenna ports is presented in [35]. The configuration of this design, which comprises two microstrip-line-fed monopoles sharing a common partial ground, is shown in Fig. 23. The sensing UWB antenna is based on an egg-shaped patch, obtained by combining a circle and an ellipse at their centers. A small tapered microstrip section is used to match the 50-Ω feed to the input impedance of the patch. The UWB response of the sensing antenna is guaranteed by the design of the patch, the partial ground plane, and the feed matching section. The reflection coefficient of the sensing antenna, and its normalized 5-GHz patterns, are shown in Fig. 24.

The communicating antenna is a microstrip line connected to a 50-Ω feed line via a matching section. Two electronic switches are incorporated along this line. By controlling the switches, the length of the antenna is changed, which leads to various resonance frequencies inside the UWB range. Three switching cases are considered: Case 1 where both switches are deactivated, Case 2 where Switch 1 is ON and Switch 2 is OFF, and Case 3 where both

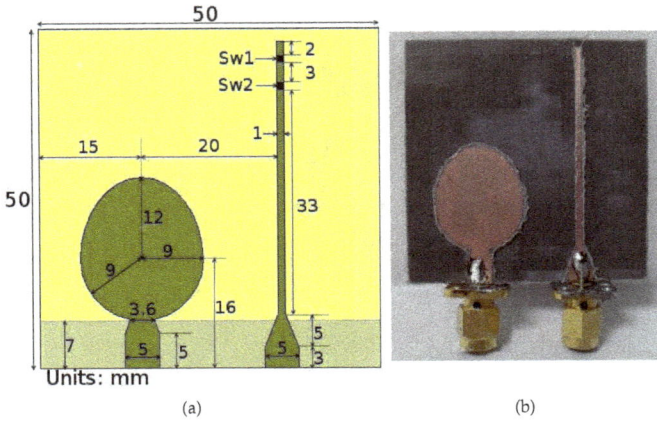

Figure 23. Dual-port UWB-NB antenna for overlay CR (a) configuration and (b) photo of a fabricated prototype [35]

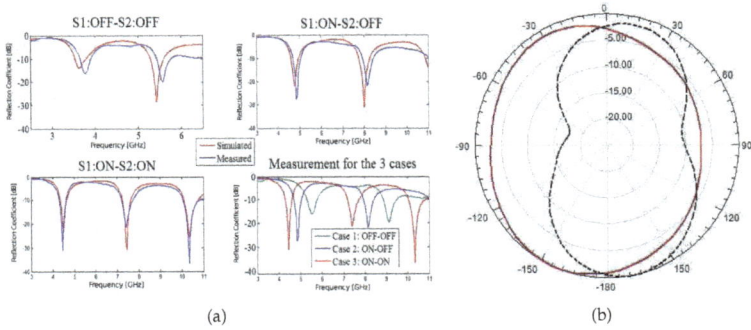

Figure 24. (a) Reflection coefficient of the sensing UWB antenna, and (b) its normalized gain patterns at 5 GHz: H-plane (solid line) and the E-plane (dashed line) [35]

switches are activated. The resulting measured reflection coefficient plots are given in Fig. 25, which shows clear frequency reconfigurability and coverage of most of the UWB range.

The transmission S_{21} at the resonance frequencies, for the three switching cases, are given in Table 2. Good isolation between the UWB and NB port is achieved, given the simplicity of the design.

	Case 1		Case 2		Case 3		
f (GHz)	5.55	9.15	4.85	8.15	4.44	7.41	10.33
S_{21} (dB)	-31.4	-14.8	-16.8	-15.5	-21.3	-22.6	-15.7

Table 2. Transmission S_{21} for the design in [35]

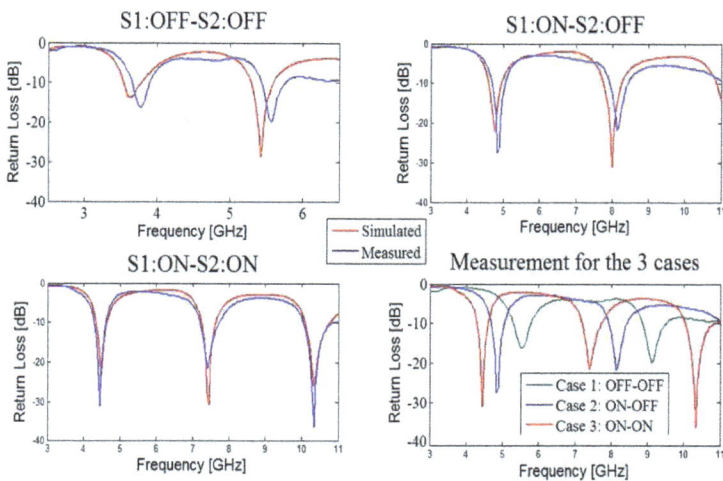

Figure 25. Measured reflection coefficient of the communicating antenna [35]

The communicating antenna has also omnidirectional patterns, but some degradation occurs due to the presence of the UWB patch.

5.2. Single-port antennas for overlay CR

Dual-port antennas enable simultaneous sensing and communicating over the channel, but have limitations in terms of their relatively large size, the coupling between the two ports, and the degraded patterns. These limitations are solved by the use of single-port antennas, but these are only suitable when the channel does not change very fast, and thus sensing and communication are possible, sequentially. Single-port CR antennas are also more challenging to design.

A reconfigurable wideband/dual-band double C-slot microstrip patch antenna is proposed in [36]. The frequency tuning is performed by switching ON and OFF two patches. The antenna operates in one of two different dual-band modes when either patch is activated, and in very wide band mode when both patches are excited.

In [37], a single-port Vivaldi antenna with added switched band functionality to operate in a wideband or narrowband mode is presented. Frequency reconfigurability in this design is attained by inserting four pairs of ring slots into the structure, and switching them using PIN diodes. A wide bandwidth mode covering the 1.0–3.2 GHz range, and three narrowband modes within this range, can be selected. A single pair of ring slots, and fifteen PIN diode switches across, are used on the single-port Vivaldi design in [38]. For this antenna, a wideband operation is obtained over the 1–3 GHz band, inside which there are six narrowband states of use. In these two Vivaldi antenna designs, the switching elements, PIN diodes in this case, are mounted on the radiating parts of the antennas. This makes the design of the DC bias circuits a complex task, as the designers have to make sure these circuits have little interference to the antenna performance.

The single-port overlay CR antenna in [39] has the switching elements mounted along its microstrip feed line, away from the radiating patch. This property has the advantage that the DC bias circuit causes limited interference to the antenna characteristics. The antenna is initially UWB, which makes it sensing-capable. A reconfigurable bandpass filter is then embedded along its feed line. When activated, the filter can transform the UWB frequency response into a reconfigurable narrowband one, which is suitable for the communication operation of the CR system. The configuration of the antenna, and a closer view of its embedded filter part, are shown in Fig. 26. It features a partial rectangular ground plane, a rectangular patch, and a curved matching section between the microstrip feed line and the patch. The filter is based on a symmetrical defected microstrip structure (DMS) implemented in the feed line of the UWB antenna. It has a T-shaped slot, which by itself, has bandstop characteristics. However, when placed between a pair of gaps, which act as capacitors, a bandpass structure results [40].

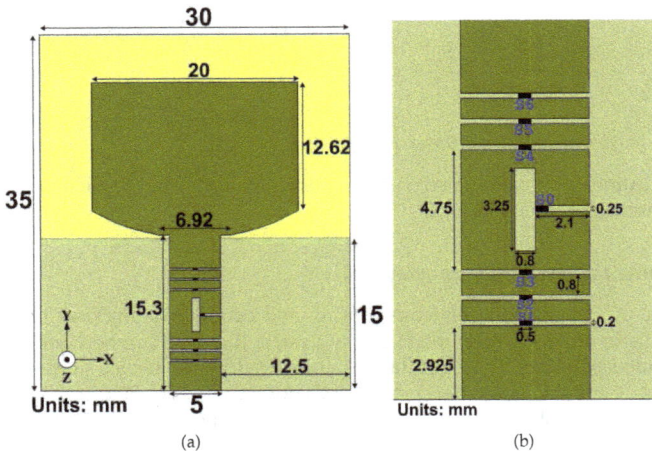

Figure 26. A reconfigurable UWB/NB filter antenna. (a) Configuration, and (b) closer view of the embedded filter [39]

For the purpose of achieving frequency reconfigurability, three pairs of gaps are symmetrically placed around the T-slot, and seven electronic switches are placed across the slots as shown. Six switching cases are considered, as indicated in Table 3. Case 0 corresponds to all the switches being ON. In this case, the effect of the filter is canceled, bringing back the UWB response of the antenna. The frequency characteristics of the filter depend on the dimensions of the slots, and on the switching state.

The computed and measured reflection coefficient plots for the six switching cases are given in Fig. 27. The operation of the antenna makes it suitable for employment in cognitive radio applications, where Case 0 could be used for sensing the channel (to determine the white spaces), and the other cases for communicating in the corresponding white space. Further resonances can be obtained by including more gaps around the T-slot and appropriately choosing their locations and widths.

Case	Switches in OFF state
0	None (all ON)
1	S0, S1, S6
2	S0, S1, S5
3	S0, S2, S5
4	S0, S3, S5
5	S0, S3, S4

Table 3. The six adopted switching cases for the single-port design in [39]

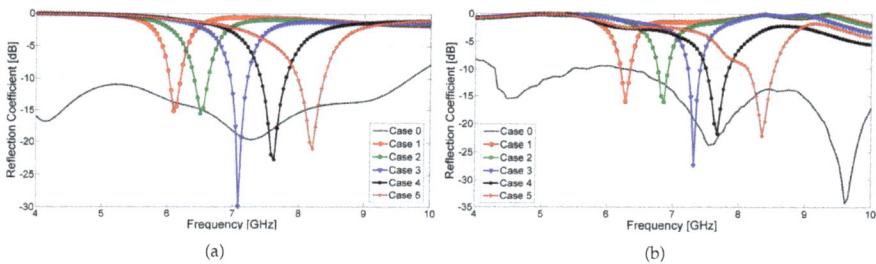

(a)   (b)

Figure 27. (a) Simulated, and (b) measured reflection coefficient [39]

5.3. A Single-port tunable filter antenna for overlay CR

Reconfigurable NB antennas can operate over a limited number of bands inside a designated frequency range. Tunable antennas, on the other hands, can be reconfigured to resonate at theoretically an infinite number of frequencies within a certain band. Tunable antennas can be used for communicating in a CR system, by tuning to a white space. They can also be employed for channel sensing, by progressively scanning small portions of the band. This depends of course on the rate at which the channel is changing, and on the tuning speed.

A tunable filter antenna is shown in Fig. 28. The initially UWB design features a tunable bandpass filter embedded along its microstrip feed line. It has a rounded patch and a partial rectangular ground plane. The Rogers RO3203 material, with $\varepsilon_r = 3.02$, is used for the 1.52mm-thick substrate. The filter is based on a T-shaped slot incorporated in the microstrip line between a pair of gaps.

For the purpose of achieving frequency tunability, a varactor is included in the design, as indicated. Changing the capacitance of the varactor changes the notch band caused by the T-slot, and as a result the narrow pass band of the overall filter. The DC lines of the varactors are connected with ease. Due to the presence of the two gaps, DC is separated from both the antenna port and the patch. Two surface-mount inductors are used over the DC lines as RF chokes. The reflection coefficient plots, obtained in Ansoft HFSS, are given in Fig. 29. They show narrowband tunability over the 4.5–7 GHz frequency range, for capacitance values between 0.3 and 7 pF.

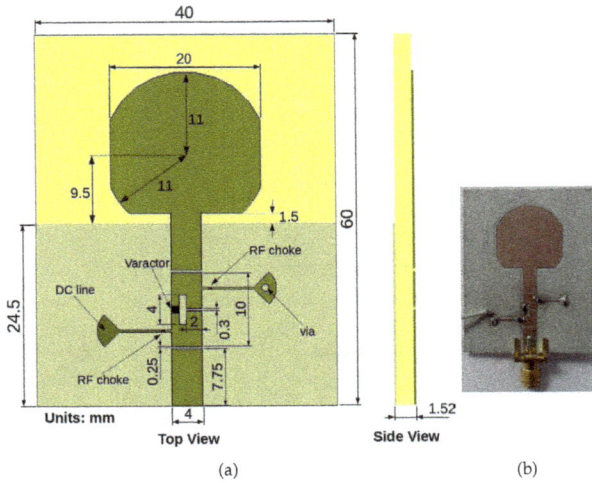

Figure 28. (a) Configuration of a tunable filter antenna, and (b) photo of a fabricated prototype. This antenna has a tunable bandpass filter embedded along its feed line.

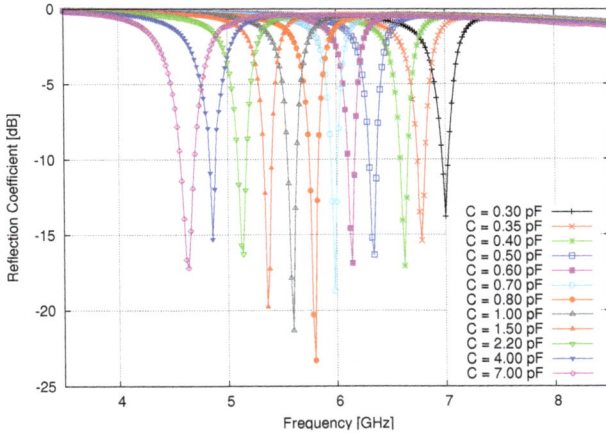

Figure 29. Reflection coefficient of the tunable filter antenna. Narrowband tunability is achieved

A UWB operation of the antenna can be made possible by installing 3 switching elements (e.g. PIN diodes) across the T-shaped slot and the two gaps. When these switches are ON, the effect of the narrowband bandpass filter is canceled, and the original UWB response of the antenna is retrieved. Tunablility would still be possible by putting the switches to the OFF state and adjusting the varactor capacitance. Extra DC biasing lines are required to control the switches, but they will be installed away from the radiating patch. This makes their design relatively simple.

6. Summary

This chapter has discussed the design of antennas for Cognitive Radio applications. CR is a revolutionary spectrum allocation technology that allows unlicensed users to access spectrum bands licensed to primary users, at the condition of avoiding interference to them. Spectrum underlay and spectrum overlay are two approaches to sharing spectrum between primary and secondary users.

UWB antennas are required for sensing in overlay CR, and for communicating in underlay CR. Modified UWB antennas with reconfigurable band notches allow to employ UWB technology in overlay CR and to achieve high-data-rate and long distances communications. Overlay CR requires reconfigurable wideband/narrowband antennas, to perform the two tasks of sensing a wide band and communicating over a narrow white space. UWB antennas, antennas with reconfigurable band rejections, and single-port/dual-port wide-narrowband and tunable antennas suitable for these approaches have been reported.

Author details

Mohammed Al-Husseini[1], Karim Y. Kabalan[1],
Ali El-Hajj[1] and Christos G. Christodoulou[2]

1 American University of Beirut, Lebanon
2 University of New Mexico, USA

References

[1] Federal Communications Commission (2002) Spectrum Policy Task Force Report, Technical Report.

[2] Chen K.-C, Prasad R. (2009) Cognitive Radio Networks. ISBN 978-0-470-69689-7. West Sussex, United Kingdom: John Wiley & Sons.

[3] Zhao Q, Sadler B.M (2007) A Survey of Dynamic Spectrum Access: Signal Processing, Networking, and Regulatory Policy. IEEE signal processing magazine. 24:3:79–89.

[4] Low Z.N, Cheong J.H, Law, C.L (2005) Low-cost PCB Antenna for UWB Applications. IEEE antennas and wireless propagation letters. 4:237–239.

[5] Mehdipour A, Mohammadpour-Aghdam K, Faraji-Dana R, Kashani-Khatib M.-R (2008) A Novel Coplanar Waveguide-fed Slot Antenna for Ultra- wideband Applications. IEEE transactions on antennas and propagation. 56:12:3857–3862.

[6] Oraizi H, Hedayati S (2011) Miniaturized UWB Monopole Microstrip Antenna Design by the Combination of Giusepe Peano and Sierpinski Carpet Fractals. IEEE antennas and wireless propagation letters. 10:67–70.

[7] Azari A (2011) A New Super Wideband Fractal Microstrip Antenna. IEEE transactions on antennas and propagation. 59:5:1724–1727.

[8] Ghaderi M.-R, Mohajeri F (2011) A Compact Hexagonal Wide-slot Antenna with Microstrip-fed Monopole for UWB Application. IEEE antennas and wireless propagation letters. 10:682–685.

[9] Liu W, Yin Y, Xu W, Zuo S (2011) Compact Open-slot Antenna with Bandwidth Enhancement. IEEE antennas and wireless propagation letters. 10:850–853.

[10] Al-Husseini M, Ramadan A, El-Hajj A, Kabalan K.Y (2008) A 1.9-13.5 GHz Low-cost Microstrip Antenna. Proceedings of the 2008 International Wireless Communications and Mobile Computing Conference (IWCMC 2008). Crete Island, Greece. 6–8 Aug 2008.

[11] Al-Husseini M, Ramadan A, El-Hajj A, Kabalan K.Y (2009) Design of a Compact and Low-cost Fractal-based UWB PCB Antenna. Proceedings of the 26th National Radio Science Conference (NRSC 2009). Cairo, Egypt. 17–19 Mar 2009.

[12] Al-Husseini M, Tawk Y, El-Hajj A, Kabalan K.Y (2009) A Low-cost Microstrip Antenna for 3G/WLAN/WiMAX and UWB Applications. Proceedings of the 2009 International Conference on Advances in Computational Tools for Engineering Applications (ACTEA 2009). Zouk Mosbeh, Lebanon. 15–17 Jul 2009. pp. 68–70.

[13] Al-Husseini M, Ramadan A, Tawk Y, El-Hajj A, Kabalan, K.Y (2009) Design and Ground Plane Consideration of a CPW-fed UWB Antenna. Proceedings of the 2009 International Conference on Electrical and Electronics Engineering (ELECO 2009). Bursa, Turkey. 5–8 Nov 2009. pp. II-151–II-153.

[14] Al-Husseini M, Ramadan A, Tawk Y, El-Hajj A, Kabalan, K.Y (2011) Design and Ground Plane Optimization of a CPW-fed UWB Antenna. Turkish journal of electrical engineering and computer sciences. 19:2:243–250.

[15] Curto S, John M, Ammann M (2007) Groundplane Dependent Performance of Printed Antenna for MB-OFDM-UWB. Proceedings of the IEEE 65th Vehicular Technology Conference. Dublin, Ireland. 22–25 Apr 2007. pp. 352–356.

[16] OOi P.C, Selvan K.T (2010) The Effect of Ground Plane on the Performance of a Square Loop CPW-fed Printed Antenna. Progress in electromagnetics research letters. 19:103–111.

[17] Arslan H, Sahin M (2007) UWB-based Cognitive Radio Networks. In: Hossain E, Bhargava V, editors. Cognitive Wireless Communication Networks. US: Springer.

[18] Zhang H, Zhou X, Chen T (2009) Ultra-wideband Cognitive Radio for Dynamic Spectrum Accessing Networks. In: Xiao Y, Hu F, editors. Cognitive Radio Networks. ISBN 978-1-4200-6420-9. Boca Raton, Florida: CRC Press. pp. 353–382.

[19] Safatly L, Al-Husseini M, El-Hajj A, Kabalan K.Y (2012) Advanced Techniques and Antenna Design for Pulse Shaping in UWB Cognitive Radio. International journal of antennas and propagation. DOI:10.1155/2012/390280.

[20] Kelly J.R, Hall P.S, Gardner P, (2011) Band-notched UWB Antenna Incorporating a Microstrip Open-loop Resonator. IEEE transactions on antennas and propagation. 59:8:3045–3048.

[21] Nguyen T.-D, Lee D.-H, Park H.-C (2011) Design and Analysis of Compact Printed Triple Band-notched UWB Antenna. IEEE antennas and wireless propagation letters. 10:403–406.

[22] Almalkawi M, Devabhaktuni V (2011) Ultrawideband Antenna with Triple Band-notched Characteristics Using Closed-loop Ring Resonators. IEEE antennas and wireless propagation letters. 10:959–962.

[23] Kim D.-O, Jo N.-I, Jang H.-A, Kim C.-Y (2011) Design of the Ultrawideband Antenna with a Quadruple-band Rejection Characteristics Using a Combination of the Complementary Split Ring Resonators. Progress in electromagnetics research, 112:93–107.

[24] Pendry J.B, Holden A.J, Robbins D.J, Stewart W.J (1999) Magnetism from Conductors and Enhanced Nonlinear Phenomena. IEEE transactions on microwave theory and techniques. 47:11:2075–2084.

[25] Falcone F, Lopetegi T, Laso M.A.G, Baena J.D, Bonache J, Beruete M, Marques R, Martin F, Sorolla M (2004) Babinet Principle Applied to the Design of Metasurfaces and Metamaterials. Physical review letters. 93:19:197401-1–197401-4.

[26] Baena J.D et al. (2005) Equivalent-circuit Models for Split-ring Resonators and Complementary Split-ring Resonators Coupled to Planar Transmission Lines. IEEE transactions on microwave theory and techniques. 53:4:1451–1461.

[27] Al-Husseini M (2012) Antenna Design for Overlay and Underlay Cognitive Radio Applications, Ph.D. Dissertation. Beirut, Lebanon: American University of Beirut.

[28] Sim C.-Y.-D, Chung W.-T, Lee C.-H (2010) Planar UWB Antenna with 5 GHz Band Rejection Switching Function at Ground Plane. Progress in electromagnetics research. 106:321–333.

[29] Perruisseau-Carrier J, Pardo-Carrera P, Miskovsky P (2010) Modeling, Design and Characterization of a Very Wideband Slot Antenna with Reconfigurable Band Rejection. IEEE transactions on antennas and propagation. 58:7:2218–2226.

[30] Al-Husseini M, Costantine J, Christodoulou C.G, Barbin S.E, El-Hajj A, Kabalan, K.Y (2010) A Reconfigurable Frequency-notched UWB Antenna with Split-ring Resonators. Proceedings of the 2010 Asia-Pacific Microwave Conference (APMC 2010). Yokohama, Japan. 7–10 Dec 2010.

[31] Al-Husseini M, Ramadan A, Tawk Y, Christodoulou C.G, El-Hajj A, Kabalan K.Y (2011) Design Based on Complementary Split-ring Resonators of an Antenna with Controllable Band Notches for UWB Cognitive Radio Applications. Proceedings of 2011 IEEE AP-S International Symposium on Antennas and Propagation (IEEE AP-S 2011). Spokane, Washington, USA. 3–8 Jul 2011.

[32] Al-Husseini M, Safatly L, El-Hajj A, Kabalan K.Y, Christodoulou C.G (2012) Reconfigurable Filter Antennas for Pulse Adaptation in UWB Cognitive Radio Systems. Progress in electromagnetics research B. 37:327–342.

[33] Ebrahimi E, Hall P.S (2009) A Dual Port Wide-narrowband Antenna for Cognitive Radio. Proceedings of the third European Conference on Antennas and Propagation (EuCAP2009). Berlin, Germany. 23–27 Mar 2009. pp. 809–812

[34] Ebrahimi E, Kelly J.R, Hall P.S (2011) Integrated Wide-narrowband Antenna for Multi-standard Radio. IEEE transactions on antennas and propagation. 59:7:2628–2635.

[35] Al-Husseini M, Tawk Y, Christodoulou C.G, El-Hajj A, Kabalan K.Y (2010) A Reconfigurable Cognitive Radio Antenna Design. Proceedings of the 2010 IEEE AP-S International Symposium on Antennas and Propagation (IEEE AP-S 2010). Toronto, ON, Canada. 11–17 Jul 2010.

[36] Abu Tarboush H.F, Khan S, NilavalanR, Al-Raweshidy H.S, Budimir D (2009) Reconfigurable Wideband Patch Antenna for Cognitive Radio. Proceedings of the 2009 Loughborough antennas and propagation conference (LAPC 2009), Loughborough, UK. 16–17 Nov 2009. pp.141–144.

[37] Hamid M.R, Gardner P, Hall P.S, Ghanem F (2011) Switched-band Vivaldi Antenna. IEEE transactions on antennas and propagation. 59:5:1472–1480.

[38] Hamid M.R, Gardner P, Hall P.S, Ghanem F (2011) Vivaldi Antenna with Integrated Switchable Band Pass Resonator. IEEE transactions on antennas and propagation. 59:11:4008–4015.

[39] Al-Husseini M, Ramadan A, Zamudio M.E, Christodoulou C.G, El-Hajj A, Kabalan K.Y (2011) A UWB Antenna Combined with a Reconfigurable Bandpass Filter for Cognitive Radio Applications. Proceedings of the 2011 IEEE-APS Topical Conference on Antennas and Propagation in Wireless Communications (IEEE APWC 2011). Torino, Italy. 12–16 Sep 2011. pp. 902–904.

[40] Kazerooni M, Cheldavi A, Kamarei M (2009) A Novel Bandpass Defected Microstrip Structure (DMS) Filter for Planar Circuits. Proceedings of the 2009 Progress in Electromagnetics Research Symposium (PIERS 2009). Moscow, Russia. 18–21 Aug 2009. pp. 1214–1217.

Permissions

The contributors of this book come from diverse backgrounds, making this book a truly international effort. This book will bring forth new frontiers with its revolutionizing research information and detailed analysis of the nascent developments around the world.

We would like to thank Ahmed Kishk, for lending his expertise to make the book truly unique. He has played a crucial role in the development of this book. Without his invaluable contribution this book wouldn't have been possible. He has made vital efforts to compile up to date information on the varied aspects of this subject to make this book a valuable addition to the collection of many professionals and students.

This book was conceptualized with the vision of imparting up-to-date information and advanced data in this field. To ensure the same, a matchless editorial board was set up. Every individual on the board went through rigorous rounds of assessment to prove their worth. After which they invested a large part of their time researching and compiling the most relevant data for our readers. Conferences and sessions were held from time to time between the editorial board and the contributing authors to present the data in the most comprehensible form. The editorial team has worked tirelessly to provide valuable and valid information to help people across the globe.

Every chapter published in this book has been scrutinized by our experts. Their significance has been extensively debated. The topics covered herein carry significant findings which will fuel the growth of the discipline. They may even be implemented as practical applications or may be referred to as a beginning point for another development. Chapters in this book were first published by InTech; hereby published with permission under the Creative Commons Attribution License or equivalent.

The editorial board has been involved in producing this book since its inception. They have spent rigorous hours researching and exploring the diverse topics which have resulted in the successful publishing of this book. They have passed on their knowledge of decades through this book. To expedite this challenging task, the publisher supported the team at every step. A small team of assistant editors was also appointed to further simplify the editing procedure and attain best results for the readers.

Our editorial team has been hand-picked from every corner of the world. Their multi-ethnicity adds dynamic inputs to the discussions which result in innovative outcomes. These outcomes are then further discussed with the researchers and contributors who give their valuable feedback and opinion regarding the same. The feedback is then collaborated with the researches and they are edited in a comprehensive manner to aid the understanding of the subject.

Apart from the editorial board, the designing team has also invested a significant amount of their time in understanding the subject and creating the most relevant covers. They scrutinized every image to scout for the most suitable representation of the subject and create an appropriate cover for the book.

The publishing team has been involved in this book since its early stages. They were actively engaged in every process, be it collecting the data, connecting with the contributors or procuring relevant information. The team has been an ardent support to the editorial, designing and production team. Their endless efforts to recruit the best for this project, has resulted in the accomplishment of this book. They are a veteran in the field of academics and their pool of knowledge is as vast as their experience in printing. Their expertise and guidance has proved useful at every step. Their uncompromising quality standards have made this book an exceptional effort. Their encouragement from time to time has been an inspiration for everyone.

The publisher and the editorial board hope that this book will prove to be a valuable piece of knowledge for researchers, students, practitioners and scholars across the globe.

List of Contributors

Daniel B. Ferreira and Cristiano B. de Paula
CPqD - Telecommunications R&D Foundation, Brazil

Daniel C. Nascimento
ITA - Technological Institute of Aeronautics, Brazil

Ouarda Barkat
Electronics Department, University of Constantine, Constantine, Algeria

Marek Bugaj and Marian Wnuk
Faculty of Electronics, Military University of Technology, Warsaw, Poland

Shun-Shi Zhong and Zhu Sun
Shanghai University, China

Ahmad Rashidy Razali
Faculty of Electrical Engineering, University of Technology MARA, Shah Alam, Malaysia

Amin M Abbosh and Marco A Antoniades
School of Information Technology and Electrical Engineering, The University of Queensland, Brisbane, Australia

Rezaul Azim and Mohammad Tariqul Islam
Institute of Space Science (ANGKASA), Universiti Kebangsaan, Malaysia
Department of Physics, University of Chittagong, Bangladesh

Osama Haraz
Electrical and Computer Engineering Department, Concordia University, Canada

Abdel-Razik Sebak
Electrical and Computer Engineering Department, Concordia University, Canada
KACST Technology Innovation Center in RFTONICS, PSATRI, King Saud University, Saudi Arabia

Gijo Augustin , Bybi P. Chacko and Tayeb A. Denidni
National Institute of Scientific Research (INRS), Montreal QC, Canada

Li Sun and Shusen Tan
Beijing Satellite Navigation Center, China

Gang Ou
College of Electronic Science and Engineering, National University of Defense Technology, China

Yilong Lu
School of Electrical and Electronic Engineering, Nanyang Technological University, Singapore

Ken G. Clark and James M. Tranquilla
EMR Microwave Technology Corporation, 64 Alison Blvd., Fredericton, NB, Canada

Hussain M. Al-Rizzo and Ayman Abbosh
Systems Engineering Department, Donaghey College of Engineering and Information Technology, University of Arkansas at Little Rock, USA

Haider Khaleel
Department of Engineering Science, Sonoma State University, Rohnert Park, CA, USA

Kazuyuki Seo
Process Development Dept., Nippon Pillar Packing Co., Ltd. Sanda City, Japan

Albert Sabban
Ort Braude College, Karmiel, Israel
Tel Aviv University, Israel
Colorado University, Boulder, USA

Haider R. Khaleel, Hussain M. Al-Rizzo and Ayman I. Abbosh
Department of Systems Engineering, University of Arkansas at Little Rock, USA

Mohammed Al-Husseini, Karim Y. Kabalan and Ali El-Hajj
American University of Beirut, Lebanon

Christos G. Christodoulou
University of New Mexico, USA

www.ingramcontent.com/pod-product-compliance
Lightning Source LLC
Chambersburg PA
CBHW072251210326
41458CB00073B/970